THE END OF OIL

Acclaim for *The End of Oil*

"Eas... ...most important book of the summer... quite ...
whole year." — *Olympia* (Washington) *Olympian*

"Roberts displays the nuanced understanding of a longtime observer of American industry and economics . . . [He] deftly writes about the passions that fueled American consumption." — *Christian Science Monitor*

"Outspoken but even-handed." — *The New Yorker*

"A chilling wake-up call." — *Tucson Citizen*

"Roberts compellingly argues that it will take more than a temporary spike at the pump . . . to wean Americans from their energy-hog habits . . . Part of the paradox of American energy use, which Roberts neatly traces, is that as we get more efficient, we respond to that trend by consuming more."
— Salon.com

"Brilliant." — *Baltimore Sun*

"Fascinating . . . Roberts offers a stinging rebuke of America's myopic, do-nothing energy policy."
— Joseph J. Romm, author of *The Hype About Hydrogen*

"[An] elegantly written tour of our energy future . . . *The End of Oil* will stoke a debate on the severity of our predicament and the paths for the future."
— David G. Victor, director of the Program on Energy and
Sustainable Development at Stanford University

"This book may very well become for fossil fuels what *Fast Food Nation* was for food." — *Publishers Weekly*

THE END OF OIL

THE DECLINE OF THE PETROLEUM ECONOMY
AND THE RISE OF A NEW ENERGY ORDER

PAUL ROBERTS

BLOOMSBURY

First published in Great Britain 2004
This paperback edition published 2005

Copyright © 2004 by Paul Roberts

The right of Paul Roberts to be identified as the author of this work
has been asserted by him in accordance with the Copyright,
Designs and Patents Act of 1988

A CIP catalogue record for this book
is available from the British Library

Bloomsbury Publishing plc, 36 Soho Square, London W1D 3QY

ISBN 978 0 7475 7081 3

19 18 17 16 15 14 13 12 11 10

All papers used by Bloomsbury Publishing are natural,
recyclable products made from wood grown in well-managed forests.
The manufacturing processes conform to the
environmental regulations of the country of origin.

Printed in Great Britain by Clays Ltd, St Ives plc

www.bloomsbury.com/paulroberts

FOR KAREN

Contents

The END
of OIL

PROLOGUE

I WAS STANDING on a sand dune in Saudi Arabia's "Empty Quarter," the vast, rust-red desert where one-quarter of the world's oil is found, when I lost my faith in the modern energy economy. It was after sundown and the sky was dark blue and the sand still warm to the touch. My Saudi hosts had just finished showing me around the colossal oil city they'd built atop an oil field called Shayba. Engineers and technicians, they were rattling off production statistics with all the bravado of proud parents, telling me how many hundreds of thousands of barrels Shayba produced every day, and how light and sweet and sought-after the oil was. Saudi oilmen are usually a taciturn bunch, guarding their data like state secrets. But this was post 9/11 and Riyadh, in full glasnost mode, was wooing Western journalists and trying to restore the Saudis' image as dependable long-term suppliers of energy — not suicidal fanatics or terrorist financiers. And it was working. I'd arrived in the kingdom filled with doubts about a global energy order based on a finite and problematic substance — oil. As we'd toured Shayba in a spotless white GMC Yukon, though, my hosts plying me with facts and figures on the world's most powerful oil enterprise, my worries faded. I'd begun to feel giddy and smug, as if I had been allowed to peek into the garden of the energy gods and found it overflowing with bounty.

Then the illusion slipped. On a whim, I asked my hosts about another, older oil field, some three hundred miles to the northwest, called Ghawar. Ghawar is the largest field ever discovered. Tapped by American engineers in 1953, its deep sandstone reservoirs at one time had held perhaps a seventh of the world's known oil reserves, and its wells produced six million barrels of oil a day — or roughly one of every twelve barrels of crude consumed on earth. In the iconography of oil, Ghawar is the eternal mother, the mythical giant that makes most other fields look puny and mortal. My hosts smiled politely, yet looked faintly annoyed — not, it seemed, because I was asking inappropriate questions, but because, probably for the thousandth time, Ghawar had stolen the limelight. Like engineers anywhere, these men took an intense pride in their own work and could not resist a few jabs at a rival operation. Pointing to the sand at our feet, one engineer boasted that Shayba was "self-pressurized" — its subterranean reservoirs were under such great natural pressure that, once they were pierced by the drill, the oil simply flowed out like a black fountain. "At Ghawar," he said, "they have to *inject* water into the field to force the oil out." By contrast, he continued, Shayba's oil contained only trace amounts of water. At Ghawar, the engineer said, the "water cut" was 30 percent.

The hairs on the back of my neck stood up. Ghawar's water injections were hardly news, but a 30 percent water cut, if true, was startling. Most new oil fields produce almost pure oil, or oil mixed with natural gas — with little water. Over time, however, as the oil is drawn out, operators must replace it with water, to keep the oil flowing — until eventually what flows from the well is almost pure water and the field is no longer worth operating. Ghawar wouldn't run dry overnight: depletion takes years and even decades; however, daily production would continue to fall steadily, and the Saudis would be forced to tap new fields, like Shayba, to maintain their status as the world's preeminent oil power. While such expansions were never a problem during the heyday of Arab oil wealth in the 1970s and early '80s, times are much tighter today for Saudi Arabia and for most other petrostates. As we drove back toward the airstrip for my flight home, my hosts bombarding me with more facts and figures, I couldn't shake the feeling that the gods of energy might not be as powerful and eternal and confident as I had imagined.

To me, Ghawar is the perfect metaphor for what is happening to the larger energy economy, a geologic cautionary tale for a complacent world accustomed to reliable infusions of cheap energy. On the face of it, our energy economy is humming along like a perpetual-motion machine. Today, billions of people enjoy an unprecedented standard of living and nations float in rivers of wealth, in large part because, around the world, the energy industry has built an enormous network of oil wells, supertankers, pipelines, coal mines, power plants, transmission lines, cars, trucks, trains, and ships — a gigantic, marvelously intricate system that almost magically converts oil and its hydrocarbon cousins, natural gas and coal, into the heat, power, and mobility that animate modern civilization. For three hundred years, this man-made wonder has performed nearly flawlessly, transforming coal, oil, and natural gas (and in much of the world, a vast volume of wood, peat, and even animal dung) into economic and political power — and nurturing the belief that the surest way to still greater prosperity and stability was simple: find more oil, coal, and natural gas.

Yet, like Ghawar, our energy economy has hit a kind of peak of its own. Each year, the world demands more and more energy, with no end point in sight. And each year, it is more and more evident that the extraordinary machine we have built to supply that demand cannot sustain itself in its present form. Not a day goes by without some new disclosure, some new bit of headline evidence that our brilliant energy success comes at great cost — air pollution and toxic waste sites, blackouts and price spikes, fraud and corruption, and even war. The industrial-strength confidence that was a by-product of our global energy economy for most of the twentieth century has slowly been replaced by anxiety.

Although, like most consumers, I've been a casual student of this energy anxiety since it began — circa 1974, with the Arab oil embargo — I began exploring the question in earnest during the boom years of the late 1990s. I was writing about America's bizarre and growing infatuation with that modern warhorse, the "sport-utility vehicle," or SUV, and its close cousin, the pickup truck. At first, the story seemed to be mainly about conspicuous consumption and automotive vanity and sheer stupidity, since very few of their owners actually took their hugely expensive SUVs off-road or loaded their pickup trucks with anything heavier than groceries or soccer balls. But the more I looked into it, the more I realized that the real story lay less in the vehicles themselves than in the oceans of oil they were burning.

As is well known by now, SUVs and pickup trucks (known collectively, and somewhat deceptively, as "light trucks") consume a great deal of gasoline: the house-sized Ford Excursion I test-drove gets something like 4.6 miles per gallon in the city, and even the more sensible models rarely do better than 18. The cumulative effect of so much unnecessary internal combustion is staggering: since the SUV craze began in 1990, the twenty-year-old trend in the United States toward improving automotive fuel efficiency not only has halted but is now sliding backward, dramatically increasing U.S. demand for oil. And here is the rub: the United States doesn't have enough of its own oil to meet that surging SUV-driven demand. After a century of full-bore drilling, oil companies are finding precious little new oil in the Lower Forty-eight, and production — the number of barrels pumped per day — is falling steadily each year. What this means is that the United States, despite being the third-largest oil-producing nation in the world, now must import even more oil from the much-maligned "foreign" producers — including many, like Iran and Saudi Arabia, whose populations regard the United States as an enemy. In one of many energy ironies, during the months leading up to the second war with Iraq (charter member of the Axis of Evil, greatest threat to the American way of life since the fall of the Soviet Union, etc.), the United States was getting more than 10 percent of its imported oil from Iraqi fields.

The United States isn't the only nation with oil issues. Europe has long been import-dependent, as has Japan. China, a rapidly industrializing giant with more than a billion people and plans to build an economy as powerful and energy-intensive as anything in the West, now uses more oil than its own fields can produce and has begun courting the same foreign producers Uncle Sam now spends so much money and time and political capital trying to control. As I charted all this rising demand for oil, I wondered where it was going to come from, and what new contradictions and hypocrisies would result.

I was certainly not the only one asking. In interviews with oil industry officials — men and a few women who are, generally, quite optimistic about their business — I heard repeatedly how oil companies were having a harder and harder time finding new oil. I learned that most of the world's oil reserves are controlled by a small number of countries whose governments are unstable and corrupt and whose dependability as suppliers is increasingly in doubt. I began to wonder whether the glorious golden age of oil might be over. How long would the supplies of oil last? What would happen to our phenomenal wealth and splendid lifestyle if oil production peaked, supplies grew scarce, and prices rose? Did world governments and energy companies have a plan to ensure a smooth, gradual shift to a new fuel or a new energy technology? Or would the end of oil catch us unprepared and send shockwaves through the global economy, touching off a dangerous race for whatever oil supplies remained?

As my research took me to places like Houston, Saudi Arabia, Azerbaijan, and other outposts of the oil empire, the more I realized the story that needed telling wasn't simply about oil, but about all energy. Oil may be the brightest star in the energy firmament, the glamorous, storied shaper of twentieth-century politics and economics, and the owner of 40 percent of the world energy market. Yet oil is only one of a triad of geological siblings known as *hydrocarbons* that have dominated the global energy economy for centuries and whose histories and destinies are hopelessly intertwined with our own. Twenty-six percent of our energy still comes from coal, a cheap, abundant mineral used to power industrial processes and generate most of the world's electricity. Twenty-four percent comes from natural gas, a versatile energy source that will soon surpass coal as the preferred fuel for heating and power generation — and quite possibly become the "bridge fuel" to some future energy system. And yet, although coal and gas are, in a sense, alternatives to oil, both impose many of the same environmental, political, and financial costs. Coal is fatally dirty. Gas is extremely hard to transport and comes with its own thicket of geopolitical snarls; a global energy economy based on either would be just as problematic as the one we have, if not more so. In other words, when I began to ask about the end of oil, I was really asking about a transformation of the entire hydrocarbon economy and the end, perhaps, of a story that is almost as old as civilization.

For most of the past six thousand years, human history has been characterized by a constant struggle to harness ever-larger quantities of energy in ever more useful ways. From the earliest experiments with animal-drawn plows in what is now Iraq, the march of material progress has been accompanied by — and, one could argue, driven by — increasingly sophisticated mastery of fuels and energy systems. Animal power made agriculture possible. Firewood let us cook our food, heat our homes, brew barley into beer, and smelt metal ores into plowshares and spearheads. The wide-scale use of coal in England set the conditions for the Industrial Revolution. A century later, oil and natural gas, followed by a plethora of "advanced" technologies ranging from nuclear to solar, completed the transformation, dragging the industrializing world into modernity and in the process fundamentally and irrevocably reordering life at every level.

We live today in a world completely dominated by energy. It is the bedrock of our wealth, our comfort, and our largely unquestioned faith in the inexorability of progress, implicit in every act and artifact of modern existence. We produce and consume energy not simply to heat and feed ourselves, to move ourselves, or to defend ourselves, but to educate and entertain ourselves, to expand our knowledge, change our destiny, construct and reconstruct our world, and fill it with stuff. Everything we buy, from a hamburger at McDonalds to a duck at a Beijing market, from plastic lawn chairs and opera tickets to computers and garbage service, from medical services and cancer drugs to farm fertilizers and Humvees, represents a measure of energy produced and then consumed.

Energy has become the currency of political and economic power, the determinant of the hierarchy of nations, a new marker, even, for success and material advancement. Access to energy has thus emerged as the overriding imperative of the twenty-first century. It is a guiding geopolitical principle for all governments, and a largely unchallenged heuristic for a global energy industry whose success is based entirely on its ability to find, produce, and distribute ever-larger volumes of coal, oil, and natural gas, and their most common by-product, electricity.

Yet even a cursory look reveals that, for all its great successes, our energy economy is fatally flawed, in nearly every respect. The oil industry is among the least stable of all business sectors, tremendously vulnerable to destructive price swings and utterly dependent on corrupt, despotic "petrostates" with uncertain futures. Natural gas, though cleaner than oil, is hugely expensive to transport, while coal, though abundant and easy to get at, produces so much pollution that it is killing millions of people every year.

Worse, it is now clear to all but a handful of ideologues and ignoramuses that our steadily increasing reliance on fossil fuels is connected in some way to subtle but significant changes in our climate. Burning hydrocarbons releases not only energy, but carbon dioxide, a compound that, when it reaches the atmosphere, acts like a planet-sized greenhouse window, trapping the sun's heat and pushing up global temperatures. If left unchecked, this so-called greenhouse effect will keep warming the earth until polar icecaps melt, oceans rise, and life as we know it becomes impossible. The only way to slow global warming (for at this late date, the process cannot be stopped) is to cease emitting carbon dioxide — a monumental and expensive task that will require us to reengineer completely the way we produce and consume energy.

Climate change is in fact widely regarded as one of the main factors driving change in the energy economy — but it is not the only one. While climatologists and environmentalists fret about the *quality* of the energy we produce, most other experts worry far more about the *quantity* of energy we can make and, more specifically, whether we can produce enough energy of *any* kind or quality to satisfy the world's present and future needs. By 2035, the world will use more than twice as much energy as it does today. Demand for oil will jump from the current 80 million barrels a day to as much as 140 million barrels. Use of natural gas will climb by over 120 percent, coal use by nearly 60 percent. Demand will be especially acute in "emerging" economies, like those of China and India, whose leaders see voracious energy consumption as the key to industrial success.

Yet while the future energy demand seems certain, no one is clear where all this energy will come from. Consider oil. Quite aside from questions of how much is left (we'll get to that matter very shortly), there is simply the matter of finding and producing enough oil, and moving it via pipeline and supertanker to the places it needs to go. The sheer scale of the task is mind-boggling: when we say that by 2035 oil demand will be 140 million barrels a day, what we mean is that by then oil companies and oil states will need to discover, produce, refine, and bring to market 140 million new barrels of oil *every twenty-four hours,* day after day, year after year, without fail. Simply building that much new production capacity (to say nothing of maintaining it or defending it) will mean spending perhaps a trillion dollars in additional capital and will require oil companies to venture into places, like the Arctic, that are extremely expensive to exploit. Repeat the exercise for gas and coal, and you begin to understand why even optimistic energy experts go gray in the face when you ask them what we will use to fill up our tanks thirty years from now.

To make matters more complicated, it is not merely a question of procuring *enough,* as our growing appetite for electricity shows. Today's boom in technology and information has made electricity the fastest-growing segment of the energy market, and a crucial resource for emerging economies. By 2020, demand for electricity could be 70 percent higher than today. Yet because most electric power is generated in gas- and coal-fired power plants, making all that new power would mean putting an even greater strain on the hydrocarbon energy economy. At the same time, moving all this new electric load will completely overwhelm the existing electrical system — from power plants and transmission lines to the emerging and problematic network of energy traders. The great blackout of 2003 and the California power crisis of 2000 (due as much to dishonest energy speculators like Enron as to any shortage of power plants) are only the most colorful examples of what we may expect to see as the need for electricity continues to outpace supply.

It is in the third world, however, where we see the energy economy breaking down entirely. In Asia today, electrical demand is growing so fast that governments in China and India have essentially declared a state of emergency, sidelining environmental concerns to build hundreds of cheap coal-fired power plants, whose emissions may make it impossible even to slow climate change. And China and India are by no means the worst cases.

Around the world, more than one and a half *billion* people — roughly one-quarter of the world — lack access to electricity or fossil fuels and thus have virtually no chance to move from a brutally poor, preindustrial existence to the kind of modern, energy-intensive life many of us in the West take for granted. Energy poverty is in fact emerging as the new killer in developing nations, the root cause of a vast number of other problems, and perhaps the deepest divide between the haves and have-nots.

My point here is not simply that the modern energy economy should be changed but that we no longer have a choice in the matter: the system is already changing, and not always for the better. Everywhere we look, we can see signs of an exhausted system giving way messily to something new: oil companies quietly reengineering themselves to sell natural gas; governments scrambling to develop, or least understand, the "hydrogen economy"; a desperate search for new oil fields; rising tensions between energy producers and importers; diplomatic skirmishes over climate policy; and the frightening energy race between countries such as Japan and China to secure access to the last "big oil" and gas in Siberia, Kazakhstan, and the Middle East.

Yet if it is obvious that the current energy economy is on its way out, no clear consensus has taken shape on what happens next, what the "next" energy economy will look like. Can existing hydrocarbon technologies be adapted to new realities, or does the world require a radical new energy technology? If so, which technology? Newspapers and magazines and political speeches are filled with descriptions of brave new energy technologies

— hydrogen fuel cells and wind farms and solar buildings and tidal generation and fantastic processes that turn grass into diesel and manure into gasoline. But are any of these truly viable? How much will they cost? Can they
be brought to bear in time?

More to the point, even if some miracle technology is developed, this
in itself is no assurance of an orderly or peaceful transition. Historically,
shifts from one energy technology to another have proved wrenching. The
leaps from wood to coal and from coal to oil caused economic disruption
and political uncertainty (sixteenth-century Englishmen nearly revolted at
having to burn sooty coal instead of wood). And these were fairly slow-
motion transitions, occurring over several decades. Given that today's energy infrastructure is even more intertwined with global economies and
politics and culture, would a fundamental change in our energy technology
be even more disruptive? How long would a transition take — a decade,
fifty years? And what would a new energy order look like? Will it be better
than the one we have, or a hastily arranged, stopgap arrangement? Will we
be richer or poorer, more powerful or more hampered, happier with our
advanced energy technologies, or bitter over our memories of a bygone
golden age? And who will be in control? Are the current world powers —
most of whom are the biggest consumers of oil — still likely to be the leaders in this brave new world? Or might a new energy order breed a new
political order as well? This book is an effort to answer these questions.

It is hard to imagine a more appropriate moment to be talking about a new
energy economy. Electrical blackouts and gasoline price spikes have reminded us of the vulnerability of our energy system and our precarious dependence on foreign producers. Europe and the United States have parted
ways over climate change and energy policy generally, with Europeans
making modest efforts to develop a post-oil economy, while American
leaders, beginning with the president, have adopted an aggressive policy of
domestic oil drilling that wishes away environmental, geopolitical, and
even geological realities. Meanwhile, OPEC, the Organization of Petroleum
Exporting Countries, the bogeyman of yesteryear, is regaining much of its
old power and is vying with an oil-rich Russia and, increasingly, the United
States for control over the world oil markets. Perhaps most tellingly, the
United States and Britain are struggling to extricate themselves from a *sec-

ond oil war in Iraq that, whether openly acknowledged or not, was clearly meant to restore Middle Eastern stability and maintain Western access to a steady supply of oil.

Moreover, if recent events are any indication, we may be entering a period of payback for a century of petro-diplomacy. Unstinting efforts by the United States, Europe, and other industrialized powers to ensure access to Middle Eastern oil — by any means necessary, and often with the help of Israel — have helped foster a perpetual state of political instability, ethnic conflict, and virulent nationalism in that oil-rich region. Even before American tanks rolled into Baghdad to secure the Iraqi Ministry of Petroleum, leaving the rest of the ancient city to burn, anti-Western resentment in the Middle East had become so intense that it was hard not to see a connection between the incessant drive for oil and the violence that has shattered Jerusalem, the West Bank, Riyadh, Jakarta, and even New York and Washington. Only days after September 11, in fact, commentators were suggesting that the attacks were not only motivated by decades of oil politics but had been financed by oil revenues from the United States.

By nearly any sane measure, then, the quest for less problematic forms of energy and more energy-efficient technologies should be a top priority for all players in the energy world. Even now, a veritable army of energy optimists — scientists, engineers, policymakers, economists, activists, and even energy company executives — is working on the next energy economy, piece by piece, each participant confident that it can be built. I have seen energy technologies that are frankly miraculous: wind farms that generate enough electricity to power a city; ultraefficient office buildings requiring no outside power; cars that get a hundred miles per gallon of gasoline or run on clean hydrogen fuel cells; refineries that turn coal into a clean-burning gasoline.

I've seen how much energy can be saved through absurdly simple efficiency measures — and how much cheaper it is to save oil or electricity than it is to go out and produce more. I have watched the world's biggest energy companies slowly emerge from a policy of flat denial and begin a cautious, calculated, yet measurable shift toward a new energy economy. I have had politicians, economists, and energy executives lay out the Realpolitik of the energy economy by showing me the money we'll need to spend, the sacrifices we'll need to make, and the political deals we will need to cut in order to launch a new, sustainable energy economy.

Yet I have also encountered phenomenal resistance. The path toward a new energy economy is fraught with political and economic risk. No one knows when or if the new technologies will be ready, or how much they will cost, or what kinds of hardships they will impose — and few countries and companies are eager to be the first to take the leap. The current energy economy, with its oil wells and pipelines, its tankers and refiners, its power plants and transmission lines, is an enormous asset, worth an estimated ten trillion dollars. No company, nor any nation, not even America, can afford to write that off — even if many of the gloomier commentators believe that doing so is the only way to slow climate change. Instead, energy companies are looking to minimize their losses, waiting till the last minute to adopt some technology so that they can squeeze the last drop of revenue from their existing hydrocarbon assets. Governments, too, fearing economic dislocation and political disadvantage, are steadily delaying any significant move away from the existing energy economy — thereby ensuring that change, when it occurs, will be all the more sudden and disruptive.

Consumers, meanwhile, seem almost oblivious. In industrialized nations, energy is so cheap and incomes are so great that consumers think nothing of buying ever larger houses, more powerful cars, more toys and appliances — increasing their energy use without even knowing it. And if people in developing nations use far less energy today, this is not by choice: they, too, want the cars, the large homes, the entertainment systems, the conditioned air, and other features of the energy-rich lifestyle enjoyed in the West. The trend seems clear: barring some economic collapse, world energy demand can do nothing but rise — and the energy industry not only intends to meet that demand but, for all its talk of novel technologies and approaches, will do so almost entirely with existing methods, fuels, and technologies — at least, for the time being.

Thus, even as it becomes more and more possible to imagine a new energy economy, the old one is switching into high gear. In places like Borneo, Kamchatka, and Nigeria, off the coast of Florida and in the South China Sea, in Alaska and Chad, multinational energy companies comb the earth and ocean beds in search for the next big oil and gas plays. And around the world, the diplomatic, economic, and military strategies of nearly every nation continue to be shaped by one overriding objective — to maintain uninterrupted access to a steady supply of energy. The goal is sacrosanct, to be pursued at all costs, regardless of the way it perverts the culture and politics of entire regions or props up corrupt governments and dictators or, ultimately, fosters the instability and resentments that have already spawned such malignant figures as Muammar Qaddafi, Saddam Hussein, and Osama bin Laden.

Yet despite the staying power of the status quo, each year that energy consumption continues unabated, the end of the current energy system not only becomes more inevitable but appears more likely to occur as a traumatic event. As energy supplies become harder to transport, as environmental effects worsen, and as energy diplomacy sows even greater geopolitical discord, the weight of the existing energy order becomes less and less bearable — and the possibility of a disruption more undeniable.

In the end, this question of disruption may be the most critical one of all — not simply for policymakers and oil sheiks, but for anyone accustomed to filling up at the gas station or switching on an air conditioner; for it is not simply change that affects us, but the *rate* of change — how quickly and cleanly one way of life is exchanged for another. A swift, chaotic shift in our energy economy almost guarantees disruption, uncertainty, economic loss, even violence. By contrast, were we somehow to manage a gradual, smooth change, phased in over time, we might be able to adapt, minimizing our losses and even allowing the more clever of our species to profit from new opportunities.

In fact, while the precise shape of our energy future remains veiled, we can already discern two distinct paths for getting there. On the one hand, we can imagine the transition as a kind of a proactive endeavor, driven by global consensus over some perceived threat, based on scientific analysis, and managed to minimize disruption and maximize economic gain. On the other, we can picture a change that is less a transition than a reaction, a patchwork of defensive programs triggered by some political or natural disaster.

Suppose, for example, that worldwide oil production hits a kind of peak and that, as at Ghawar, the amount of oil that oil companies and oil states can pull out of the ground plateaus or even begins to decline — a not altogether inconceivable scenario. Oil is finite, and although vast oceans of it remain underground, waiting to be pumped out and refined into gasoline for your Winnebago, this is *old* oil, in fields that have been known about for years or even decades. By contrast, the amount of *new* oil that is being discovered each year is declining; the peak year was 1960, and it has been downhill ever since. Given that oil cannot be produced without first being discovered, it is inevitable that, at some point, worldwide oil production must peak and begin declining as well — less than ideal circumstances for a global economy that depends on cheap oil for about 40 percent of its energy needs (not to mention 90 percent of its transportation fuel) and is nowhere even close to having alternative energy sources.

The last three times oil production dropped off a cliff — the Arab oil embargo of 1974, the Iranian revolution in 1979, and the 1991 Persian Gulf War — the resulting price spikes pushed the world into recession. And these disruptions were *temporary*. Presumably, the effects of a long-term permanent disruption would be far more gruesome. As prices rose, consumers would quickly shift to other fuels, such as natural gas or coal, but soon enough, those supplies would also tighten and their prices would rise. An inflationary ripple effect would set in. As energy became more expensive, so would such energy-dependent activities as manufacturing and transportation. Commercial activity would slow, and segments of the global economy especially dependent on rapid growth — which is to say, pretty much everything these days — would tip into recession. The cost of goods and services would rise, ultimately depressing economic demand and throwing the entire economy into an enduring depression that would make 1929 look like a dress rehearsal and could touch off a desperate and probably violent contest for whatever oil supplies remained.

When such a production peak will occur is, as we shall see, a Very Big Question. Optimists like the U.S. government believe that a peak in oil production cannot occur before 2035 or so and that would give the world plenty of time to find something else to burn. Pessimists, by contrast, a group whose members include geologists, industry analysts, and a surprising number of oil industry and government officials, believe that a peak may come much sooner — perhaps as soon as 2005. (Indeed, a small but vocal minority believes that the peak has already occurred and that this is why oil companies like Shell and BP are struggling to find untapped sources of oil to replace all the barrels they produce.)

Granted, such a wide range of dates is not particularly helpful for anyone wanting to know when to start hoarding diesel, light out for the hills, or invest in oil company stocks. But lest you think it's about time to buy a larger SUV, it is worth noting that even the oil optimists concede, usually privately, that the important oil — that is, the oil that exists outside the

control of the eleven-country OPEC oil cartel — will in all likelihood peak
between 2015 and 2020. We call this "important oil" because, once it peaks,
the free world will have to rely more each year on oil controlled by the
likes of Saudi Arabia, Venezuela, and Iran — governments that cannot be
counted on to bear the best interests of the West in mind in setting pricing
policy.

That brings us back to the question of smooth or sudden change. Admittedly, even if the world knew exactly when non-OPEC oil was going to
peak, only so much could be done to prepare, given the size of the existing
oil infrastructure and the complacency of the average consumer. Yet it's
also true that were Western governments to begin taking steps to reduce oil
demand, or at least to slow the rate at which it is growing (by, say, raising
fuel efficiency standards for cars), the impact of such a peak would be less-
ened dramatically — and the world would gain all the benefits of using
something other than oil.

At the same time, if the consuming world instead continues in its current mode — known by energy economists and other worriers as "business
as usual" — oil demand will be so high by 2015 that a peak (or any big disruption, such as a civil war in Saudi Arabia or a massive climate-related
disaster that kills thousands and forces politicians to cut the use of oil and
other hydrocarbons in a hurry) could be an unmitigated disaster. Thus, the
real question, for anyone truly concerned about our future, is not *whether*
change is going to come, but whether the shift will be peaceful and orderly
or chaotic and violent because we waited too long to begin planning for it.

In writing this book, I have focused on all aspects of the energy economy
— the past and present of energy, the technology and business of energy,
and the major players. I've studied the big energy producers, like Saudi
Arabia and Russia, who control most of the world's oil reserves and who
will play a critical role in the transition to a post-oil economy. I've looked
in depth at China and India, two energy paupers whose enormous populations and growing economies will nonetheless make them the biggest energy players of the twenty-first century. I have examined Japan and Germany, countries that, lacking their own domestic oil supplies, have adopted
energy-efficient policies and have fostered a culture that accepts if not em-
braces a low-energy way of life.

But by necessity, much of this book will focus on the United States.

For all that the new energy economy is an international issue, no nation will play a greater role in the evolution of that economy than ours. Americans are the most profligate users of energy in the history of the world: a country with less than 5 percent of the world's population burns through 25 percent of the world's total energy. Some of this discrepancy is owing to the American economy, which is bigger than anyone else's and therefore uses more energy. But it is also true that the American lifestyle is twice as energy-intensive as that in Europe and Japan, and about ten times the global average. The United States is thus the most important of all energy players: its enormous demand makes it an essential customer for the big energy states like Saudi Arabia and Russia. Its large imports hold the global energy market in thrall. (Indeed, the tiniest change in the U.S. energy economy — a colder winter, an increase in driving, a change in tax law — can send world markets into a tailspin.) And because American power flows from its dominance over a global economy that in turn depends mainly on oil and other fossil fuels, the United States sees itself as having no choice but to defend the global energy infrastructure from any threat and by nearly any means available — economic, diplomatic, even military.

The result of this simultaneous might and dependency is that the United States is, and will be, the preeminent force in the shaping of the new energy economy. The United States is the only country with the economic muscle, the technological expertise, and the international standing truly to mold the next energy system. If the U.S. government and its citizens decided to launch a new energy system and have it in place within twenty years, not only would the energy system be built, but the rest of the world would be forced to follow along. Instead, American policymakers are too paralyzed to act, terrified that to change U.S. energy patterns would threaten the nation's economy and geopolitical status — not to mention outrage tens of millions of American voters. Where Europe has taken small but important steps toward regulating carbon dioxide (steps modeled, paradoxically, on an American pollution law), the United States has made only theatrical gestures over alternative fuels, improved efficiency, or policies that would harness the markets to reduce carbon. As a result, the energy superpower has not only surrendered its once-awesome edge in such energy technologies as solar and wind to competitors in Europe and Japan but made it less and less likely that an effective solution for climate change will be deployed in time to make a difference.

Critics place much of the blame on a political system corrupted by big energy interests — companies desperate to protect billions of dollars in existing energy technologies and infrastructure. An equal measure of blame, however, must fall on the "average" American consumer, who each year seems to know less, and care less, about how much energy he or she uses, where it comes from, or what its true costs are. Americans, it seems, suffer profoundly from what may soon be known as energy illiteracy: most of us understand so little about our energy economy that we have no idea that it has begun falling apart.

The End of Oil is a dramatic narrative in three parts. In the first five chapters, I set the stage for the current crisis, by explaining how and why energy has become so vital a part of our existence. Chapter 1 offers a short history of energy, describing the long, slow rise from muscle power and sweat to a sprawling, hydrocarbon-powered economy. In Chapter 2, we tackle the question of how much oil is left and see firsthand how difficult the search for oil has become. Chapter 3 takes a sharp look at one of oil's most talked-about challengers — the hydrogen fuel cell — highlighting that technology's awesome potential, yet showing just how far it has to go. Chapter 4 discusses the connections between energy and power and outlines the role energy plays in domestic and international politics, trade, and even war. This first part closes with a chapter on global climate change — a complex phenomenon that is both the consequence of our current energy economy and, perhaps, the most important impetus for building a new one.

In Part Two, we look at the mechanics of the energy order. In Chapter 6, we examine energy consumption and see how our evolving use of oil, electricity, and other forms of energy has become one of the most powerful economic and political forces on the planet. In Chapter 7, we meet the producers of oil and gas, and learn how the energy business is undergoing a radical and potentially disastrous transformation. Chapter 8 takes us on a tour of the options for that new system — the alternative fuels and systems, their potential for changing the world, and the many obstacles they face. Chapter 9 introduces the important yet often-neglected concept of energy conservation and shows how a radical improvement in energy efficiency will be essential to any new and sustainable energy economy.

In Part Three, we chart the promise and the peril of our energy future. Chapter 10 describes how the existing energy system is already failing to meet even current needs — and shows how the race to develop "clean" energy must compete with the more basic need to produce enough energy of any kind. Chapter 11 describes the colossal inertia of the current energy order, and the way it has influenced, shaped, and, too often, corrupted economies and entire nations. Chapter 12 lays out the terms of the coming struggle, as defenders of the energy status quo go up against a new generation of players. Chapter 13 offers a speculative account of the transition to a new energy economy, in extrapolating current trends to show how a new system might actually emerge.

I am under no illusions that this book addresses all the important aspects of the evolving energy economy, or even most of them. Energy is a vast topic, with millions of components interwoven in a complex and ever-changing pattern that defies quick answers or simple truths. Instead, my hope is to provide an introduction, a way for nonexperts to begin to think about what experts have long known: that energy is the single most important resource, that our current energy system is failing, and that the shape of the next energy economy is being decided right now — with or without our input. Ideally, readers of this book will acquire a better understanding of what is coming, and perhaps a better chance of making a difference in that future.

PART I

THE FREE RIDE

1

LIGHTING
THE FIRE

ONLY THE BAREST details remain from the day Thomas Newcomen
saved the Industrial Revolution from collapse and launched the great race
for energy that has defined civilization ever since. But we can reconstruct
the scene. The year is 1712, the month probably March. The setting is the
Coneygree Coal Works in Staffordshire, on the site of England's greatest
coalfield. Inside a neat two-story brick building, we find a middle-aged
man clambering around a large, upright contraption of brick, iron pipes,
and brass that rises thirty feet from the floor and protrudes into the chamber above. He is Thomas Newcomen, a forty-nine-year-old metal smith
and Baptist preacher turned inventor. The contraption is his "heat engine,"
a coal-powered, "self-acting" device that has taken ten years to perfect and
which, if all goes according to plan, will soon be pumping water from a
flooded mineshaft 160 feet below.

A private, guarded man by habit, Newcomen has today thrown open
the doors to his engine room. Around him, a small crowd has gathered —
coal mine officials, a handful of investors, perhaps an attorney or two representing Newcomen's many creditors. As the visitors gawk at the engine,
we can picture their upturned faces, their expressions of alternating doubt
and desire. Certainly, they will have heard the criticism from leading scientists who believe that such a contraption cannot work — especially one
built by a mere tinker like Newcomen. They will have heard that the early
prototypes for this engine, hand-built devices that pushed at the limits of
existing technology, have all failed. Those assembled will know of Newcomen's mounting debts.

But anyone in the cramped, smoky room that day in March will also
have been keenly aware of the stakes if Newcomen succeeds. England is in a

fuel crisis. The rapidly industrializing country has used up most of its firewood and is now utterly dependent on coal. Coal powers the thousands of factories and foundries that are popping up like dark mushrooms all over the English landscape. Coal provides heat and fuels the cook fires for the hundreds of thousands of Englishmen who now live in the cities. London alone, the world's largest metropolis and its commercial center, teeming with more than six hundred thousand people, consumes a thousand tons of coal a day — and each year needs more.

Yet more isn't coming out of the ground, not fast enough. In Wales, the Midlands, and other coal regions, British miners have already hollowed out the easy coal seams on the surface and must now delve deeper. Unfortunately, the new shafts are constantly flooding with groundwater. Many mines have installed crude horse-driven pumps, but the contraptions are slow, inefficient, and prohibitively expensive. Around the country, mine after mine has lost productivity or shut down entirely. The nation is desperate. "Drainage" has become the topic of the year, and it is clear that if a solution can be found, it will not only save the day but, as one observer puts it, prove "most lucrative to the inventor."

Newcomen motions to an assistant to shovel more coal into the firebox beneath the huge brick boiler. The inventor turns a valve, directing steam into an eight-foot-tall brass cylinder. Smoke and a great hissing and clanging fill the air, and inside the cylinder a massive piston begins to rise and fall, once every twelve seconds. The spectators look up. High above them in the rafters, a twenty-eight-foot horizontal beam has begun to rock up and down, like a giant teeter-totter, raising and lowering an iron chain that drops through a hole to the mine below. For a long moment, nothing happens. The onlookers fidget, shifting from foot to foot, clearing their throats. Then, from an open pipe outside comes the gurgle of rushing water, followed by a great spurt of blackish liquid — water from the mineshaft far below. Twelve seconds later, another gush, and another. The mine is draining. The investors cheer. Newcomen has just made them very rich.

In fact, on that day in Staffordshire, Thomas Newcomen had improved the fortunes not simply of a few local capitalists, but of all humankind. The Newcomen engine may have been expensive, noisy, and comically inefficient (more than 99 percent of the coal's heat energy was wasted, owing to poor design). The engine may have burned through more than a ton and a half of coal a day. But even at that, the new device was considerably cheaper than the alternatives. One engine could replace a pumping operation employing fifty horses, thereby cutting operating costs by 85 percent. Within twenty years, more than a hundred Newcomen engines would be clanking away across England and Continental Europe, bringing mine after flooded mine back into production, and contributing to a spectacular increase in coal production. In Britain alone, yearly coal output jumped from around three million tons in 1712 to nearly double that by 1750. At the end of the century, England was producing ten million tons, making the island the undisputed king of coal, and the world's first modern energy economy.

The consequences went well beyond an increase in coal production, however. Newcomen's engine was, after all, an *engine*, one of the world's first — an automatic, or "self-acting," device that transformed chemical energy from coal into physical energy — work — and did so more efficiently than the horses and men it replaced. In so doing, Newcomen's engine gave us our first real mastery over energy and set humanity on a course that would change the world forever. True, our ancestors had been running machines with energy from water mills and windmills for centuries, but these crude devices often lacked the power or rotating speed needed to drive complex machines like pumps or mechanized looms. More to the point, water mills and windmills worked only in certain places, such as riverbanks or spots where the wind was constant, and mills could be idled by low water or a calm day: Nature, in other words, still meted out the energy. The steam engine had speed to spare. It could be installed anywhere and would run continuously — assuming you had a continuous supply of coal. For the first time, human beings had the potential to harness energy in quantities far greater than previously imagined, and the impact would be enormous.

Within a century of Newcomen's successful demonstration, the world was being remade by coal energy. Although wood and other types of "biomass" would remain important sources of fuel through the nineteenth century, coal and the power it supplied transformed Western commerce and society, by increasing productivity and wealth and accelerating the great shift from agriculture to industry. First in mining, then in textile manufacturing, and then in transportation, the rapid advances in the mastery of energy allowed people to produce more goods, faster and more efficiently, and transport them to more distant customers, at lower cost, than had ever before been even conceivable. The potential of coal-fired steam power seemed boundless. Released from the age-old limitations of muscle, wind, and water, Industrial Man was poised on the brink of limitless wealth and material progress — just as long as he could keep the coal coming.

In many ways, Newcomen's engine marked the culmination of human-kind's ten-thousand-year march toward what might be called energy con-sciousness. From the moment humans sought to control their environ-ment, success and material progress have been intimately bound up with the ability to find and exploit greater, more concentrated sources of energy. Early tribes of hunter-gatherers, for example, probably had no phrase for "energy cost-benefit analysis," but they knew which roots and berries had the highest caloric content and thus offered the richest energy returns for a given investment of energy. Cro-Magnon hunting parties learned to target larger prey in part because the energy costs — the calories required to track and chase — were more readily reimbursed by a larger chunk of meat than by a smaller one (even if the larger beast might be more difficult and dan-gerous to bring down).[1]

This primitive energy calculus became more essential when farming began, somewhere in present-day Iraq around ten thousand years ago. Clearing land and tilling soil are brutally hard work. They suck up more energy, in the form of food calories, than does foraging for nuts and ber-ries. In return for those extra calories, though, our ancestors received sub-stantial benefits, including the ability to produce more food on a far more reliable schedule. Similarly, when draft animals came into widespread use, probably around 4000 B.C., energy requirements kept climbing, but so did productivity. An ox might require a great deal of forage and water; but tied to a plow, that four-legged tractor would allow its human master to till three times as much land as he could by hand with a hoe.[2]

Did early humans see the connection between the extra energy costs and extra benefits? We'll never know. Yet visible or not, the advantages stemming from increased energy use — greater productivity, the new reli-ability of the food supply, and so on — were real. They were also critical in encouraging the shift from hunting and gathering to agriculture and ur-banism, as small, mobile tribes now had the tools and capabilities to sup-port larger, sedentary farming societies. And notice the trend: as humans became dependent on the increased production, and consequently greater reliability, of food, what they were actually becoming dependent on was the underlying increase in energy consumption. The link between energy and progress would become much more obvious as people began to live in larger, more concentrated communities — although the focus would now expand from calories to fuel. Whereas nomadic hunter-gatherers could dependably find wood, grass, dung, or other heating and cooking fuel wherever they roamed (indeed, the availability of fuel may have helped determine where they roamed), that didn't work for city dwellers. No longer could a person simply wander out at sunrise and pick up a few sticks for a fire to cook breakfast. Urbanites were stuck in one place, competing with neighbors for fuel — in even shorter supply now that early craftsmen had begun to fire pottery or smelt metals. By some estimates, every man, woman, and child in these early cities required a half ton of firewood a year, a requirement that put an enormous strain on local forests. Factor in the added energy demands from a primitive industry like copper smelting — a ton of firewood was needed to smelt ten pounds of metal[3] — and you have the beginnings of the earliest energy crunch. For perhaps the first time in history, humans could see the threat that lay in the gap between fuel demand and fuel supply. They had, in other words, achieved a rudimentary energy consciousness.

To bridge this gap between demand and supply, the world's first energy economy arose. In a pattern that would be repeated centuries later with coal and then with oil, our enterprising ancestors invented a system to find and distribute fuel as efficiently and cheaply as possible. In forested regions, this would have meant organizing wood-gathering parties to comb the neighboring countryside. Logs or branches had to be collected, hauled to the city, stored in stacks to dry properly, then sold, traded, or given to those in need. As the energy economy evolved, these tasks became distinct specialties: those of the woodcutter, the warehouse owner, the wood-seller. Social, political, and even legal questions had to be addressed: Who owned a particular forest? How quickly could it be cut down? Gathering costs needed to be accounted for, especially as the forests closer to town were chopped down and foragers were forced to range farther afield, and conservation schemes were likely to be considered. At some point, the possibility of running out of fuel came to be feared as much as any other large-scale disaster — war, drought, or plague.

In short, energy had become a strategic resource — a factor in the rise, and fall, of economies and civilizations.[4] Catastrophic fuel shortages were probably quite frequent. Indeed, the march of human progress may well have been marked by a series of energy crises that either killed off a particular civilization or helped push it to the next level of technological and economic development.

This, certainly, was the case in Europe, which by medieval times had worked itself into an energy shortage as severe as anything we moderns can complain about. As elsewhere, Europe's agricultural revolution had depleted its primary fuel source: firewood. Not only were farmers clearing forests to plant more crops, but the additional crops supported a larger population with even greater needs for wood fuel (not to mention lumber for buildings and ships), which only took more of a toll on already decimated forests. New industries also depleted precious fuel supplies. The expanding manufacture of glass, dyes, ale, lime, salt, and bricks, though key to improved living conditions, consumed entire forests of firewood. Iron making was especially fuel-intensive, requiring a ton of firewood to produce twenty pounds of metal. To operate year-round, a single iron smelter needed more than four hundred square miles of forest. Something had to give.[5]

England, with its limited forests and comparatively advanced industry, suffered acutely. By the thirteenth century, wood shortages were so severe that English officials were shutting down metal forges and forbidding the cutting of any Crown forests. The crisis eased briefly in the fourteenth century, when the plague killed off a third of Europe's population and allowed forests time to grow back. But by the fifteenth century, the recovering population had wiped out any wood surplus, and firewood became an expensive luxury, available to only the wealthiest citizens.[6]

Coal was the obvious alternative, and the transition from wood to that fossil fuel would utterly transform the economy, culture, and politics of the world and spark what we now understand to be the energy revolution. Yet early on, few Englishmen regarded the move to coal as positive. The soft brown lignite then being mined in England and Europe was chock full of sulfur and other impurities: when burned, it produced an acrid, choking smoke that stung the eyes and lungs and blackened walls and clothes. More fundamentally, coal simply did not work with an energy technology designed for wood. Brewers and bakers refused to use coal, for it fouled the taste of food and drink. And because sulfur interferes with the chemistry of iron, coal could not be used in smelting furnaces. In nearly all cases, wood was the preferred fuel, when it was available.

Still, even the most grudging Englishmen had to acknowledge that coal had important advantages over wood. Just as oil and gas would out-perform coal two centuries later, coal was simply superior to wood economically. First, the great abundance of coal made it much cheaper. Second, coal provided a better energy payoff: not only did it take less energy to dig out a pound of coal than to cut a pound of wood, but that pound of coal, when burned, released up to five times as much energy. Coal's higher *energy density* made it far more economical to produce, cheaper to transport over long distances (from mines in the north to London in the south), and easier to store. Unlike a wood lot, which depends on a scattered and in-efficient fuel-gathering process, a coalfield concentrates a massive volume of chemical energy more or less in one physical location. Production can be centralized and efficient, and therefore much more profitable — a critical prerequisite to the birth of any industry, especially in a time when labor is scarce.

Nor, finally, were all types of coal so messy. Medieval Europeans had been using poor-quality coal, but there were better varieties. Coal is a fossil fuel: it forms when layers of dead trees and leaves and other organic materials, known as peat, become buried and, over the course of millions of years, compressed and heated into a carbon-bearing mineral. The quality of a particular coal depends on its carbon content: high-carbon coals burn hot and relatively cleanly; low-carbon coals do not. Generally, the longer coal stays buried, the harder it becomes, but also the deeper one must dig to ex-tract it. Not surprisingly, the first coal that Europeans found was the softer, more recently formed lignite. But as miners emptied the surface seams and were forced to dig deeper, they found the harder, cleaner-burning bitumen and anthracite. And with Newcomen's engine pumping out the mine shafts when they flooded, the coal age had truly begun.

⚬⟊⟊⟊⟋

The significance of the Newcomen phenomenon wasn't just that it allowed us to produce more energy, but that it changed the way we used the energy. Until that point, coal, wood, and other fuels were simply sources of heat: their chemical energy was converted, through burning, into heat energy used mainly for cooking or heating. But Newcomen's engine took the pro-cess one step further, by converting the heat energy from combustion into the physical, or *mechanical,* energy of work. In a sense, Newcomen's engine was achieving with coal what men, horses, and oxen already did naturally with calories, but far more efficiently, and with no apparent limitations. If you wanted more power, you simply built a bigger engine and fed it more fuel. All that was necessary was an ever-larger supply of coal, which Newcomen's engine itself seemed to ensure.

The remarkable thing was that the more coal England produced, the more coal England burned. Because coal was now so readily available, industrial users could not only expand existing operations but develop entirely new uses for the abundant fuel. This phenomenon was most dramatically evident in the iron industry, where manufacturers, having developed a sulfur-free form of coal, known as coke, could now use coal to smelt iron. It is impossible to overstate the impact of this partnership between energy and iron, the twin building blocks of the industrial age. Cheap, abundant iron touched off an acceleration in the manufacture of machines, including steam engines. Factories of all kinds sprang up, using steam engines to drive looms, lathes, presses, and every other conceivable kind of device. And naturally, each advance in technology only increased the demands on the coal industry. In the early 1800s, coal-fired steam locomotives began carrying people and cargo, including coal. As rail transport expanded, so did demand for iron rails, creating yet further demand for coal at the iron foundries. Then, as the rail system grew, the greater number of locomotives necessitated higher coal production, as did England's move from a wind-powered to a steam-powered navy. This interdependence — between production and consumption, between supply and demand — was perhaps the most critical element in the success of the Industrial Revolution, and it remains a fundamental aspect of the modern energy economy.

Inevitably, so much new demand brought in new producers. Germany, France, and Belgium developed their coal industries, followed somewhat belatedly by the United States. American coal reserves are massive (the largest in the world, in fact), but the young nation had so much forest that wood remained the dominant fuel until the end of the Civil War. Before 1850, most Americans didn't even know coal could be burned. Yet by 1900, U.S. mines were outproducing those in England and contributing to a world production total of nearly a billion tons — more than ten times the volume of just fifty years before.[7]

Beneath this staggering increase in volume, an even more interesting transformation was taking place. In 1701, the average Englishman used less than half a ton of coal a year. By 1850, he was using nearly three tons, and by 1900, more than four tons. Similar changes were occurring in industrializing Europe and the United States.[8]

What were people using all the extra energy for? More heating and cooking, to be sure, and more travel in steamships and trains; but mainly, people were manufacturing more things: more textiles, more machines, more food and ale, more paper. The pattern was clear: the more you produced, the more energy you needed. And conversely, the more energy you used, the more things you produced — and the wealthier you or, more likely, your employer or the state, became.

One might just as well relabel the expanding Industrial Revolution the energy revolution, because the industrial economies of the nineteenth century simply could not have developed without the parallel emergence of energy economies to sustain them. And as industrialization spread, country by country, region by region, so did demand for energy.

To meet this rising tide of demand, the energy industry itself had to change, becoming not only one of the largest businesses in the world, but among the most sophisticated and enterprising, and certainly the most widespread. Coal technology advanced quickly. Mines themselves became enormous underground factories, served by hundreds of miles of tunnels, rail systems, and subterranean canals. To transport the coal to the cities and burgeoning industrial areas, extensive networks of canals and then railways were built; in many cases, the iron foundries and other factories simply relocated to the coal fields in the north of England, the Ruhr Valley in Germany, and the coal regions of Pennsylvania and Ohio.

At the other end of the supply chain, an entire system of distribution and marketing arose to sell the coal to industrial and residential users and to promote new uses and new demand. Coal-fired boilers were engineered to fit inside every factory, every office building. Coal-burning ovens and heaters were refined for home use, thereby increasing domestic consumption and, just as important, teaching consumers to expect better, easier lives and more "convenience" through greater energy consumption.

The energy industry grew so rapidly that traditional business practices could not keep pace. It was in the coal business that consolidation became an established practice, as hundreds of small, inefficient coal mines in England, Europe, and the United States were rolled up into massive corporate entities fundamentally different from anything that had existed before. This new breed of organization required new approaches to everything from production and delivery to accounting, cost control, labor management, and, above all, finance. Industrial-era coal mining was one of the first truly capital-intensive industries. In 1800, the start-up costs for a large coal mine could easily run to tens of thousands of pounds — a vast sum in those days — and force owner-entrepreneurs to come up with new ways to attract capital investment and ultimately created a web of interdependence between the energy industry and the financial community that still exists today.[9]

As with the modern energy sector, the coal business was enormously risky. Return on investment, the one number that mattered in so capital-intensive an industry, could be destroyed by any number of things: price fluctuations, a glut in supply, production bottlenecks, mining disasters, and eventually the temerity of the coal miners themselves.

Yet lest we grow too sympathetic, it should be understood that the early energy business was, on the whole, exceedingly profitable. Demand was rising and coal companies became expert at protecting their position. They invested in new technologies and practices. They lobbied government for favorable laws, including laws preventing miners from striking for better working conditions. And increasingly, coal companies simply cheated.[10] Just as OPEC would several centuries later, coal-mining companies in England, Europe, and later the United States joined in great regional monopolies, colluding shamelessly to limit production and thus keep prices high, then peacefully dividing up the big urban markets in London, Paris, Berlin, and New York to avoid price competition. Consumers complained bitterly to government, but periodic reform efforts had little effect, because any new laws were rarely enforced. Coal companies were simply too politically influential, and few in government wanted to interfere with so important an industry.

Price gouging was far from the only problem coal presented. Labor woes continued to mount, as miners protested horrific working conditions. Cave-ins and gas explosions, which claimed hundreds of lives, provided reform-minded writers like Zola with powerful material and constantly reminded the public of the cost of a coal-based economy.[11] For that matter, coal had hardly turned out to be the ideal fuel, after all. Though it contained more energy than wood, coal was still too bulky to be a completely efficient fuel. Long-haul steamships, for example, required such large coal bunkers that they had little room left over for cargo or passengers. And even the best coal didn't burn hot enough for many of the new industrial processes. Nor had the soot problem been solved. By the end of the nineteenth century, the air was so black in London, Pittsburgh, Berlin, and other industrialized cities that trees died, marble facades dissolved, and respiratory ailments became epidemic.

For the moment, however, there was nothing to be done: coal had become something no person or business or country could live without. By 1900, the coal industry stood at the very center of the industrial world, interconnected with, and supporting, every other sector, and generating a substantial proportion of the national wealth, jobs, and export income for the producing countries.

Perhaps more important, by the dawn of the twentieth century, coal had created something more lasting: a new kind of economy, or perhaps more accurately, new kind of economic order. This new order had engendered a powerful system of production practices and distribution networks, tailored to the reciprocal dynamic of supply and demand. It included a corporate business model designed for massive economies of scale, a financial structure to manage the large capital requirements, and political relationships to protect these investments. Just as significantly, around the new energy order had arisen a culture of energy consumption and a social and political awareness of the critical role that energy played in rising living standards and wealth, in national success and international power. Coal might be dirty and dangerous, and the coal economy might be monopolistic and corrupt. But coal was without question the basis for the industrial world's burgeoning prosperity. As one English observer noted, "Coal stands not beside but entirely above all other commodities. It is the material source of the energy of the country — the universal aid — the factor in everything we do."[12]

Coal production would continue to grow for decades. Yet by the end of the nineteenth century — to the great dismay of English coal barons, U.S. coal miners, and the centuries-old coal industry — the energy order had become far too large and global to be dominated by a single country, or even a single fuel.

✿

The end of the coal age began on the morning of January 10, 1901, just outside Beaumont, Texas, on a small hill called Spindletop. It was half past ten, and a frustrated man named Al Hammil had just stepped away from the well he had been drilling to tell his brother, Curt, that there was in fact no oil here, when the sandstone some 1,100 feet below his feet proved him wrong in spectacular fashion. With a deafening blast and a great howling roar, thick clouds of methane gas jetted from the hole. Then came the liquid, a column of it, six inches wide and brownish green. It rocketed hundreds of feet into the winter sky before falling back to earth as a dark rain. It soaked the wrecked drilling derrick, the red Texas earth, and the Hammil brothers, who were now dancing for joy, for there *was* oil here — more than the Hammils, or anyone else, for that matter, had ever seen before. Most oil wells of that time were yielding fifty to a hundred barrels a day. The record breakers, like those in Russia, produced maybe five thousand barrels a day. But Spindletop was pumping out five thousand barrels every *hour* — one hundred thousand barrels a day — more than the combined production of every other well on earth.

Spindletop's plume was visible from downtown Beaumont, four miles away, and within hours of the strike, townspeople had flocked to the site to stare at the gusher, which was now creating a lake of oil. Tourists began arriving from Houston the next day, followed by journalists and a few skeptical geologists, some of whom had helped advance the view, widely held at the time, that oil simply did not exist in such quantities. This was no negligible point. Although oil had been known about for thousands of years and produced commercially since the 1850s, the world oil business in 1901 was comparatively small and centered mainly on the refining of oil into kerosene fuel for lamps. Cheaper, cleaner, and safer than all other lamp fuels, kerosene had been a godsend to a rapidly industrializing world desperate to light its homes, libraries, factories, and office buildings. But now, with the advent of the newfangled *electric* light, oil's future in the illumination market looked dim. True, some scientists believed oil would work as an engine fuel, like coal — only better, for oil burned cleaner than coal and had a higher energy content. At that point, however, world oil production was only a trickle, and geologists said oil could never be produced in great enough volume to compete with King Coal. The supply of oil, in other words, would never meet the world's demand for energy.

Now, however, as Spindletop poured forth its dark river, the skeptics felt their theories eroding. A few diehards pronounced the flow of oil too large to be sustainable: Spindletop, they said, was a geological fluke, soon to be depleted, never to be repeated. But in March the Hammils drilled a second well, which also produced at the same unearthly rate of one hundred thousand barrels a day, as did a third, a fourth, and a fifth well. The skeptics withdrew, and the speculators and opportunists and investors rushed in. The age of oil had begun.

Until Spindletop, oil had been regarded as something of a sideshow in the energy economy. The ancients had come across it mainly by accident, in natural oil springs, or "seeps," and had used it sparingly, in caulking, glue, and liniment, as well as weapons. (Flaming arrows were popular in warfare, as was Greek fire, a liquid incendiary that could be flung, via catapult, at enemy ships and armies — the world's first weapon of mass destruction.) First-century Persians learned to distill oil into lamp fuel that was prized throughout the Middle East and Europe, but oil remained scarce. Even during the mid-1800s, after the discovery of oil fields on the Caspian Sea near Baku and in Pennsylvania, production was still small — partly because the early oil barons did not drill deep enough, but mainly because they had no clue what oil was or where it came from.

Oil, like coal, is an ancient substance. The crude that gushed from Spindletop on that January morning was the product of a process begun fifty million years before, when Beaumont and much of eastern Texas lay submerged beneath a much wider Gulf of Mexico. The warm waters were ideal for great, state-sized blooms of plankton and other microscopic life forms, whose tiny bodies rained down like a rich dust to form an organic mat on the muddy gulf floor. Over millions of years, this mat hardened into a layer of nutrient-rich rock — geologists call it source rock — which was slowly buried beneath megatons of sandy sediment that poured out from the mouths of nearby rivers. The sandy sediments gradually compacted, in turn, into a layer of sandstone five miles thick. The weight of so much stone atop the source rock, coupled with the naturally high subterranean temperatures, pressure-cooked all those tiny fossilized bodies and chemically transformed the biological molecules made of hydrogen and carbon into a complex hydrocarbon brew known as petroleum.

The creation of petroleum is similar to that of coal, with a key difference: whereas coal derives mainly from dead plants, petroleum's raw ingredient is mainly animal. Animals contain more fat than plants do, and fat contains more hydrogen; this extra hydrogen yields a hydrocarbon that is far more fluid than coal. In fact, what we call petroleum is actually a blend of hydrocarbons: liquids like kerosene and gasoline and semisolids like asphalt (which we call collectively crude oil) mixed up with gaseous hydrocarbons, such as propane, butane, and methane (or natural gas), whose presence, in the form of billions of tiny bubbles, makes the petroleum even more fluid.

This fluidity means that, whereas coal is content to sit underground until the end of time, petroleum does everything it can to escape the underworld. From the moment petroleum is pressure-cooked into existence, the trapped bubbles of gas expand violently, shattering the source rock. Then, because it is lighter than the surrounding groundwater, the gas-oil mix begins migrating upward, pushing through microscopic pores in the sandstone, rising through any cracks or fissures it comes to, spreading toward the surface like a blot of ink through a giant sponge.

Often, the oil and gas (which separate from each other as they rise) reach the surface and simply leak away, as uncountable trillions of barrels of oil and natural gas have done over the millennia.[13] In some cases, however, the migrating oil and gas encounter some kind of obstacle or "trap," which in the case of Spindletop was a two-hundred-foot-thick layer of dried sea salt, left behind when the waters of the ancient gulf receded. Salt, though impermeable by oil, has only temporary powers of containment. Under the enormous subterranean pressures, the salt layer buckled and folded, until finally a massive finger of salt, miles tall and perhaps half a mile wide, was extruded upward, smashing through the sandstone sediments above it like an enormous battering ram — and carrying a great volume of the trapped oil and gas in its wake. As the salt column neared the surface, it pushed up a mound of topsoil, known as a salt dome, which white settlers would later dub Spindletop. The tagalong oil and gas, meanwhile, tens of billions of barrels' worth, came to a halt beneath a superhard layer of limestone, one thousand feet from the surface. Here, in the pores of the sandstone, a petroleum reservoir formed: a layering of hydrocarbons with groundwater at the bottom, oil in the middle, and on top a cap of gas, trapped, pressing up against the limestone, and thus pressurizing the entire reservoir like a can of soda that has been shaken but not yet opened.

A thousand feet is far deeper than anyone had ever drilled for oil before. The wells in Baku and Pennsylvania, for example, rarely went down more than a few hundred feet, because oilmen were stuck with an obsolete drill technology that essentially pounded a sharpened bit down through the dirt and rock like a jackhammer. But at Spindletop, the Hammils tried something new — a rotary drill. Powered by a small steam engine (coal-fired, no doubt), the rotary drill not only could go deeper into the earth but could pierce the kind of hard limestone rock that had guarded Spindletop's dark treasure for so many million years — and which, geologists now suspected, might be guarding similar oil fields elsewhere in the world.

⊚§§~

Just as the Newcomen engine had helped ignite the Industrial Revolution by making coal cheap and abundant, the rotary drill and the new science of oil geology now made it possible at last to satisfy years of latent oil demand. Spindletop and the subsequent discoveries of even larger fields in Texas, Oklahoma, Mexico, and Venezuela unleashed tens of millions of barrels of oil, flooding the market and giving the nascent industry the boost it needed in order to break into the energy economy. As oil prices fell, coal users began switching in droves to the more efficient oil. Railroads converted their coal-fired locomotives to burn cheap Texas crude. Shipping companies, quickly recognizing that oil made their ships go faster — and also that it took up less storage room onboard than coal did — refitted cargo vessels to run on oil.

It was the gasoline-powered internal-combustion engine, however, that sealed oil's dominance. Although early automakers had tried steam engines and electric motors, by the time Henry Ford introduced his Model A in 1903, the gasoline engine had demonstrated its greater power and range. By 1913, more than a million cars and trucks were racing across America and Europe, and most of them ran on gasoline or diesel.[14] With the advent of automobiles, oil gained a virtual monopoly. Whereas preceding generations had been able to choose between coal, oil, and even wood for their transportation fuel, by the age of the automobile, the choice had been made: the internal-combustion engine ran on oil-based fuels. If people wanted to drive, they had no alternative: oil was it.

As clichéd as it has become to say that the oil-fueled engine utterly remade modern life, the transformation was undeniably profound. The popularity of the automobile made possible a host of new lifestyles and social forms, including commuting, suburban living, geographically dispersed families, and, of course, the motor holiday. A larger and more important transformation was occurring in the commercial sphere. Not only did the transportation industry itself now represent a huge chunk of the national economy in America and Europe (the U.S. auto industry alone would one day account for nearly 5 percent of the U.S. gross domestic product, or GDP[15]), but, more significantly, the new modes of oil-fueled transportation — the ships, trains, aircraft, and especially cars, trucks, and buses — were themselves essential to the new global economy. With cheaper, more reliable transportation, companies could move farther and faster, reaching more customers, delivering more products, exploiting more markets, and responding to competitors far more quickly than before; with oil, in other words, companies could succeed in an economy that favored speed, flexibility, and above all unceasing growth. Even more than had been true of coal, oil was essential to economic success. Between 1895 and 1915, per capita energy consumption in America and other industrialized countries nearly doubled, and much of that growth was in oil.

As with coal in the 1700s, the reciprocal mechanism of supply and demand took root in the oil economy. Greater supply fostered new uses for oil, which in turn spurred even greater demand — and forced industry to reinvent itself. Oil companies that had previously focused on making and selling lamp fuel now had to grow larger and more sophisticated to supply a world economy that increasingly fueled itself with oil. Companies such as Standard Oil, Royal Dutch–Shell, and British Petroleum scrambled to erect a new system of oil wells, pipelines, tankers, and storage depots.[16] Drilling technologies improved. Exploration teams learned to "look" for oil deep underground with a technology called seismology. Companies became adept at refining oil, thanks to processes that could efficiently separate the various "fractions" in the crude — gasoline and kerosene, as well as the heavier asphalts and heating oils — for sale to newly segmented markets.

As worldwide oil demand rose — from a mere 500,000 barrels a day in 1900 to 1.25 million barrels a day in 1915 to 4 million by 1929 — oil companies looked farther afield for new supplies. British Petroleum wangled a deal with the shah of Persia to exploit huge deposits in what is now southern Iran, while Royal Dutch–Shell found even larger fields in neighboring Iraq. (Saudi Arabia, paradoxically, was dismissed by geologists as a poor oil prospect.) Meanwhile, huge discoveries in Texas, Louisiana, Oklahoma, and California ensured that the United States remained the world's dominant producer.

Inevitably, so much oil on the market caused problems. New fields would be exploited at top volume, subjecting the market to devastating gluts and driving down prices — only to evaporate when the reservoirs ran dry. Refineries were either starved for supplies of crude or were drowning in it. Oil prices whipsawed violently. Many refiners and producers tried to protect themselves from the volatility with "futures" contracts,[17] but the larger problem was simply that the two ends of the business — the "upstream" of oil-field production and the "downstream" of refining and marketing, were woefully out of balance.

The chaos and uncertainty of the oil boom gave rise to a new corporate model that strove to bring the upstream and the downstream back into sync. The pioneer of this new corporate ideal was John D. Rockefeller, whose Standard Oil would become the largest oil company in the world and the template for the modern energy giant. It was Rockefeller who envisioned a *vertically integrated* industry, in which upstream and downstream meshed in perfect harmony. Whereas other oil companies limited themselves to a piece of the business — production or refining or marketing — Rockefeller and his descendants wanted to control the *entire* oil "stream," from oil well to gasoline pump. Thus, Standard acquired not just oil fields, but tankers and pipelines, refineries and filling stations.

A man well ahead of his time, Rockefeller grasped the importance of technology and was constantly searching out ways to increase productivity while cutting costs. Above all, he perfected the now-standard strategy of being the lowest-cost producer, making his profits through ever-larger sales volumes, while mercilessly, and often illegally, undercutting his competitors. In market after market, Standard would set up a front company, slash prices so low that most competitors were driven into bankruptcy, then demand that any surviving refiners sell out to Standard. "If you refuse to sell," Rockefeller once explained to a defiant refiner, "it will end with your being crushed."[18] At one point, Standard controlled 90 percent of the U.S. market and much of the international market as well.

Ultimately, Rockefeller's great success ran afoul of U.S. antimonopoly laws: in 1914, Standard was forcibly broken into dozens of smaller companies. In a sense, however, Rockefeller's legacy never died. Most of Standard's corporate shards have since been reconstituted into the handful of giants that now control a large chunk of the international oil business; in fact, two Standard spinoffs, Exxon and Mobil, recently merged to form the largest oil company in the world. More to the point, the business model Rockefeller pioneered — that of the giant multinational corporation, capable of operating in any market or sector, but dependent for its profits on ever-greater oil production — remains the standard in the energy business.

Even before the fall of Standard Oil, it had become clear that the oil business was more than a business. Although the entire world production was controlled by a small number of private oil corporations, the sense among governments was that oil was too important to be left in private hands — or even trusted to the laws of supply and demand. Even more than coal before it, oil had become so central to the economic well-being of nations that its value went beyond economics: oil was a political commodity, subject not simply to the laws of supply and demand, but to the national agendas. In 1908, less than seven years after Spindletop, Britain took the bold step of converting its entire navy from coal- to oil-powered ships. The intent was to gain an advantage over the coal-fired navy of Germany, then girding for the first of two world wars. But the move was a huge gamble: Britain had plenty of coal but not a drop of oil domestically. By switching to oil, the English were making themselves dependent on a resource that was by definition undependable. "Security of supply" was no longer guaranteed. Britain would now need to protect access to Middle Eastern oil supplies, which meant keeping a navy in the Mediterranean (much as the United States keeps the Fifth Fleet there today). Henceforth, national security would be tied to the ability to maintain access to foreign oil.

In a remarkably short time, oil had moved to the very epicenter of geopolitics. Just as nineteenth-century imperial powers had competed for the colonies with the best sugar and tea and slaves, the industrial powers of the twentieth century maneuvered for the choicest oil regions. Driven by the ravenous demand for oil, Western governments and their able assistants, the international oil companies, vied for control over the hapless oil states of Venezuela, Mexico, Sumatra, Borneo, and especially the Middle East, where European and U.S. diplomats redrew the map to maximize access to oil. As one French diplomat declared during a period of particularly frenzied boundary drawing, "He who owns the oil will own the world."[19]

Not every oil colony appreciated these new masters. Western "oil imperialism" — by which we mean the collaborative effort between industrial governments and international oil companies to control the oil resources of various less advanced countries — was igniting political fires around the globe that would smolder for decades. In 1938, a resentful Mexico went so far as to kick out Shell, Standard, and other Western oil companies and nationalize their assets. Oil executives, rightly afraid that this "socialist" infection would spread to other oil colonies, lobbied Washington to intervene militarily and make an example of Mexico. But Washington had other fish to fry. In a move that presaged its modern-day appeasement of oil sheiks, Washington refrained from scolding the Mexicans for fear that Mexico might ally itself — and, more important yet, its oil — with Japan and Germany, then well along the path to another world war.

Inevitably, as oil became inseparably tied to diplomacy, it became inseparably linked with war as well. Not only did industrialized nations need oil to wage war (the modern army was now a "mechanized" force, with tanks, ships, and planes), but countries increasingly went to war *for* oil. This was especially true of the Second World War. Lacking domestic oil fields to fuel their industrial and military ambitions, both Nazi Germany and Imperial Japan faced a stark choice: curb those ambitions, or find oil elsewhere. Both chose the latter. In Germany, Adolph Hitler knew his only hope of victory lay in taking the oil fields of the Middle East and Russia (despite a pledge of loyalty to Stalin). In Tokyo, meanwhile, Hirohito's vision of an Asian empire depended heavily on gaining control of the oil-rich East Indies. In fact, when the Japanese bombed Pearl Harbor in December 1941, a primary objective was to sink any U.S. warships that might otherwise have prevented Japanese tankers from reaching Indonesia.

Oil soon became the main war supply, as critical as munitions or labor supply. France and Britain quickly exhausted their own oil supplies and, as they had during the First World War, turned to the United States for help. The United States responded in typical Yankee fashion, opening the taps of the huge fields in Texas and Oklahoma and making sure Allied armies were never without fuel. Desperate to stem this flow of oil, Germany dispatched its deadly U-boats to torpedo U.S. oil tankers as they delivered Texas crude to the eastern seaboard. In the first five months of war, German subs sank fifty-five tankers and littered American beaches with oil slicks and dead sailors.[20] Nevertheless, the tide of American oil was unstoppable. Just as England had dominated the energy order in the age of coal, the United

States, the energy superpower of the twentieth century, was feeding not only its own enormous appetite, but the world's as well.

Japan and Germany, meanwhile, were not so lucky. Hitler's desperate lunge for the Russian oil fields ended with a catastrophic defeat at Stalingrad. In the Pacific, Japanese oil tankers became sitting ducks for U.S. warships. By the time an American B-29 dropped an atomic bomb on Hiroshima in August 1945, Japan's air force was completely out of fuel. The war was over, and though many factors had contributed, once again, the winners had been those best able to keep the oil flowing. "The Allies had floated to victory on a wave of oil," declared a British official at the end of the First World War. Twenty years later, his assessment was even truer.

With the end of the Second World War, any question about the supremacy of oil in the energy order, or of the role that oil would play in the postwar global economy, had been put to rest. Crucial in wartime, oil was now the linchpin for postwar prosperity, the true currency of geopolitical power. Coal might still produce more total energy, but oil fueled the ships and aircraft, the freight trains and automobiles on which military and commercial dominance were increasingly based.

The oil industry itself reflected this ascendance. Enlarged by the demands of war, the oil sector that emerged was more sophisticated, with more fields, pipelines, tankers and terminals, and refining capacity. Oil company research led to a myriad of new oil-based products, from plastics to synthetic rubber, further contributing to the demand for oil. Between 1945 and 1960, as the war-ravaged economies of Europe and Asia were resurrected, worldwide consumption of oil rose sharply, from six million barrels a day to twenty-one million barrels.[21] And although some production had been nationalized (oil in Mexico and the USSR was controlled by the state), the lion's share of world production was in the hands of a tiny number of companies — Exxon, British Petroleum, Shell, Texaco, Chevron, Gulf, and Mobil — the "majors," along with a few dozen smaller outfits that somewhat defiantly called themselves independents.

Oil was, for all intents and purposes, *the* fuel of the twentieth century. Although coal would retain a huge market share in heating and power generation, it would never have oil's political or economic importance or its star status as the world's first geopolitical commodity: to be a world power, a nation needed either oil or the money to buy it. Countries like Britain, which lacked domestic supplies, recovered only partially from the war. Producers like Mexico, Venezuela, and Russia enjoyed increasing power in the world economy, while Saudi Arabia, now understood to possess the largest oil reserves of the world, was no longer dismissed as a nation of Bedouin princes and camel drivers.

At the top of this new energy order stood the United States. By 1960, it was producing seven million barrels a day — one of every three barrels pumped. Just as important, the United States, and U.S. companies, enjoyed increasing influence in oil-rich regions elsewhere in the world, most notably, Saudi Arabia, to which Washington had tacitly agreed to offer military protection in exchange for drilling rights for U.S. companies.

Where the United States truly dominated the world of energy, however, was in consumption. By 1955, the country was using more than a third of all energy produced in the world. Per capita consumption was six times as high as any other nation's. We were using that energy to produce more goods and wealth, to be sure, but we were also simply using more energy, to heat our homes, cool our offices, and, above all, drive our cars. In the decade after the war, the number of passenger cars in America nearly doubled, from twenty-five million to forty-eight million,[22] and gasoline consumption doubled as well.[23] The age of the automobile was in full swing. Cities like Los Angeles became famous for their car culture, highways, and traffic jams, as well as for the sprawling suburbs and bedroom communities that the automobile culture encouraged.

Yet within this rosy picture of robust energy preeminence, serious problems were emerging. The oil economy, and American dominance in it, had always been predicated on ready supply and on the ability to meet the ceaselessly rising demand simply by pumping more oil or going out and finding more fields. Yet now this paradigm was failing. "Security of supply" was no longer certain. By 1946, America was consuming more oil than it could produce domestically, and for the first time in its history, it became a net oil importer.[24] The ramifications were enormous. After fueling the world through two wars, the United States, as one historian noted, "had actually become an importing nation whose East Coast would freeze in winter were it not for the liquid warmth of Venezuela and Arabia."[25] Americans would now understand firsthand the anxiety and insecurity that had long afflicted Britain, Europe, and Japan. America would now become that great twentieth-century paradox — an economic and military giant whose life-blood was controlled in other parts of the world.

And as if to emphasize the precariousness of the new circumstances, "foreign" oil suddenly seemed far less dependable. The anger that had led Mexico to nationalize in 1938 had indeed spread to other oil colonies. As oil's importance swelled, and as it became clear that oil held the key to future power and wealth, foreign producers began to demand a larger share of both. Venezuela raised the price for its oil and began making diplomatic overtures to its oil allies in the Middle East. A far more serious consequence was that Arab nations, enraged by the creation of Israel in 1948, threatened to embargo oil to the United States or any other nation that supported Israel.

Three years later, in yet another sign of things to come, Iran nationalized its oil industry, throwing out the English and American majors. Other oil-rich countries followed suit, and by 1961 they had formed the world's first oil cartel, the Organization of Petroleum Exporting Countries (OPEC). All at once, it seemed that the world oil map was also the map of political instability — nowhere more so than in the Middle East, which was now understood to possess well over half of all the world's oil. In a few short years, a global industry that had largely been controlled by a handful of international oil companies was now mostly in the hands of a new kind of oil entity, the petrostate, as Saudi Arabia, Venezuela, and other oil-rich nations were now called. In a stark reversal of fortune, the majors found themselves fighting for the scraps of world production — the so-called non-OPEC oil — and increasingly that meant looking for oil in ever more remote, ever more challenging places.

It was not simply the business and politics of oil that had become risky. Like coal before it, oil had begun to display its downsides. The production and refining of oil contaminated rivers and lakes, while the exhaust from millions of cars and trucks was creating serious air pollution problems. During the war, Los Angeles had suffered its first "smog" alerts, and by the 1960s, smog was being blamed for poor visibility, health problems, and property damage, even forcing some residents to leave the city.[26] Mexico City, London, and Tokyo reported similar problems.

There was another complication as well. In 1970, U.S. oil production hit its peak. The flow from the big U.S. fields began to taper off, and the number of barrels that the majors could bring out began to fall. Imports, already a necessity in the U.S. energy economy, suddenly surged. As the reality of America's energy dependence set in, many government and industry officials began to wonder whether a similar trend might not also affect the world supply of oil. Was it possible, despite the oceans of oil then on the market, that production might also peak and decline worldwide? Almost overnight, oil had changed from a factor in economic success to a source of economic and political vulnerability. The age of oil, it now was clear, would be just as susceptible to anxiety as any that had gone before.

2

The Last of the Easy Oil

Two miles off the coast of Azerbaijan, on a windswept strip of land called Sand Island, the former glories of the Soviet oil empire rust away beneath a relentless Eurasian sun. Twenty years ago, this three-hundred-acre island was the toast of the Soviet oil industry, with row after row of gushing wells and thick pipelines crossing the water to refineries in Baku. Then oil production hit its natural peak, the flow subsided, and Sand Island fell into the kind of profound industrial decay that Hollywood spends millions trying to replicate. Rusting pipelines line the roads. Empty buildings, some still sporting the red Soviet star, lean at odd angles. Old barrels, bits of broken machinery, and permanently parked trucks litter the grounds, while just offshore a line of gigantic rust-colored oil derricks, most of them abandoned, marches away toward the horizon. "No pollution," insists Sahib Siradjev, my translator from Azerbaijan's State Oil Company, for maybe the tenth time since we drove onto the island. "You can fish here."

Inside an old administrative building, we are met by Hüseynov Vaqif, general manager of Sand Island — a big wedge of a man with a beefy face, beautifully coiffed silver hair, and a reputation in Caspian oil circles as something of a star. When he was brought to Sand Island in 1996, the easy oil was long since gone. The entire operation was producing barely 1,500 barrels a day — hardly enough to pay the salaries of its 1,600 employees. Vaqif swung into action. In booming, rapid-fire Russian, he tells me how he retrofitted old wells and sank dozens of new ones, some as deep as two miles, eventually increasing the flow on Sand Island by a factor of nearly three — and all this despite limited resources and a fraction of the technology Western oil companies take for granted. "Put me anywhere, and I can get you the last drop of oil," boasts Vaqif, standing beneath a huge, hero-of-the-people portrait of Heydar Aliyev, the former Azeri president. "Even if they put me on the moon."

Vaqif, unlike his counterparts at Western oil companies, actually seems pleased to have a media person on the island. Untroubled by the dilapidation around him, he barks out orders that I be driven around the facility, then hosts me at a sumptuous luncheon of borscht, sturgeon, and vodka. Later, over tiny glasses of sweet Azeri tea, Vaqif presents me with autographed copies of an engineering manual he himself penned and gently lectures me on the superiority of Soviet-trained oil engineers. Western oilmen are "too specialized," says Vaqif, pausing briefly to shout into his desk intercom at some distant underling. In the former Soviet Union, Vaqif continues, oilmen were trained to rely on their instincts. "You must work with an oil well as you would with a lady," he tells me. "That way, she won't refuse you."

As Vaqif walks us back to the car, bouncing along in jaunty good humor, it seems the wrong moment to point out that his "ladies" have in fact been refusing him ever since he arrived on Sand Island, and fairly assertively. Although production has indeed nearly tripled here, current output is still barely a sixth of what it was during the time of peak production, in 1986. At this rate, the flow of oil will slow to a trickle within a few years, and Sand Island will permanently enter the ranks of the abandoned fields that now surround the city of Baku, the fading former capital of the old Soviet oil empire.

Sand Island is little different from thousands of other former boomtowns in Texas, in Pennsylvania, on Borneo, and elsewhere — richly endowed oil frontiers where the industry came in, erected an enormous and expensive infrastructure, and then, when most of the oil was gone, packed up and moved on to the next big strike. At one level, Sand Island will hardly be missed. Historically, oil companies have been so adept at finding new oil fields that the loss of a single operation is a microscopic blip in global oil production, which has surged relentlessly from half a million barrels a day in 1900 to around eighty-two million barrels today.

Yet as the unfortunate Vaqif knows quite well, what goes up must come down. Oil is a finite substance, and at some point, just as Sand Island's volumes have fallen off, all the oil being discovered around the world will no longer replace the oil that has been produced, and global production will peak. Oil companies and oil states will find it harder and harder to maintain current production levels, much less keep up with rising consumption. Demand will again outstrip supply, and prices will rise.

Worse, although the term "peak" suggests a neat curve with production rising slowly to the halfway point, then tapering off gradually to zero, in the real world, the landing will not be soft. As we approach the peak in production, soaring prices — seventy, eighty, even a hundred dollars a barrel — will encourage oil companies and oil states to scour the planet for oil. For a time, they will succeed, finding enough to keep production flat, stretching out the peak into a kind of plateau and perhaps temporarily easing fears. But in truth, this manic, postpeak production will simply deplete remaining reserves all the more quickly, thereby ensuring that the eventual decline is far steeper and far more sudden. As one U.S. geologist put it, "the edge of a plateau looks a lot like a cliff."[1]

In short, oil depletion is arguably the most serious crisis ever to face industrial society. And yet, according to Colin Campbell, a former Amoco oil geologist and currently the éminence grise of the so-called oil pessimists, "governments remain pathetically ill informed and unprepared."[2] For years, the official line of the big importing nations, the big exporting countries, and the big international oil companies, with few exceptions, has resembled that of an annoyed parent dealing with an overly curious child. Yes, yes, yes, we're told, in tones of exasperation and condescension, oil production *will* peak — eventually; but that isn't something humanity needs to worry its pretty little head about anytime soon. Not only are the known reserves of oil enormous, we're told, but oil scientists, engineers, and other clever types are getting better at finding *new* oil in unexpected places — in the North Sea (in the 1960s), for instance, or off the shore of Angola (in the 1990s).

Factor in the indescribably vast reserves of so-called unconventional oil — whether in the form of the molasseslike "heavy oil" in Venezuela, for instance, or the oil-bearing tar sands in Alberta — plus all the known reserves of natural gas (which can be processed into synthetic gasoline and diesel) — and, say optimists, the world won't reach a peak in production for fifty or sixty or a hundred years. Such a buffer, optimists say, leaves us plenty of time to develop new energy technologies and ensure an orderly transition to a post-hydrocarbon order without having to take rash or costly emergency measures, or even to upset ourselves thinking about it.

To the extent that governments and energy companies even mention long-term oil prospects publicly, it is almost entirely in a political "if only" context: we could have as much oil as we needed *if only* OPEC would stop limiting the supply; or *if only* oil companies were allowed to drill in the Arctic National Wildlife Refuge; or *if only* malcontents in Iraq would stop blowing up their own oil pipelines. According to this view, any concerns relating to long-term oil can be addressed through legislative, diplomatic, or, on occasion, military means. Long-term oil supply is, in other words, really a question of political will, of deciding how much oil we need and then going and getting it.

In truth, however, as some energy companies and government agencies tacitly acknowledge, the optimists' rosy picture is far from accurate. Though vast quantities of oil still remain in the ground, most is what might be called *theoretical* oil — it may exist, but in highly uncertain and even problematic environments: deep below the Arctic ice, for example, or in small African regimes wracked by civil war. Or, most important, inside the oil fortress known as OPEC, whose political machinations will affect long-term supply more powerfully than any geology. Thus, our ability to get at this theoretical oil, and to use it, depends on a myriad of variables — technological, economic, financial, and political — that are, at this point, hard to predict and even harder to control.

In other words, although we will not run out of oil tomorrow, we are nearing the end of what might be called the easy oil. Even in the best of circumstances, the oil that remains will be more costly to find and produce and less dependable than the oil we are using today. This fact means not only higher prices, but more volatile prices, which will make it harder to see how fast oil supplies are being depleted, and harder still to know when we'll need to start looking for something new.

So when do we peak? In theory, the production of oil reaches a peak when half the original supply has been pumped from the ground. This holds true whether you're talking about a single oil well or the collective behavior of all oil wells on the planet: with half the supply gone, it simply gets harder and harder to maintain the same levels of production — the same number of barrels per day — and eventually, production falls.

Presumably, if we know the total volume of oil the world had to begin with, as well as the amount of oil we've already used and the amount we will use in the future (calculated from forecast energy demand), we can predict the arrival at a depletion "midpoint" and thus the production peak; but of course, we don't know the total volume. Although we are reasonably sure how much oil we've used since the dawn of the oil age — around 875 billion barrels — estimates of the amount of oil still in the ground are tremendously suspect, and therein lies the crux of the problem.

Generally, when we ask how much oil is left in the ground, we're talking about two kinds of oil — proven and undiscovered. "Proven" is the term used for oil in fields that have already been discovered but not yet pumped out. Proven reserves are essentially the inventories held by oil companies like ExxonMobil and oil states like Saudi Arabia or Norway. According to the U.S. Geological Survey (USGS), one of the most respected and widely quoted oil agencies in the world and a leader among the so-called oil optimists, the world's proven reserves stand at 1.7 trillion barrels, over half of which are in the Middle East.[3]

"Undiscovered" oil, by contrast, is oil whose existence has not yet been confirmed by the drill but is strongly indicated by various geological markers. Undiscovered oil is the exciting oil — the stuff of romantic stories about hardy John Wayne types, "wildcatters," who risk their lives searching steamy jungles and barren steppes in hopes of striking a gusher (even if oil exploration is essentially automated these days). In theory, undiscovered oil fields are scattered around the world, although certain regions appear favored — among them, Siberia, western Africa, eastern South America, and the Caspian. According to the USGS, undiscovered oil amounts to around 900 billion barrels. Adding proven and undiscovered oil deposits together, we get a total of 2.6 trillion barrels. Assuming that world oil consumption, now 80 million barrels a day, continues to grow at the rate of 2 percent per year, a 2.6 trillion-barrel reserve has us hitting our peak somewhere around 2030 — or even later if world oil consumption slows.[4]

The problem is that both numbers, those for proven and for undiscovered, are doubtful. Estimates of proven reserves, for example, are routinely exaggerated for economic and political gain. The classic case came in the late 1980s, when the six big OPEC producers — Kuwait, the United Arab Emirates, Iran, Iraq, Venezuela, and Saudi Arabia — collectively added more than 300 billion barrels to their stated reserves. The move nearly doubled reserve numbers that had been on the books for years and, in one stroke, "delayed" a peak in world production by nearly a decade. Saudi Arabia alone, owner of the largest oil reserves in the world, raised its estimate from 167 billion barrels to a breathtaking 257 billion barrels, overnight.

Why is this figure suspect? Generally speaking, oil producers revise their reserve estimates in only two situations: when discoveries are made or when some new assessment methodology reveals that they have more (or less) oil in existing reserves than previously stated. But none of the six OPEC countries had announced any significant new discoveries during the 1980s or 1990s, nor had assessment technologies suddenly improved. The six countries themselves claimed to be correcting for past mistakes: the Western oil companies that founded Middle Eastern oil operations had routinely underreported the size of their reserves.[5] Yet although some correction was in order, it is worth noting that the upward revisions just happened to coincide with a 1985 OPEC edict stipulating that the higher a member's stated reserves, the more oil that country could export and thus the more revenues it could earn. Going country by country, says Campbell, the Amoco geologist-turned-pessimist, it becomes apparent that the revisions were largely bogus. "It is obviously absurd to imagine that Iraq, for example, has increased its reserves fourfold since 1980," says Campbell, "when much of the time it was at war or embargoed."[6]

This is classic "pessimist" rhetoric, and it tends to reinforce the image of the oil pessimists as conspiracy nuts who believe that the energy-industrial complex is trying to conceal the imminence of the peak in oil production. Campbell, in particular, a stout, square-faced Englishman with a serious tone and a penetrating, gloomy stare, has earned the undying enmity of oil executives everywhere by repeatedly declaring their reserve estimates to be bald-faced lies. "If you get to meet with Shell and BP," Campbell warned me once, "I recommend that you admit to no contact with myself, whom they apparently regard as a terrorist."

Yet for all their dark theories and occasional paranoia, oil pessimists are right to challenge the oil numbers being tossed around today, because in many cases those numbers simply don't make sense. Take the estimates for "undiscovered" oil. Many optimists, including the USGS, believe that a huge amount of oil remains to be found — anywhere from 1 trillion to 1.5 trillion barrels. The problem is, few places on earth remain where all that oil could be hiding but where oil companies have not already looked. Oil is not a random geological event, something that can occur just anywhere. It is the product of complex geological processes that take place only in certain quite specific conditions. As we saw in the story of Spindletop, you must first have *source rock* — the deeply buried sediments rich in organic matter. It is also necessary to have a migration pathway — cracks or porous rock through which the newly formed petroleum can escape toward the surface. Finally, a layer of impermeable stone or clay or salt is required, to trap the petroleum and create a reservoir, or field.

This three-part source-reservoir-trap configuration — "a petroleum system," in geologists' terminology — constitutes an underground hydrocarbon machine that generates, transports, and stores oil and gas. Petroleum systems exist all over the world and comprise anything from smallish entities producing just a few hundred barrels a day to the four massive systems in the Middle East that together account for half of the world's known oil reserves. Yet for all their variety, all petroleum systems operate according to a set of rigid natural rules. The source rocks, for example, must contain enough organic material to generate usable volumes of oil and gas. The migration rock must be sufficiently permeable, or the oil won't flow freely through it. The cap rock must be sufficiently impermeable, or the oil will simply leak away.

Above all, the timing must be perfect. To become oil, the organic material in the source rock must be heated to a certain temperature for a certain period of time. Typically, this happens when the source rock gets buried and, over millions of years, is pushed downward, into what oil scientists call the "kitchen" — a geological zone between ten thousand and thirteen thousand feet below sea level where temperatures are high enough (100 to 135°C) to boil organic matter into petroleum. Petroleum forms only in the kitchen. Source rock that isn't pushed low enough will not be cooked, whereas source rock that is pushed too far, past the kitchen, becomes too hot, and the petroleum is either "cracked" into gas or simply destroyed. There is no halfway: the conditions for oil either exist or they don't. Nor are there any guarantees: even if a system meets all these criteria, it may still reveal itself to be empty when oil companies pierce it with their drills. Many older petroleum systems have produced oceans of oil in the distant past, only to have it leak away millions of years before humans even knew what oil was.

Oil, in other words, is a relatively rare phenomenon, produced only in certain geological spaces, under certain conditions, and within a shallow zone just below the surface of the earth. Worldwide, there exist approximately six hundred petroleum systems capable of producing commercial volumes of oil and gas. Of these, approximately four hundred have been explored. The remainder lie in places like the Arctic or in deep offshore waters — remote, hard-to-reach areas that oil companies have turned to only after exploiting the more accessible oil.

This state of affairs helps explain why oil exploration has become so much more difficult in recent decades. Not only are the remaining "undiscovered" systems harder to reach, but they are likely to be smaller: historically, larger systems, being easier to find than smaller ones, have tended to be discovered first. What is more, oil companies prefer to develop the large discoveries first and put off exploring the smaller, less profitable fields until later. "In any region, the large fields are the biggest targets and are usually discovered first," says petroleum geologist Joseph Riva, a former oil analyst with the U.S. Congressional Research Service (CRS). "As exploration progresses, the average size of the fields discovered decreases, as does the amount of oil found per unit of exploratory drilling."[7] Or in plain English, remaining undiscovered fields not only will be smaller but are likely to yield ever-smaller volumes of petroleum.[8]

In fact, when one charts the average volume of oil that has been discovered each year since the beginning of the century, it becomes clear that new oil is indeed getting harder to find. Year by year, the volume of newly discovered oil — that is, the number of barrels found each year and recorded in the books as known or discovered reserves — climbs steadily upward from 1860 until around 1961, when it peaks. Since then, oil companies have found, on average, a little less oil each year — with the exception of a small blip in the late 1990s, as big finds were announced in the Caspian, off the shore of West Africa, and in the Gulf of Mexico. In fact, since 1995, the world has used at least 24 billion barrels of oil a year but has found, on average, just 9.6 billion barrels of new oil annually. According to a study by Wood Mackenzie Consultants, industry is finding less than 40 percent of the new oil it needs to keep the base of known reserves from shrinking.

Barring some fairly spectacular disruption to historical patterns, there is little reason to expect anything to alter the downward trajectory of discovery. "We've been drilling holes all over the world since the early 1900s," says Les Magoon, a geologist with the USGS who has mapped world petroleum fields for three decades and does not share his employer's optimism. "Statistically, it's unlikely that there is all this 'hidden resource,' waiting to be found; [it] is pretty hard to support scientifically."[9]

Indeed, according to pessimists, when we use these more realistic forecasts of future oil discoveries, our estimates for the world's total remaining oil — proven and undiscovered — drops to a trillion barrels (not 2.6 trillion, as the USGS claims) and puts the peak at around 2010. That doesn't leave us a lot of time — certainly not enough time to prepare for the kind of consequences that a peak is expected to unleash. Even if we assume that the peak would actually be a plateau, with the "cliff" pushed out till, say, 2016, the deadline is still fairly imminent, given the size and value of the oil-based infrastructure — the tankers, the pipelines, the refineries, 747s, Greyhound buses, and, above all, cars — that would need to be upgraded or replaced outright. "The point to remember about production isn't that it peaks, but that it declines rapidly afterward, at a time when the world demand would be moving rapidly in the opposite direction," Joe Romm, former acting U.S. assistant energy secretary for the Clinton administration, told me last year. Once a decline begins, Romm says, "there is very little time for the U.S. to react."[10]

On the road back from Sand Island, crammed into the backseat of the oil ministry's tiny Lada sedan, Sahib has finally stopped reminding me about the cleanliness of the water. We are bouncing along through yet another dreary petro-landscape — treeless brown hills, rusty pipelines, oily lakes, and mile upon mile of oil derricks — interrupted periodically by an olive grove or a flock of dust-covered children. As we approach Baku, the hills sink lower, sprouting apartment blocks and refugee shantytowns, before giving way to a broad plain of oil refineries, factories, and soot known as the Black City — the heart of the old Soviet oil empire. We pass a shipyard, where a massive oil-drilling rig lies on its side, its bright new paint in stark contrast to the surrounding decrepitude. In the seat next to me, Sahib seems to revive. He taps on the window and tells me how the platform will soon be taken by barge out to the new fields in the Azeri sector of the Caspian Sea, "where the really big oil is."

In the 1990s, in what was dubbed the deal of the century, a consortium of Western oil companies paid Azerbaijan eight billion dollars for the right to look for oil in the seabed a few miles off Baku, far from Sand Island and Azerbaijan's other antiquated oil fields. Soviet engineers had long suspected the presence of huge oil reserves beneath the deep Caspian waters yet lacked the technology to prove the oil existed, to say nothing of actually pumping it out of the ground. Within a few months of signing the deal, Western operators, armed with the latest seismic technology and deep-water drills, struck pay dirt at Chirag-Azeri-Gunshali, a supergigantic formation initially believed to hold some 3 billion barrels of oil — enough to earn Azerbaijan anywhere from eighteen billion to thirty billion dollars a year in oil revenues, depending on oil prices.

The promise of such oil wealth has become a national obsession in Azerbaijan. In speeches and ads and on billboards and huge banners, Azeri politicians lose few opportunities to remind voters how bright their future is. Sahib, who worked on some of the first oil company negotiations, recalls them proudly — not least because they involved lavish trips to the United States. "I have been to Houston many times," he tells me, with the nonchalance of a world traveler. "I have stayed at the Hilton, the Hyatt. I also went to New Orleans. What a town. '*Show us your teets!*'"

Many in the oil business share Sahib's enthusiasm. Western analysts described the Azeri deal, the first of many between Western oil companies and various Caspian governments, as a geopolitical win-win: a desperate former Soviet republic gets oil revenues; the industrialized world gets an alternative to Middle Eastern OPEC oil. Thus far, Azerbaijan has been slow to build up its offshore operations, and the massive Chirag-Azeri-Gunshali and other fields have generated only a modest flow of oil — less than a million barrels a day. But officials with the State Oil Company of the Azerbaijan Republic, or SOCAR, insist that production is growing and that the "big oil" is just around the corner. By 2010, I am told, Azeri production will reach 2 million barrels a day. And in the meantime, estimates of Chirag-Azeri-Gunshali's total size are updated almost monthly, as new test wells are sunk and operators find new oil: as recently as 2003, Azeri officials claimed the field contained at least 4.7 billion barrels of oil.

To many in the oil industry, stories like this help justify the optimistic scenarios for future oil discoveries — and go a long way toward dispelling the more pessimistic predictions about the end of oil that have lingered since the oil shocks of the 1970s. Thirty years ago, the world truly did seem to be running out of oil. In 1971, U.S. production had peaked. After serving as the world's oil pump for nearly a century, America could no longer cover global oil shortages, much less meet its own domestic needs, simply by opening the taps in Texas and Oklahoma. Henceforth, those taps would remain wide open pretty much twenty-four hours a day, and still it would not yield enough. America's reign as the dominant oil power was over, and newer oil producers were quick to take advantage of it. In 1973, when Arab oil states like Saudi Arabia and Iraq cut off oil shipments to America, the United States watched helplessly as oil prices tripled and the world economy plunged into deep recession.

Fearing that a similar peak might be imminent for world oil, energy analysts scrambled to assess global oil supplies. The early forecasts were not encouraging. By most estimates, the world's proven reserves stood at around 1.3 trillion barrels, which, at the then-current rate of consumption, would not last very long. Esso (later Exxon, then ExxonMobil) predicted a peak in 2000, as did Britain's Department of Energy. Royal Dutch–Shell said production would plateau by 2005.[11]

By the time revolutionaries shut down Iran's oil fields in 1979, sending oil prices to their highest level in history, oil pessimism had become the reigning paradigm in Western political and economic circles and a fixture of popular opinion. Convinced that oil depletion was imminent, Western governments and consumers embraced energy conservation with patriotic fervor, while environmentalists and energy activists welcomed the opportunity for a new, cleaner energy order. Meanwhile, an army of experts, many of them former oil company geologists, devoted themselves to calculating the date of the peak and creating highly detailed and gruesome postpeak scenarios, most involving worldwide recession, political chaos, and the military conquest of the Middle East by desperate industrialized nations.

But as in times past, depletion anxiety was quickly replaced by a surge of oil optimism. In 1975, spurred on by the high prices caused by the Arab oil embargo, oil companies began producing enormous volumes of oil from the North Sea, a deep-sea frontier previously dismissed as too technically challenging to develop economically. Two years later, huge volumes began to flow from extensive fields on Alaska's equally inhospitable North Slope.

Optimists say that these successes and the many more since highlight a major flaw in the pessimists' theory: namely, their failure to credit the oil industry for becoming much cleverer since the gloomy 1970s. Barred from access to "easy" Middle Eastern oil, oil companies were forced to reinvent how they looked for and produced oil, and the results have been astonishing. Drills today can now reach ten miles underground, move in any direction — even horizontally — and electronically detect oil and gas. Operators employ powerful supercomputers to create stunning three-dimensional seismic images of underground structures, showing precisely where oil- and gas-bearing rocks are and even identifying the best routes for drilling.

For the industry, this explosion of technological advances has had three major effects. First, companies can now work in nearly any climate or environment, from permanently frozen tundra to a floating platform anchored two miles above the ocean floor — places previously dismissed as technically or economically impractical, like the Caspian or even frigid Siberia, which is widely regarded as the "next" oil frontier. Thus, each year oil that was regarded as unreachable — or "unconventional" — becomes conventional. For example, new production technologies are even allowing oil companies to produce previously unusable oil, such as the molasseslike "heavy" oil of Venezuela and the massive reserves of tar sands in Alberta, Canada; indeed, the government of Alberta now claims to have "reserves" equivalent to more than a *trillion* barrels of oil.

Second, companies have dramatically increased the amount of oil they get from a given field. As recently as the 1970s, drillers were lucky to extract 30 percent of the oil from a field, while effectively leaving 70 percent in the ground as "unrecoverable."[12] Even today, in less-developed oil regions, like Saudi Arabia, recovery rates are said to average just 25 percent. But with new mapping and drilling technology, operators can see where the remaining oil lies within a reservoir, and then drop in a precisely targeted new well to reach it. Such techniques have raised recovery rates to as high as 80 percent — a success that not only has boosted yields at new fields but is allowing companies to revive declining and even abandoned fields.

Worldwide, according to the USGS, enhanced recovery technologies will add another seven hundred *billion* barrels of oil to the world's tally of remaining oil — and delay by years the peak in production. Dan Butler, an analyst at the Energy Information Administration, the very optimistic forecasting arm of the U.S. Energy Department, says some of the biggest potential for improving recovery is in the Middle East. "The Saudis have very primitive operations," says Butler. "They just let the oil gush out. But if you could get another 5 percent out of Saudi Arabia and the rest of the Middle East, you would up your reserve base by at least a hundred billion barrels."

Third, companies are much smarter at knowing where to look for oil. New geological understandings — for example, that oil can form anywhere within dozens of miles of a river delta, even in superdeep waters — have led to a welter of new discoveries in unexpected places, like the deep waters off the coast of West Africa. Deep-water oil is touted as the real frontier of the future and is the place where most oil companies and many analysts expect to find the bulk of the undiscovered oil. Excitement is particularly keen over "deltaic" prospects in the deep-water Gulf of Mexico, off the coast of Africa and Brazil, as well as in the Arctic provinces of Canada and Greenland, Norway, and Siberia, where seismic surveys reveal subterranean structures identical to those beneath the oil-rich North Sea, but far larger. "The Arctic is going to be the next big play," promises Tom Ahlbrandt, the director of the USGS world assessment project and a prominent oil optimist. "We feel that more than half of all undiscovered resources are in the deep offshore, of which half are in the Arctic. And we've looked at only seven Arctic provinces; there are twenty-eight more we need to look at. We haven't even begun to discover all the oil that is out there."[13]

But even the USGS is not the last word in oil optimism. When U.S. policymakers want the most positive energy forecast, they turn to the U.S. Energy Information Administration (EIA). Whereas USGS forecasts take into account only oil that could be extracted with today's technology and at today's oil prices, the EIA assumes substantial improvements in both — with encouraging results. So, for example, while most optimists believe that the Caspian region might hold 100 billion barrels, EIA numbers show a staggering 292 billion barrels of "ultimately recoverable reserves" in Kazakhstan, Azerbaijan, and other "-stans." The EIA further believes that newly discovered fields off West Africa and South America may, when combined, come close to rivaling those of some Middle Eastern states. "It's probably not a new Saudi Arabia," says EIA's Butler, smiling faintly, "but certainly enough to push the world production peak to 2035."[14]

Butler's comment about a "new Saudi Arabia" bears closer scrutiny. No matter how good we get at finding new oil, the world oil map remains fundamentally unchanged. We may find reserves in Africa, in Siberia, and elsewhere. Sooner or later, though, we must come back to the fact that the lion's share of world oil is in the Middle East, controlled by OPEC, a cartel of unfriendly, unstable regimes that already exercises too much control over world oil prices and will gain even more sway once oil fields outside the OPEC countries have begun running out. Thus, although it will eventually be important to know when *total* world oil production will peak, for now, when governments and oil companies and pessimists ask about a peak, what they really want to know is, When can we expect a peak in *non-OPEC* oil, the free oil, the oil we have a chance at exploiting?

This is where the depletion picture really gets ugly. Clever though we may be at finding new oil, the fact remains that there is simply less of it to be found in the regions outside OPEC control. Yes, exploration technologies have improved dramatically. The supercomputers that companies brought in during the 1980s to help map out new fields and zero in on oil did in fact yield a burst of discoveries. But neither supercomputers nor anything else has been able to halt the long-term decline in new discoveries outside OPEC, where oil producers and international oil companies alike continue to pump out more oil than they can replace through exploration.

Where the non-OPEC world's troubles are most evident is in the decline of the supergigantic oil fields — those massive, multibillion-barrel behemoths that could change a third-world nation into an oil empire but which now rarely come to light. The two largest fields exploited in the last thirty years have been Kazakhstan's Kashagan field, with an estimated fifty-five billion barrels, and Kuwait's Kra al Maru, reportedly of similar size. And while this *is* a lot of oil — enough to keep the world humming for about four years — we should note that it is not non-OPEC oil. Kra al Maru is in Kuwait, which is part of OPEC.[15] Kashagan is in Kazakhstan, which, though technically outside OPEC, was similarly off-limits to Western exploration methods until the early 1990s. (When international oil companies were allowed into Kazakhstan, they found Kashagan in about thirty-six minutes.)

If we want to see the last monstrous non-OPEC fields, we have to go back all the way to South America's Canatrell, discovered in 1976, and Prudhoe Bay in Alaska, found in 1968. Giant fields are still being uncovered outside the OPEC countries, but mainly in the one-billion-to-three-billion-barrel range — again, substantial volumes, in absolute terms, but piddling by comparison with superstars of yore. The trend is clear: in places where international oil companies have been allowed to look — that is, places that OPEC does not control — the industry is finding smaller and smaller fields.

To be sure, this reality tends to get lost amid all the hullabaloo over the "next hot prospect." A decade ago, the deep-water Gulf of Mexico was supposed to be the new El Dorado, although after a string of successes, it has disappointed. British Petroleum's biggest find — the 1.5-billion-barrel Thunderhorse field[16] — barely qualifies as a supergiant, and other companies have been similarly frustrated. ExxonMobil's chairman and CEO Lee Raymond has gone so far as to complain that "the best thing ExxonMobil could have done after it drilled its first well in the Gulf was to never drill another again."[17]

Declining field size is one reason that many of the large oil companies have recently been missing their growth targets and are struggling to "replenish" reserves — that is, to discover a new barrel of oil for each one they produce. Adds analyst Fadel Gheit, "The low hanging fruit has already been picked. There *is* more fruit, but it's harder to pick."[18]

The story is the same whether we're talking about oil companies or entire oil provinces. Despite billions of dollars in investment by the industry, production in oil fields in Alaska, the Western Basin of Canada, and Britain's North Sea — once-prolific regions that provided the oil economy with a bulwark against OPEC — is today in steep decline. In the North Sea, for example, oil companies recently celebrated the discovery of the 1.1-billion-barrel Buzzard field, but it was not enough to keep the United Kingdom's production from peaking in 2002 at 2.3 million barrels a day and falling to 1.8 million barrels a day the next year.[19]

Depletion is rampant. Mexico, the sometime ally of the West and a loyal supplier to the United States, could reach its peak as early as 2005. Nigeria, which the United States is trying to woo away from OPEC, could peak by 2007. Worse, Norway, whose state-owned oil company, Statoil, exports three million barrels a day and consequently ranks as the third-largest exporter, behind Saudi Arabia and Russia, is likely to see a production peak in 2004. Even the mighty gush of Russian oil is beginning to look temporary. Since the fall of the Iron Curtain, Russian oil production has come roaring back, and today every major Western oil company with a passport is in Moscow, bidding for a share of Russia's near-mythic petroleum riches. U.S. diplomats, meanwhile, are wooing Moscow to be America's chummiest (non-Arab) oil supplier. Yet although Russia does have a great deal of oil — perhaps as much as 200 billion barrels, according to the congenitally optimistic EIA — that is peanuts by comparison with the nearly 850 billion barrels believed to be held by Saudi Arabia and other Arab states. And whereas most OPEC states are restricting their production (in an effort to keep world supplies tight and prices high), Russian oil companies are producing at full throttle, and many experts expect a Russian peak no later than 2015.

This, then, is the final act in the oil saga. According to even optimistic projections that take Russian oil into account, non-OPEC oil production could peak by 2015 — at which point, the world's big importing nations will be forced to turn to the one supplier they trust least: OPEC. OPEC, of course, faces a peak of its own — probably sometime in 2025. Yet as long as OPEC's peak comes later, the effect is the same: world oil supply will come increasingly under the control of a cartel with a history of rash behavior and dubious sympathy for the West. By some estimates, as early as 2010, even before a non-OPEC peak, the countries of OPEC will be supplying approximately 40 percent of the world's oil, up from around 28 percent today. Presumably, its share will rise dramatically as non-OPEC oil production falls. What this will mean for the oil markets, and for energy geopolitics generally, is impossible to say. But to judge by deteriorating relations between the oil-consuming West (read: the United States) and many players in the Arab Middle East (read: Saudi Arabia), few of the possible scenarios are very encouraging. At the very least, OPEC countries would be fairly free to push prices higher than they are now, without fear of competition from non-OPEC producers. The last time OPEC had such control over oil prices, during the 1974 Arab oil embargo, Western powers came only the tervening militarily and simply *taking* the oil. By so doing so, and threat of a counterstrike by the Soviet Un

that deterrent no longer exist then, oil optimists ask, has panic not sumably, if traders got even a hint of a more permanent disruption in oil supplies — a peak in world oil supply, for example — they would scramble to buy up as much oil as they could, in hopes of selling it for a higher price later. Indeed, the scramble to buy would send up prices now, well in advance of an actual shortage, and the higher prices would then provide what economists call a signal to consumers and politicians, telling them to either conserve or find an alternative — as happened during the oil shocks of the 1970s. The fact that this is not happening — despite occasional spikes, oil prices have averaged twenty dollars a barrel for decades — is proof, optimists say, that a peak is by no means imminent.

There are, of course, a few flaws in this reassuring argument. First, one reason we haven't seen a price signal is that we couldn't: there has been too much slack in the oil markets. Although non-OPEC producers — the international oil companies, plus countries like the United States and Russia — have been pumping at their maximum, the situation in many OPEC countries is different. In fact, countries like Saudi Arabia, Kuwait, and Venezuela have historically held back: they have extra wells, pumps, and pipelines that are not being used but can be brought into use in fairly short order.

Until recently, OPEC had as much as three million barrels a day above world demand, and this surplus capacity has come in handy, allowing OPEC to fill in supply gaps when Iraq or Venezuela suddenly stops producing. Unfortunately, spare capacity, or "overhang," also serves to cloud the supply-depletion picture, because it muffles any signs of production difficulties. If non-OPEC production begins to fall, OPEC countries can call on their spare capacity before markets get too tight and prices rise too high.

As a _ can ima_ no price signal is sent. Thus, with this overhang in place, one one would _ vestment bank_ rio in which non-OPEC oil could actually peak but no tion on energy issu_ time. Or as Matt Simmons, an oil industry inalready happened, and _ _pert who advises the Bush administrauntil we see it 'in our rear _ _oil and gas will occur, if it has not Second, in order for pri_ _ warn us whether depletion is_ _n the event has happened _ be relatively free, which for_ _ was accessible to whoever _ _curately and _oduce the easiest, most acc_ _uesless to do so. As that easy oil was depleted, companies would turn to the increasingly expensive oil, which would gradually push up the price and simultaneously send a timely signal to consumers to start using less oil. For this marvelous mechanism to function, though, oil companies must have access to that cheaper oil, so they can use it up first, before moving on to the expensive stuff. In the real world, however, just the opposite occurs. Because OPEC owns most of the cheap "easy" oil and limits how much is produced (and who can produce it), Western oil companies are essentially forced to produce the *expensive* oil first, and so must charge more for it — around twenty to twenty-five dollars a barrel — to cover their higher production costs. (This dynamic in turn allows OPEC to charge the same price for its oil, even though OPEC oil is much, much cheaper to produce.)

This market inversion, according to many analysts, has effectively kept the world oil price double what it would be on the free market, a situation that not only encourages a production overhang but masks many changes in long-term supply. As a result, says Alfred Cavallo, a Princeton-based energy consultant who has studied depletion, "the price warning that consumers expect to have as resources are being exhausted is totally obscured."[21]

Ideally, when markets fail like this, governments are supposed to intervene — in this case, by giving some indication that they have doubts about long-term oil. In reality, no intervention takes place. Having witnessed the political damage and panic caused by the bleak forecasts of the 1970s, today's consuming nations tread ultracautiously when speaking officially about future supply. To suggest that something was amiss — that non-OPEC oil production might peak as early as 2015, for example — would not only spook the markets and give bargaining power to OPEC but run counter to the Western mantra of nonstop economic growth. As Joe Romm, the former U.S. assistant energy secretary, put it, "if the U.S. government even brought up the possibility that global oil production might peak in, say, 2020, not only would that have an enormous and very negative impact on the markets, but it would essentially force the United States abruptly to change its energy policy to one that emphasized energy efficiency and alternative energy."[22]

Thus, despite the widely understood fact that all oil estimates are highly speculative — statistical extrapolations based on data from known oil fields[23] — such forecasting agencies as the USGS, the EIA, and Europe's

International Energy Agency are under intense political pressure to err on the side of wild optimism. And err they do. During the 1990s, for example, a USGS report giving a low figure for oil reserves in the Arctic National Wildlife Refuge was withdrawn under pressure from pro-oil lawmakers in Alaska and rewritten with a more optimistic conclusion.

According to industry and government officials, this Panglossian dynamic occurs in every forecasting bureaucracy and does little to encourage policymakers even to consider the issue of oil depletion. "It would be a huge mistake to base U.S. energy policy on what the USGS thinks about future oil supplies," says one former high-ranking U.S. energy official, "and the Energy Information Administration has put out such overblown numbers, and done it with such arrogance, that it should be statutorily barred from answering questions about oil."[24]

The State Oil Company of the Azerbaijan Republic is headquartered in a huge Georgian mansion overlooking the Baku waterfront, about an hour away from Sand Island. Built by an oil millionaire during the city's first oil boom a century ago, it's a juxtaposition of old and new that a Western oil company would have seized upon as a marketing bonanza, but which here looks to have been purely accidental. The building is rundown. The grand old roofline has been disfigured by a row of massive blue letters that spell "SOCAR," as if the Azeri oil bureaucracy were some kind of Hollywood icon. Inside, most of the spacious rooms have been chopped up into tiny offices, and the fine old parquet floors look as if they've been driven over repeatedly by tractors.

Still, this being the former Soviet Union, some elegance has been preserved for senior officers. In one especially grand corner space, with a large conference table and sweeping vistas of the refineries and tankers in Baku Harbor, Natig Aliyev, SOCAR's president, smokes a slim cigarette and dismisses the disappointing exploration results from one of the country's much-hyped offshore fields. The previous summer, ExxonMobil, the biggest and most successful of all Western oil companies, drilled a test well in a new formation called Nakhchivan, not far from the giant Chirag-Azeri-Gunshali. When the first round of drilling revealed no "commercial" volumes of oil or gas at Nakhchivan, the well was deepened. When oil was still not forthcoming, ExxonMobil again deepened the well, this time to around twenty-two thousand feet — a record in Caspian drilling. Still no oil was found, and ExxonMobil announced that the well would be "plugged and abandoned."[25]

Natig insists that the fault lies with ExxonMobil, not the field. "We have made an analysis," he tells me through a translator, "and we concluded that the well drilled by our foreign partner was outside the oil-bearing structures." In other words, ExxonMobil simply *missed* the oil. Satisfied, apparently, with this explanation, Aliyev rises and walks over to his desk to retrieve another cigarette. Slender and darkly handsome in an elegant dark gray suit, which he carefully protects from cigarette ash, he looks nothing like the traditional Soviet oilman. "Only one well was drilled," he continues. "It is impossible to judge reserves by just one well."

That may be true. But around Baku these days, at least outside the offices of SOCAR, talk is decidedly less optimistic than it used to be. Although production from the big Chirag-Azeri-Gunshali field has increased steadily, if slowly, Azeri "big oil" otherwise has not only failed to materialize but seems to be shrinking. The Nakhchivan failure is actually the second disappointment for ExxonMobil, which recently came up short in another field, the much-touted Oguz formation. Nor is ExxonMobil the only Western partner to strike out. Despite numerous test wells throughout the Azeri sector of the Caspian, four other majors — Eni Agip of Italy, TotalFinaElf of France, ChevronTexaco, and BP — have all failed to find "commercial volumes" of hydrocarbons.[26] Firms have quietly tried to break their contracts with the government of Azerbaijan, and it is no secret that many now wish they had bet less heavily on Azeri oil and more heavily on the north Caspian, where the massive Kazakh fields are capturing all the headlines — and most of the Western oil investment. "Azerbaijan has only confirmed what people always knew," complains one senior Western oil executive. "Only a tenth of explored structures usually turn into real fields."[27] And even Kazakhstan is losing its luster. Just recently, BP, Norway's Statoil, and British Gas have sold their interest in the mighty Kashagan field. As one oil analyst quipped, "maybe they were just embarrassed at the prospect of so much wealth. Or maybe they'd begun to suspect that Kashagan wasn't the largest field ever found."

Such bad luck fits into the larger pattern of very mixed exploration results worldwide. Although the new technology is unquestionably uncovering new fields, it has not reversed the trend of declining discoveries. In 2002, for example, worldwide discoveries fell to six billion barrels of new oil — far less than the historic average and well below the twenty-seven billion barrels that the market sucked up. Most of the easy oil — the huge oil reserves in easy-to-reach fields — has already been discovered and in many cases, especially outside OPEC, pumped out. The oil that remains will be riskier to extract, and the likelihood of unexpected costs, missed production targets, and outright failure will be greater. The more oil we produce, the greater the risks associated with what remains.

The Arctic, for example, may indeed hold huge untapped reserves. Yet as even many optimists acknowledge, drilling and producing oil in deep, ice-covered waters, thousands of miles from any tanker port, pose enormous technical challenges. Special equipment and highly trained crews must be brought in and protected in a harsh environment. Thousands of engineering and technical hurdles must be overcome simply to bring the oil to the surface — to say nothing of building the thousands of miles of pipeline that must be laid to get the oil to market. What is more, according to some geologists, once oil companies finally do tap into the Arctic, the formations are far more likely to hold gas than oil.

In addition, the Arctic is among the more fragile ecosystems on the planet, one that environmental groups have been willing to fight hard to protect. For nearly twenty years, Greens have effectively kept oil companies from tapping into a reserve estimated at fifteen billion barrels that lies beneath the Arctic National Wildlife Refuge in Alaska, despite decades of well-financed oil industry lobbying. Signs of similar resistance to exploration in Greenland and Arctic Scandinavia are already in evidence.[28] Likewise, many analysts are already raising questions about plans to produce synthetic oil from Alberta's tar sands and from other heavy oils: the refining process produces massive emissions of carbon dioxide, the main suspect in climate change.

Meanwhile, among some analysts, confidence is fading for the supplies of OPEC oil as well. After analyzing more than one hundred technical production reports written by Saudi oil engineers, Simmons, the Bush energy adviser, believes that the Saudis themselves fear that Saudi Arabia "has very likely gone over its peak. If that's true, then it's a certainty that planet earth has passed its peak of production."[29]

❦

The picture for long-term oil is not encouraging. Even if you don't subscribe to the fear that oil will run out tomorrow, it is clearly going to become riskier by the year — technically, geologically, environmentally, and ultimately economically and politically. Yet thus far, governments, and the populations that elect them, seem to be in a state of denial about petroleum. It is true that efforts have been made to develop alternative fuels or shift the energy economy to natural gas, but such programs will cost trillions of dollars and require decades to carry out. Thus, the real question is not whether oil is going to run out (it will) but whether we have the capacity, the political will, to *see* that outcome soon enough to prepare ourselves for it. Even though the peak is probably further away than many pessimists argue, its arrival may be difficult to detect, given such masking factors as supply overhang and price manipulation. Worse, because depletion will probably *accelerate* in a postpeak environment, as companies strive to capitalize on higher prices, world markets — and the political systems that depend on those markets — could deteriorate with surprising speed once it becomes widely known that a peak has occurred. "The experts and politicians have no Plan B to fall back on," complains Simmons.[30] Adds Romm, "I do not share the alarmists' point of view [about the imminence of a peak], but I am increasingly of the opinion that when it does peak, it will be too late to do anything about it."[31]

In Azerbaijan today, two years after I visited the country, the future still lies very much with oil, at least officially. Despite new setbacks — among them, a failed test well that a Japanese oil consortium drilled in 2002 — the government continues to pin its hopes on the coming "big oil." Around town, motorists can still see banners bearing slogans like THE OIL INDUSTRY IS THE POWER OF THE PEOPLE. At SOCAR, talk has turned to new formations even farther offshore, in deeper water, although some outside geologists are skeptical. As the USGS's Gregory Ulmishek observes: "The source rock is there, the structures are there, the reservoirs are there. The question is whether the source rock is as good in the deep-water part of the basin, as it was in some of the shallow areas. And whatever I say, we just won't know until the first well is drilled. Until you try it with a bit, you just can't know."[32]

Back at SOCAR headquarters, Natig continues to dismiss any such skepticism. "When will the oil run out?" he says. "Thirty years is what we hear, but who knows. We have no idea where we will be in thirty years, or even twenty." He smiles. "We have only just begun. The first wells were very shallow. We will just go deeper and deeper."

3
THE FUTURE'S
SO BRIGHT

IN EARLY MARCH 2000, a short, wiry man with a trim beard and a weary expression hurried into the offices of Ballard Power Systems in suburban Vancouver. For two weeks, Paul Lancaster, Ballard's vice president of finance, had been in constant motion, visiting bankers and investment analysts in New York, London, Zurich, and a dozen other cities, looking for buyers for three million new shares of stock that the Canadian technology company would soon be offering. It was a brutal schedule, but Lancaster and his colleagues could see that it was clearly the right time to be selling. Not only were tech stocks red-hot, but Ballard's main product — an amazing device known as the hydrogen fuel cell — had been looking more and more like the power source of the future.

A kind of battery that never needs recharging, fuel cells mix hydrogen and oxygen to produce electric current. They make little noise, emit nothing more troublesome than water vapor, and have long been touted as the key to a clean energy future. Since the 1980s, Ballard engineers had worked feverishly, often in the face of skepticism and outright hostility, to make fuel cells a commercial reality, and now those efforts were paying off. Just weeks before, the Coleman-Powermate appliance company had announced plans for a portable home generating unit, built around a Ballard fuel cell. At the Detroit auto show, Ford unveiled its new TH!NK, a four-door family sedan powered by a Ballard fuel cell. The timing couldn't have been better. The previous December, Ballard's stock had been trading at an anemic $25. Now, as Lancaster arrived at Ballard's headquarters, shares were nearing $120 and showing every sign of climbing.

Ballard hadn't invented the fuel cell. The concept of generating current by mixing hydrogen and oxygen had been around for nearly two centuries, and it had been marketed for almost as long as the way to move beyond a hydrocarbon energy economy. Nor had Ballard been the first to suggest using the electricity from a fuel cell to power the wheels on a car — essentially creating an "electric" car than never needed plugging in. But the company founded by Geoffrey Ballard had taken two critical steps toward making that vision real. First, Ballard had found a way to shrink the fuel cell while simultaneously boosting its power: its Mark 900 unit, for example, is the size of a suitcase, yet cranks out enough horsepower to move a midsized sedan. Just as important was a superalliance Ballard had created with Ford and DaimlerChrysler, leveraging their enormous expertise to build a new generation of ultraefficient, nonpolluting cars that, by many accounts, would end oil's hundred-year monopoly over transportation and usher in the "hydrogen economy." No less a figure than William Clay Ford, chairman of Ford, had hailed the fuel cell as representing the end of the internal-combustion engine. "It is time to replace fossil fuels," declared Ferdinand Panik, director of DaimlerChrysler's fuel cell division. "Hydrogen offers the best opportunity to do that, and I don't see anything else coming along with the same potential."[1]

Indeed, as far as investors were concerned, fossil fuels were already dead. When the Ballard stock offering closed a few days later, the company had raised $340.7 million — nearly twice what analysts had predicted. Staff and management were jubilant. After nearly fifteen years in the role of an R & D shop, Ballard had the bucks to take its technology to the marketplace.

To be sure, much remained to be done before a fuel cell car would truly compete with the ICE, as the internal-combustion engine is called. Fuel cells were still vastly more expensive than ICEs, although Ballard was sure the Mark 900 would be cost-competitive by the end of 2000.[2] Moreover, fuel cells run on hydrogen, a fuel that isn't sold at service stations, or anywhere else a consumer might go. Even so, the sense at Ballard, and, indeed, at other fuel cell companies, was that fuel cell technology had finally achieved critical mass. A few months later, an ebullient Lancaster told a reporter that hydrogen technology had crossed a key threshold. "In the early days, fuel cells were just a curiosity," he said. "Now all of the major obstacles have been overcome."[3]

◦§§~

Stories like Ballard's capture both the anxiety and the excitement inherent in the transformation of the energy economy — and nowhere more dramatically than in transportation. For more than a century, our mobility has been utterly dependent on oil and the internal-combustion engine. Of the 750 million cars, trucks, and other vehicles now roaming the planet (and the number grows by 50 million a year[4]) some 90 percent use oil — not because of some vast oil company conspiracy, but because, by conventional measures, oil-fueled ICEs generate more power, more efficiency, more bang for the energy dollar, than any other fuel-technology pair. Until something economically more appealing comes along, the oil-powered internal-combustion engine will be the automotive technology of choice.

Yet as we have seen, oil is approaching a threshold, a tipping point, in its dominance. Questions about long-term supply, pollution, and political stability now pose a permanent challenge to the apparently eternal, unchanging oil economy. The petroleum monolith is showing hairline cracks — and these fissures are being exploited by an army of technologies that promise energy without oil's risks, as well as rewards we can scarcely imagine. On any given day, in thousands of machine shops, research facilities, and conference rooms around the globe, exceedingly bright people are refining a full range of startling and wonderful alternative energy technologies, from solar panels to wind power and biomass — any one of which could lay the foundation for a post-oil energy economy.

At the vanguard of this energy insurrection is the hydrogen fuel cell, a 150-year-old energy technology that is clean, quiet, and nearly three times as energy-efficient as even the best internal-combustion engine. Just as coal replaced wood and as oil replaced coal, the hydrogen fuel cell may at last offer the economic proposition that could end oil's hundred-year monopoly over transportation and revolutionize the economics and politics of energy.

And the revolution won't stop there. Because fuel cells can be built to any scale, they can be used to power just about anything, from cell phones and cars to city buses and office buildings. Ultimately, fuel cells may provide the foundation not simply for a new mobility, but for an entirely new energy economy. In place of our sprawling and inefficient hodgepodge of pipelines, refineries, and polluting power plants, we would have thousands of interconnected yet independent microsystems, each powered by a mix of alternative fuels and technologies, including fuel cells, and each generating energy cleanly, cheaply, and locally. Equipped with a backyard or basement fuel cell system, consumers and businesses could achieve a kind of energy independence, fueling their cars and powering their lights and machinery without having to worry about rolling blackouts, manipulative power traders, or monopolistic utilities. After centuries of an increasingly centralized energy economy, controlled by a tiny elite of corporations and investors and protected by government, energy might again become a very local matter.

Such enormous potential is finally gaining an audience. In the past decade, interest in fuel cells and the so-called hydrogen economy has grown exponentially. All the major automakers have fuel cell programs. Policymakers, pundits, and advocates of alternative energy now routinely refer to hydrogen as the end game in the post-oil energy sweepstakes. Even the most recalcitrant oil companies pay lip service to the idea that at some point hydrogen will be the fuel of choice. What remains to be seen is when this revolution will start, how long it will take, and how much the entire thing will cost.

There is a certain poetic elegance to the notion of a future built on hydrogen. The hydrogen atom is the smallest, simplest, and most ancient known. It was the first type of matter to be created after the explosive birth of our universe, the building block from which all other elements were ultimately constructed; and hydrogen remains the most abundant of all elements, forming 75 percent of the mass in the universe. For all its profusion in the cosmos, however, here on earth hydrogen is hard to find in its pure state. The hydrogen atom is highly reactive: it abhors solitude and binds readily with other elements; in fact, it is almost always found in a hybrid form. The most famous hydrogen compound, of course, is water, or hydrogen plus oxygen — but there are many others. Hydrogen and nitrogen make up ammonia. Hydrogen and carbon form the all-important *organic* compounds, the basis of all earthly life and, more to our point, the root of all fossil fuels, or *hydrocarbons* — oil, gas, and coal.

It is in making and breaking these bonds that hydrogen stores and releases the energy for which it has become so famous. We can watch this dynamic at work during photosynthesis, the process by which green plants transform water, air, and sunlight into sugar. Photosynthesis begins when solar energy falls on a leaf and causes a water molecule inside to split into oxygen and hydrogen. This sundering is not easily accomplished; water is a very stable compound: its oxygen and hydrogen atoms are tightly bound.

Splitting them apart requires a great deal of energy — in this case, a burst of solar energy, which, in essence, attaches itself to the hydrogen atom. The cargo of solar energy makes the newly liberated hydrogen atom highly unstable. To regain stability, the hydrogen must now share its extra energy by binding with a new partner — in this case, an atom of carbon, to create a new compound, *carbohydrate*, or sugar. This is why sugars are high-energy compounds: their bonds contain the solar energy brought over by the hydrogen. Sugars, in other words, are a chemical storage for energy from the sun. And hydrogen is the energy carrier.

Once stored, the chemical energy can be released any number of ways. If an ox eats the leaf, the hydrogen-carbon bond is broken by metabolism, which is essentially photosynthesis in reverse. The hydrogen splits from the carbon and reunites with oxygen (from the ox's lungs), thereby creating a new molecule of water. But — and here is the important part — in order to rebind with oxygen, the hydrogen must surrender its cargo of solar energy. In metabolism, this surrendered energy takes the form of heat, which warms the ox, and of chemical-electrical energy, which drives muscle movement and tissue growth. (The liberated carbon, meanwhile, also rebinds with oxygen to produce carbon dioxide, which, as it is exhaled, releases its own share of stored solar energy.) This is the marvelous thing about energy: it can take on any number of forms — solar, chemical, mechanical, or electrical.

This process of rejoining with oxygen and releasing energy is *oxidation,* which is a fancy way of saying "burning." When something burns, it simply means that some energy carrier, such as hydrogen, has bonded with oxygen and is releasing its stored energy. Metabolism is essentially a controlled kind of burning, one that converts pent-up solar energy, stored as carbohydrate, into heat and mechanical energy. Something very similar happens, albeit in a less controlled fashion, when a leaf actually catches fire. Once again, the carbohydrate is split into carbon and hydrogen. The hydrogen instantly re-forms with oxygen (oxidizes), thereby producing water (in the form of steam) and releasing its stored solar energy as heat and light. In either metabolism or actual combustion, then, the energy-carrying cycle is essentially the same: the hydrogen takes on solar energy at the beginning, releases it through oxidization at the end, then reverts to water, in which form it is ready to take on yet another load of solar energy in the next round of photosynthesis.

From the standpoint of human civilization, the truly amazing thing about hydrogen as an energy carrier is that it can store energy for a very long time. Suppose that our leaf isn't eaten or burned, but instead falls into a bog and eventually becomes buried at a great depth, and, over millions of years, is pressure-cooked into coal. When we later burn the coal, we reverse the photosynthetic process, producing water and carbon dioxide and releasing this stored — and very old — solar energy.

In this way, nature has essentially transformed uncounted trillions of kilowatt-hours of solar power into highly concentrated and exceedingly useful forms — coal, oil, or gas. Granted, nature's method of storing solar energy as hydrocarbons is not terribly efficient: the average leaf converts less than 1 percent of the solar energy it receives into chemical energy in carbohydrate form, and more than 90 percent of that stored energy is lost during the long process by which carbohydrate is later cooked into coal. Oil and gas are even less efficient: less than a tenth of 1 percent of the energy contained in the original ocean plankton winds up in the oil or gas we extract from the ground. As a consequence, it takes many hundreds of thousands of watts of solar energy, accumulating over many years, to produce the energy stored in a gallon of gasoline.[5] Still, even grossly inefficient systems can, over hundreds of millions of years, put away a great deal of energy, fortunately for us: if humans hadn't found such an accessible and concentrated form of energy — if we had been forced to rely on wood or water or wind instead — our industrialized civilization could never have come so far so fast.

Yet as we have seen, getting our energy this way entails disadvantages. First, the supply of hydrocarbons is finite; in less than 150 years, we have managed to use up much, if not most, of an energy source that took several hundred million years to store. Second, burning hydrocarbons produces a whole host of noxious substances, ranging from sulfur, which destroys forests, to carbon dioxide, which has serious climatic consequences.

Third, hydrocarbons, for all their concentration and convenience, are not the most efficient energy carriers. The problem, it turns out, is the carbon. Recall that carbon, too, binds with oxygen and releases energy during oxidization. But pound for pound, carbon actually carries less stored energy than hydrogen does. Thus, when we burn hydrocarbons, such as gas, oil, and especially carbon-rich coal, the high energy content of the hydrogen is partly offset by the lower energy content of the carbon. The more carbon a fossil fuel contains, the less energy it can release. Coal, in which carbon and hydrogen atoms exist in a roughly one-to-one ratio, has the lowest energy content of all fossil fuels. Oil, with one carbon atom for every two hydrogen atoms, can release more energy, and methane, or natural gas, with only one carbon to every four hydrogen atoms, releases the most.

If, however, we dispense with carbon altogether and instead burn pure hydrogen, we suffer no carbon offset at all. Hydrogen is thus a far more energy-intense fuel than, say, oil or even gasoline. Burned in an internal-combustion engine, hydrogen produces nearly three times the energy as the same weight of gasoline, and far fewer emissions.[6]

The downside, as noted, is that pure hydrogen does not exist in nature, but must be produced. One method is to break, or "re-form," a fossil fuel like methane or gasoline, by splitting off the carbon atoms from the hydrogen — but other ways exist. If you run an electrical current through a container of pure water, the electrical energy causes the hydrogen atoms to split off from the oxygen and form new bonds, this time with other hydrogen atoms. These new hydrogen pairs carry the energy from the electrical current. The process, *electrolysis,* was the earliest method for producing hydrogen, and it remains the preferred method when very pure hydrogen is needed for industrial processes. The real usefulness of electrolysis, though, is that it can be reversed. In 1839, a British scientist named William Grove discovered that under certain conditions, if hydrogen and oxygen were recombined to form water, the stored energy of the hydrogen would be released as electrical current, plus a small amount of heat. In short, Grove had invented the fuel cell.

෴

More than a century and a half later, Grove's innovation remains largely unchanged. In its simplest form, a fuel cell is little more than a box divided into two chambers. Pure hydrogen, in the form of pairs of hydrogen atoms, is pumped into one chamber. Oxygen is pumped into the other. In its chamber, the hydrogen comes in contact with a special metal known as a catalyst. The catalyst, usually platinum, causes a chemical reaction that splits the hydrogen pairs back into single hydrogen atoms. Single hydrogen atoms, however, are unstable; their natural inclination is to bond quickly with something else, and they are strongly attracted to the oxygen in the opposite chamber.

However, the fuel cell is designed to make this reunion difficult. Between the two chambers lies a substance called an electrolyte. The electrolyte is a strangely selective barrier: it will allow only the core of the hydrogen atom, known as the *proton,* to pass through to the other side to join the oxygen. By contrast, the *electron,* the tiny, electrically charged particle that normally orbits the proton, is stripped away from the hydrogen core and drawn out via a metal wire.

This is the key to the fuel cell. The electron is in effect a discrete burst of electrical current, a "piece" of electricity. Once drawn up the wire, the electron can be made to do all sorts of work, such as illuminating a light bulb or making an electric motor turn. Each time a hydrogen atom crosses the electrolyte barrier, another electron is sent up the wire, until we have a veritable flow of current. The electrons aren't lost forever. After lighting bulbs or turning motors, the electrons return to the fuel cell via another wire, but this time on the other side of the electrolyte barrier. Here they rejoin the hydrogen proton, and both then reunite with oxygen to form water vapor, plus a relatively small amount of heat. Scientists call this process *cold oxidization:* the hydrogen is still oxidizing, but most of the energy is released as electricity, not as heat.

Cold oxidization is the main reason the fuel cell is so much more efficient than the internal-combustion engine. In an ICE, the burning fuel expands, pushing pistons, which turn wheels: in other words, chemical energy is converted into mechanical energy to do work. Most of the stored energy, though, is converted into heat, which is taken away (and wasted) by the engine's cooling system. Thus, where today's ICE cars average twenty-seven miles to the gallon of gasoline, a vehicle running on a fuel cell can theoretically get eighty-one miles to a kilogram of hydrogen — which contains roughly the same amount of energy as a gallon of gasoline — ergo, nearly triple the efficiency. Throw in the fact that hydrogen fuel cells are quiet, produce no vibrations, start instantly, and emit only steam, and you begin to see why this technology has been hailed as the harbinger of a completely new energy order.

⦿⦚⦚⦚⦚⦚⦚

The headquarters of Ballard Power System sits in a sprawling 1970s-era business park just outside Vancouver, and when I visited in September of 2002, the atmosphere was of a company that considered itself to be on a mission from God. In the spacious main lobby, serious-looking men and women wearing blue lab smocks and safety goggles strode purposefully in and out. Visitors from around the world queued patiently at the reception desk or quietly studied the shrinelike exhibit of fuel cells. On a table in the center of the room sat a mock-up of the Mark 900, a black, suitcase-sized box covered with valves and plugs. Nearby hung a recruiting poster with a picture of a young boy in a homemade Superman costume: goggles, boots, and a gold cape. The poster read, "Remember when you dreamed of being a hero and saving the world?"

Upstairs, in a small, unadorned conference room near the main manufacturing floor, Paul Lancaster was explaining just how the world is going to be saved. An intense man with penetrating eyes, puckish features, and a reputation for fiscal brilliance, Lancaster was also one of the industry's most energetic spokesmen. He would spend countless hours on the road or the phone, working with reporters, industry groups, and activists, pitching the promise of a clean, efficient hydrogen economy — and Ballard's place in it. Two years after the big sale of shares, Lancaster seemed to have lost none of his enthusiasm, despite a recent plunge in share price. When I asked him to predict the role of fuel cells in the evolution of energy, he quickly offered two scenarios. "Some people will tell you that the fuel cell is the bridging step, in terms of energy conversion, that will enable the hydrogen economy to actually happen," Lancaster said. "But that, in my opinion, is the conservative view. The not-so-conservative view is that the fuel cell will revolutionize how we think about energy, in the same way that microprocessors transformed how we think about electronics."

The outlines of this revolution were set out in 1923 by a British scientist named John Haldane. In a now-famous lecture at Cambridge University, Haldane described a civilization powered entirely by hydrogen, electrolyzed from electricity generated by huge windmills. The hydrogen would be liquefied and stored in massive underground tanks and then, on days when the wind wasn't blowing, converted back to electricity, either in combustion-driven generators or "more probably in oxidation [fuel] cells." Such a system, Haldane declared, would decentralize energy production and do away with air pollution. Moreover, creating hydrogen via electrolysis would "enable wind energy to be stored" for later use.

By the 1970s, Haldane's idea had evolved into an entire vision for a hydrogen-based economy. Advocates saw hydrogen as the perfect energy form — an energy "currency" that could be produced from any energy source, stored like money in the bank, and then withdrawn as needed to produce electricity in fuel cells, or burned as fuel in car motors, power plants, even jet engines. Hydrogen seemed particularly well suited for storing *surplus* electricity. Most power plants — whether nuclear, coal, or hydroelectric — have a generating capacity that is only fully tapped during relatively short periods of *peak* demand; the rest of the time, this capacity goes unused. Hydrogen advocates asked: Why not run the plants at full capacity all the time, storing the excess power as liquid hydrogen? In the same way, hydrogen could be used to store energy from a variety of alternative energy sources, like solar and wind power — two "renewable" energy sources whose intermittent nature — solar panels, for example, work only when the sun shines — renders them undependable.

Above all, hydrogen could be used in automobiles, either in fuel cells or in ICEs converted to burn hydrogen. Not only do cars account for the lion's share of all oil use and oil-based pollutants, but they are far less amenable to alternative-energy technologies. Whereas homes and businesses run on electricity, which can be made from a variety of sources — coal, oil, natural gas — cars need *liquid* fuel, and until recently the only liquid fuel came from fossil fuels. With a hydrogen economy, transportation could finally become a benign activity.

Predictably, the hydrogen economy has been slow to emerge. Although the oil shocks of the 1970s led oil-dependent America, Japan, and Europe to invest heavily in fuel cell research, the sense of urgency faded as oil prices fell, and hydrogen technology made only slow advances. By the mid-1980s, fuel cells capable of producing enough power for an automobile were far too large and heavy to fit into a car. Yet glimmers of hope were discernible. General Electric had been working on a compact fuel cell, known as Proton Exchange Membrane (PEM), for use in the U.S. space program. In a PEM cell, the bulky electrolytes are replaced by a thin polymer film, which, just like the traditional electrolyte, strips away electrons from protons, yet it is smaller, lighter, and far more stable. Unfortunately, PEM fuel cells were hugely expensive, owing to the high cost of the membrane material and the heavy platinum catalysts.

Worse, PEM cells were not very powerful. Power, in technical terms, is the amount of energy you can bring to bear during a given amount of time. Two energy carriers might contain the same amount of energy, but because one carrier releases it faster, we say it produces more power.[7] Power also describes the rate at which energy is consumed. A large electric fan needs more power than a small one, because it is performing more work — turning a heavier fan blade — during the same period of time. Power is measured in watts. A standard flashlight uses about 1 watt of power. A household light bulb uses anywhere from 60 watts to 150 watts. A space heater uses around 1,000 watts, or one *kilowatt* (kW). The average house in an industrialized country needs from 5kW to 20kW of power, depending on appliances and usage. The average car needs 90kW. To put this in perspective, in the 1980s, General Electric's best fuel cells produced 300 watts — not even enough power to toast bread.

Enter Geoffrey Ballard, a former petroleum geologist turned energy expert with a plan to reinvent the energy economy. By all accounts brilliant, driven, and arrogant, Ballard had worked as an energy conservation expert for the U.S. government during the 1970s oil shocks but had grown cynical about America's obsession with saving energy. In Ballard's view, energy salvation lay not in conserving existing resources but in developing new ones, especially solar energy. He also saw need for better technologies to store the abundant solar electricity and, after quitting government, founded a company to develop a rechargeable lithium battery.

In the mid-1980s, Ballard learned that the Canadian government wanted to fund a small PEM cell demonstration project. Although he knew little about fuel cells, Ballard won the contract and set up shop in Vancouver, British Columbia. In what must rank as one of the century's most brazen entrepreneurial tales, Ballard and a small team of engineers, electricians, and machinists took a crash course in PEM technology and, after combing the literature and dissecting existing models, concluded that the potential for improved power output was enormous.[8]

The main problem, as far as Ballard's team could see, was that the PEM cells weren't getting enough hydrogen. By increasing the inflow of fuel, engineers were able to bump up the production of power substantially. They also found new ways to reduce the amount of platinum in the catalyst and located a kind of polymer membrane that not only was cheaper and more efficient but boosted power. Advance followed advance, and power output soared — in some cases faster than the Ballard engineers expected. One night in 1986, a new design yielded such a massive jump in power output that the electrical cable melted in half.

Jubilant, Ballard officials flew to New Mexico to share their results with colleagues at Los Alamos National Laboratory, then the leading fuel cell research center in the United States. Since the 1970s, Los Alamos scientists had been studying whether compact fuel cells could be used in transportation but had largely dismissed them as hopelessly underpowered. Now, however, as a Ballard engineer described the test results, the astonished Los Alamos scientists realized that the game had changed completely. Pulling aside a Ballard manager, one Los Alamos researcher said, "I don't think you appreciate what it is you people have done."[9] Fuel cell cars were not only possible but probably inevitable. After more than a century, the end of the oil economy had finally begun.

This was heady stuff. For decades, industrial economies had been searching for an alternative to the ICE and oil. Electric cars had initially seemed a likely candidate, but batteries were still so heavy and inefficient that the vehicles were tiny, with a short driving range and a long recharge period. With fuel cells, however, that problem seemed solved. Onboard batteries would still be needed to power the electric motors that drove each wheel, but it would no longer be necessary to recharge the battery by plugging it in: the fuel cell would provide a steady supply of electricity, continuously "recharging" the battery, which could therefore be smaller and lighter. In short, the fuel cell car had all the benefits of an electric car — for example, quick, quiet acceleration — but no power limitations or recharging requirements. Indeed, while fuel cells might ultimately power everything from laptops to office buildings, the largest and most lucrative application would be the automobile. Developing a lightweight, compact fuel cell with the power of an internal-combustion engine had become the Holy Grail.

Predictably, skepticism ran high. The average car needs at least 100 horsepower, equal to 75kW of electrical power, or eight times the power of Ballard's best cell in 1992. Few auto companies believed such a gap could be closed anytime soon. When Geoffrey Ballard talked at technical conferences about powering city buses with fuel cells, he was openly laughed at. "The consensus," he explained in one account, "was that if [the technology] was any good, the big companies would be doing it."[10]

Yet as power output kept doubling every two years — to 25kW in 1994

— skepticism turned to antipathy. According to Tom Koppel, a Canadian journalist who has chronicled the Ballard story, once it was clear that Ballard was within striking distance of an automotive fuel cell, U.S. automakers and oil companies became openly hostile. Companies took out advertisements ridiculing the fuel cell. At trade shows and conferences, auto executives derided fuel cell advocates and their research.

In one particularly illuminating moment, Ballard says he was warned by a former high-level oil company executive that "the oil companies are ganging up on you."[11] The reason, the executive explained, was fear: oil companies did not want to lose gasoline's monopoly over the transportation market. Gasoline can be made from only one source: oil; but hydrogen could be made from numerous sources — oil, natural gas, and gasoline, but also solar energy, wind energy, and even methanol from fermented manure. Once an automotive hydrogen fuel cell became feasible, the executive told Ballard, oil companies would lose their control over the lucrative transportation and energy markets.

No doubt such paranoia was rampant: early in the twentieth century, oil companies had helped destroy the nation's system of electric trolleys in order to increase the market for gasoline. Yet despite a history of ruthless suppression of competition, the industry's reluctance to embrace a "hydrogen economy" is somewhat more complex. Gasoline's great advantage has been its cheapness, its high energy content, and, above all, its liquid nature, which makes it relatively easy to store, transport, and dispense. Hydrogen, by contrast, shares none of these features. It is expensive — more than twice the cost of gasoline — and very hard to handle. It leaks from nearly any container, on account of the tiny size of its molecules, and is highly flammable, though less so than gasoline. Hydrogen gas is also exceedingly dispersed; a kilogram of hydrogen may have three times the energy of a kilogram of gasoline, but it also takes up considerably more volume. To work as a consumer fuel, hydrogen needs to be concentrated, either by compressing it under extremely high pressure, or by condensing it, via refrigeration, into a supercold liquid — which, for fuel cell cars, would require specially designed fuel tanks. Likewise, fueling stations would need special pumps and nozzles that could handle hydrogen safely and efficiently, yet without challenging consumers raised on easy-to-use gasoline.

It was not clear either where all this hydrogen would come from. Just as the technical superiority of oil over coal had been negated by an inadequate supply of oil, many advantages of hydrogen have been offset by its own short supply. Today, demand for hydrogen is relatively small, and refiners make most of what is needed from natural gas. But the large volumes of hydrogen required for a real hydrogen economy would require multibillion-dollar investments by energy companies in new refineries, a new distribution system, and a new source of natural gas.

For that matter, other "clean" alternatives to gasoline are considerably cheaper and easier to use than hydrogen is, at least at present. Ethanol, for example, a fuel currently brewed from corn, is already added to gasoline to control emissions. Methanol, another high-energy "biofuel," can be made from fermenting grain, crop waste, and other organic matter. Both ethanol and methanol burn more cleanly than gasoline, though they contain less energy. More to the point, both exist naturally in a liquid state, making them much easier to handle than hydrogen. Today, ethanol and methanol are too expensive to compete with gasoline, but researchers have developed more cost-effective refining methods as well as specialized fuel crops, such as switchgrass, that grow fast, require no fertilizer and little water, and are easy to process into so-called biofuel. By 2020, says Lee Lynd, a researcher at Dartmouth College and one of the top biofuels experts in the world, biofuels produced from marginal croplands could replace a fifth of U.S. transportation fuel. The point, says Lynd, is that if the world is looking for an alternative to gasoline, hydrogen is not the only candidate.

Automakers, too, despite their own long-documented hostility toward fuel efficiency and alternative technologies, have traditionally had many good reasons to doubt the fuel cell. Here was a novel technology that was still years from commercial feasibility, which, even in the best case, would be costlier to build than the gasoline-powered vehicle (thus yielding a smaller profit margin), and whose consumer appeal was wholly unknown. And in exchange for this uncertain technology, automakers were being asked to abandon an existing technology — the internal-combustion engine — that was proven, highly efficient, and consumer-friendly and that already had a convenient fueling structure in place.

In private, auto industry officials admitted that the existing gasoline engine was becoming obsolete. Energy efficiencies were embarrassingly low — less than 20 percent of the energy in the gasoline actually reaches the wheels — and emissions are still higher than they need to be. Yet rather than throw out the basic technology entirely, automakers argued, we should simply improve the existing technology to achieve the efficiency and lower emissions energy advocates were demanding. One such improvement was the so-called hybrid concept, which married an electric motor with a small, ultraefficient gasoline or diesel engine to dramatically improve fuel efficiency and lower emissions. Like fuel cell cars, hybrids never need plugging in: the engine charges the battery. Unlike fuel cell cars, however, hybrids use existing oil-based fuels, which already have a global fueling infrastructure. Also, if gasoline or diesel become too environmentally or politically problematic, hybrids can be reconfigured to burn natural gas, ethanol, and methanol — even hydrogen, for that matter.

In the meantime, it was beginning to look as if Detroit needed to do nothing more than *talk* about the future, since it was far from clear that consumers — who ultimately decide whether any energy technology succeeds — even wanted fuel-efficient cars. These were the roaring nineties, and while fuel efficiency might appeal in Europe and Japan, where heavily taxed gasoline sold for four dollars a gallon, in the United States, the world's biggest and most influential car market, consumers were happily shelling out thirty thousand dollars for fuel-chugging pickup trucks and the new sport utility vehicles. The future, it seemed, would never arrive.

෴

In some sense, the problem with the fuel cell was bad timing. Here was a promising alternative technology that had shown up too early, before the market knew it even needed an alternative. Oil might have myriad problems, but in the current economy none of these problems had registered where it mattered: in consumers' checkbooks. Things might change in five or ten or twenty years, at which point hydrogen might displace oil just as oil had finally displaced coal at the turn of the last century. As far as today's market was concerned, however, that point had not yet been reached.

But this was about to change. If the economics of the oil-powered internal-combustion engine seemed unassailable in the early 1990s, politically a transformation was under way that would shift the advantage toward the fuel cell. Around the world, strict antipollution laws were doing what markets would not, forcing automobile companies to build cleaner cars — and nowhere would this have a greater impact on the industry than in the United States. In 1990, the U.S. amended its clean-air laws to require car companies to cut emissions by 60 percent by 1996. California, birthplace of the American car culture, immediately enacted its own clean-air laws, insisting that by 1998, 2 percent of all new cars sold in the state be zero-emission vehicles (ZEV), a figure that would jump to 10 percent by 2003. In a stroke, automotive economics were changed: because California is the largest, most lucrative auto market in the world, car companies were essentially forced to reconsider alternative cars. Aghast, American automakers lobbied hard to defeat the state initiatives and then began — reluctantly — developing electric cars in order to meet the new requirements. As far as fuel cell cars went, Detroit's public position remained as hostile as ever.[12]

This time, the unlikely savior of the fuel cell was a foreign automaker, Daimler-Benz. In contrast to its American rivals, the German maker of Mercedes and other brands *did* see a market for fuel efficiency and alternative fuels, both in Europe and, in the long term, in the United States. Since the 1980s, the company had been testing an internal-combustion engine that burned hydrogen. But after seeing how quickly Ballard's fuel cells were improving, Daimler decided to build a fuel cell car and, in 1993, formed a partnership with Ballard.

The alliance was a strategic coup for Ballard, and a turning point in the financial fortunes of the automotive fuel cell. Daimler brought to the table money, marketing muscle, and an eye-opening ambition to have a fuel cell vehicle on the road within the decade.[13] Daimler's first fuel cell car — the 1993 NECAR I — was hardly a dazzling success. The van was so crammed with fuel cells and hydrogen fuel tanks that it had room for only two passengers, a top speed of sixty miles per hour, and a cruising range of fifty-five miles between refuelings. But in 1996, Daimler-Benz revealed the NECAR II, a normal-looking minivan powered by only two Ballard 25kW fuel cells. The NECAR II had room for six passengers, a maximum speed of seventy miles per hour, and a range of a hundred and fifty miles between fuel stops.

All at once, the fuel cell began to look like the Next Big Thing, and analysts and investors practically swooned. A "breakthrough in pollution-free motoring," said one. A "giant step forward for Daimler and Ballard," proclaimed another. "In one bold stroke, Daimler has accelerated the race to perfect fuel cells," marveled the usually skeptical *Business Week:* "Rival carmakers must suddenly speed up their own fuel-cell research — or risk being left in the dust."[14]

Suddenly, car companies that had dismissed the fuel cells as too risky even to consider saw them instead as too risky to ignore. Within the year,

General Motors, Toyota, Honda, and Nissan all announced major fuel cell initiatives; Ford, meanwhile, actually joined the Ballard-Daimler alliance and committed $420 million to the project, thereby bringing the alliance's budget to nearly $1 billion. "We reckoned it would be well after 2020 before these vehicles would be hitting the road," said one research official at the recently merged DaimlerChrysler; "however, given the present state of the technology, it might now be as early as 2010, if not a lot sooner."[15]

One after another, automakers rolled out demonstration fuel cell vehicles, as well as a growing number of electric cars and gas-electric hybrids. Great strides were also being made in refueling. Engineers were developing machines that would produce hydrogen by breaking down, or "reforming," gasoline — potentially a huge breakthrough. By placing gasoline reformers at existing service stations, fuel companies could produce and sell hydrogen without abandoning the existing fuel infrastructure. In 1997, Chrysler announced plans to put a gasoline reformer on the car itself, thereby allowing the driver to fuel up at any service station. Even the oil industry broke its silence on hydrogen. In 1999, Shell launched a hydrogen division, and British Petroleum, now calling itself BP, for "Beyond Petroleum," soon followed.

With the entrance of the oil companies, fuel cells had achieved a kind of critical mass. Although fuel cells were still vastly more expensive than their ICE competitors, it was understood that the costs would come down as the technology advanced and, just as important, as fuel cell cars entered into mass production. The key would be a phased transition. Fuel cell cars would be rolled out gradually, to allow a market to build by increments. First would come the small "demonstration" fleets for government agencies, to develop public awareness and let companies road-test the technology. Next would come sales to so-called fleet owners, such as delivery companies or taxi services or perhaps some large government customer, like the U.S. Postal Service — organizations that would already be using centralized fueling and maintenance, and whose drivers could easily be trained in the new technology.

As fleet sales grew, economies of scale would kick in, bringing down manufacturing costs. The next step would be limited consumer rollouts, which auto companies and oil companies would carefully coordinate between themselves in selected markets, like Southern California, New York, London, and Tokyo, where the cars would arrive simultaneously with a small number of strategically located hydrogen fueling stations. These roll-outs would be targeted at the so-called early adopters — activists, Holly-wood actors, and ecologically inclined millionaires — those willing to pay a premium in price and performance for the status effect. As manufactur-ing numbers climbed, costs would fall further.

Just as important, even as prices fell, the cars themselves would be im-proving rapidly, and not simply in regard to fuel efficiency or environmen-tal friendliness. With fuel cells, it is possible to build an entirely new kind of car — one that is not only as fast, safe, and comfortable as its internal-com-bustion predecessor, but actually better in all respects. Fuel cell cars would be quieter. They would be easier to handle: whereas conventional vehicles rely on mechanical controls for steering and braking, fuel cell controls, be-ing electronic, would allow for far lighter, more precise handling. Fuel cell vehicles would also be much roomier. Conventional vehicles sacrifice space to a bulky gasoline engine and a drive train (which in running from the front of the car to the back, create the infamous floor hump). In the fuel cell car, by contrast, the fuel cell and electric motors could all fit into the floor.

In one particularly eye-catching prototype, General Motors' Hy-wire vehicle, the entire drive package — fuel cell, fuel tanks, wheel motors, and driving controls — all fit into a four-wheeled, skateboardlike platform, which is then bolted to the passenger cabin. The configuration not only gets rid of the bulky engine compartment and drive train "hump" but al-lows drivers to swap cabin types — sedan, for example, or van — depend-ing on need or interest.

These features, fuel cell advocates argued, would sell the new kind of cars. "Forget emissions — forget fuel efficiency," argued Ballard's Lancaster when I spoke with him in November of 2002. "It will turn out that with fuel cell technology automakers will be able to make a better car than the ones they have been manufacturing. Better performance, better comfort, better convenience, and better cost. It will turn out that no automaker can afford *not* to be in fuel cells, and those who do will be left behind and run the risk of never catching up."

⊙⟩⟩⟩~

As the end of the century approached, the auto industry seemed on the verge of a transformation. Encouraged by the lightning-fast advances in fuel cells during the early 1990s, many companies now laid out bold schedules for introducing the vehicles. Ballard, which had warned against releasing a consumer product too soon, simply in order to have something on the market, now firmly believed that a fuel cell competitively priced with the internal-combustion engine could be available by the end of 2000. DaimlerChrysler was so confident that it promised to spend $1.4 billion on the hydrogen fuel cell and vowed to have as many as forty thousand fuel cell cars on the market by 2004 — and production runs of as many as a hundred thousand by 2006.[16] "We are not aiming at a niche market," declared DaimlerChrysler's Ferdinand Panik. "The objective is really to concentrate on mass production. We want to compete against the internal-combustion engine."[17] Other analysts talked about a million fuel cell cars and annual sales of $10 billion by 2010.

Such optimism touched off a giddy, gold-rush mentality. After Ballard's stunning $340-million stock sale in 2000, venture capitalists scrambled to find the "next Ballard," just as they had scrambled a decade before to find the next Microsoft or Intel. They launched dozens of new fuel cell companies; established companies spun off their own fuel cell divisions. With an intensity and confidence reminiscent of the Internet boom, a new generation of entrepreneurs laid out ambitious business plans for a whole host of hydrogen products, from automotive engines to power generators for home use to industrial-scale units. Fuel cells became the next big technology stock. By 2001, an estimated $600 million had been invested in a sector that was newer than the software industry and had yet to produce a single "deliverable."

Hydrogen mania spread into the political sphere. Environmental groups championed the fuel cell as the key to a cleaner future and began berating oil companies for not immediately building new hydrogen fueling stations. Politicians did not lag far behind. After largely abandoning hydrogen and fuel cells in the 1980s, governments in Europe and North America launched or enlarged dozens of hydrogen programs. In Europe, major cities made plans to purchase buses with fuel cells. In the United States, federal and state officials kicked off programs to commercialize hydrogen technology. Iceland vowed to become the world's first hydrogen economy. "One can see the dream of a hydrogen-based economy becoming a reality," observed U.S. Senator Tom Harkin, a long-standing proponent of hydrogen, in early 2001. "I am confident that I will one day walk from my hydrogen-heated office through clean air to my hydrogen fuel cell car."[18]

Not everyone was so sanguine. Many old-school hydrogen advocates and researchers found the burgeoning enthusiasm disquieting and even disingenuous. It seemed clear that some oil and auto companies, as well as politicians, were using hydrogen to avoid having to improve fuel efficiency in gasoline-powered cars; for example, even as DaimlerChrysler was championing the fuel cell, its business strategy continued to focus heavily on sales of huge pickup trucks and SUVs, resulting in a new model fleet whose average fuel efficiency was the lowest in the industry.

Perhaps more fundamentally, veteran hydrogen advocates were disturbed by what they saw as stupidly high expectations. Though excited by the long-term prospects of the fuel cell, hydrogen experts were horrified by suggestions that the advent of fuel cell cars was imminent — in large part because many issues of cost, performance, and fueling had not yet been solved. "Fuel cells are not like software," says Karen Miller of the U.S.-based National Hydrogen Association. "It takes years and even decades to do this right."[19]

But "years and even decades" was not the time frame that many of the new hydroentrepreneurs had in mind. Whereas Ballard Power routinely spent five years or more developing and testing its products, some newer companies boasted timetables that were far more aggressive — and to some observers' way of thinking, far less realistic.[20] Several market analysts I interviewed could recall being told by fuel cell company executives that they would have a new fuel cell product designed, tested, and brought to market in less than two years — despite the fact that nothing like that had ever been done before. As one analyst told me, everyone was "assuming that you could get the costs down where they needed to be, and get the functionality and reliability you need, and not run into some huge R & D roadblock." In short, he says, "a lot of these companies were counting on some 'eureka moment' — a miraculous engineering breakthrough, and that's a risk government labs should be taking, not public companies."

But the market, apparently, didn't care. "There was a lot of exuberance," adds Peter Hoffmann, an industry observer and publisher of *The Hydrogen & Fuel Cell Letter*. In many ways, Hoffmann says, the hydrogen boom "emulated the dot.com bubble."[21] Another analyst put it this way: many fuel cell entrepreneurs "took advantage of a market mentality that had been built on the software boom. The difference was, fuel cells are *hardware*. Their performance is measureable: they either work or they don't. At some point, you can't just finesse your way" with wild claims.

In 2001, the hydrogen bubble burst. Over the next year and a half, one fuel cell company after another delayed or even canceled promised products; even Ballard's assurance of a cost-competitive automotive cell by 2001 had to be withdrawn. Share prices for the whole hydrogen sector nosedived, and companies watched as their once-golden shares were transformed into near-worthless penny stocks. Many of these companies closed their doors or were bought up for a fraction of their bubble values. By 2003, analysts had written off two out of every three ventures. Among the hardest hit was Ballard. Even though its technology and management are still seen as light-years ahead of the competition, the company suffered a huge loss of investor confidence. The plunge in the stock price from $140 a share to as low as $6 scotched plans to finance new research and forced drastic cuts. Just before Christmas of 2002, Ballard laid off 25 percent of its work force, including the ebullient Lancaster, and significantly scaled back its project development schedules.

Since then, hydrogen mania has cooled considerably, most notably in the automotive arena. Although car companies are continuing with their own internal fuel cell research programs (energy experts say General Motor's fuel cells are as advanced as Ballard's or anyone else's), the companies have grown far more cautious in their forecasts for a fuel cell revolution. Although nearly all the major automakers had demonstration fuel cell vehicles out by 2003, manufacturers have dramatically pushed back the large-scale rollouts promised only a few years ago. Toyota leased just twenty of its prototypes — including a fuel cell SUV — to California governmental agencies in 2003. DaimlerChrysler, which once promised forty thousand fuel cell vehicles by 2004, now projects tiny test fleets of twenty units or so. Overall, projections for future large-scale rollouts are more on the order of fifty thousand to a hundred thousand units, by 2010 at the earliest. Everyone still insists that larger output will come, but no one is willing to publicly predict when that might be. "GM says they want to be the first to sell a million fuel cell vehicles," says Hoffmann. "But they don't say when and they don't say how."[22]

Some energy advocates have been quick to blame automakers for never having committed to fuel cells and for using them mainly as a brilliant PR tool. Yet even if oil and car companies have exploited fuel cells for

the maximum political gain, it is also true that companies, as well as investors, are encountering fundamental obstacles to a full-scale launch of fuel
cell cars. Despite rapid technological advances, questions abound regarding
the reliability and durability of fuel cell cars, especially in extreme environments. Whereas the standard ICE car will last for 150,000 miles, today's fuel
cell vehicles must struggle to run longer than 30,000 miles — hardly something that will entice the average buyer.[23]

More seriously, the fueling issue has not been solved. Despite an apparent
consensus among developers, automakers, and energy companies that fuel
cells will use compressed hydrogen, not methanol or reformulated gasoline, onboard fuel storage is still problematic. Even when compressed, hydrogen remains far less energy-dense than gasoline. A large fuel tank,
barely big enough to fit into the trunk of a car, would still yield a cruising
range of well under two hundred miles, as compared with the three-hun-
dred- to four-hundred-mile range that consumers now expect from standard ICE vehicles — and apparently feel they can't live without.

More fundamentally, no hydrogen-fueling infrastructure has emerged
— or even moved beyond the discussion phase. Although it is possible to
produce hydrogen using solar or wind power, both these methods are enormously expensive. The most cost-effective means would be to use natural
gas, but doing so would require significant new supplies of natural gas at
a time when the North American market is already tight. Even assuming
that adequate gas supplies were on hand, building a fueling infrastructure
would cost billions of dollars. Thus, despite constant talk about phased
rollouts, and pledges of interest from Shell, BP, and now ChevronTexaco,
only a handful of demonstration stations have been built, and many of
these by the automakers or companies specializing in hydrogen. Shell, having become more circumspect in its forecasts, insists in its long-term energy scenarios that hydrogen could develop a mainstream market but admits that it could just as easily remain only a niche market. Meanwhile,
ExxonMobil, the largest oil company in the world, and in the late 1990s one
of the more vociferous critics of fuel cells, still refuses to endorse hydrogen
fully as having any kind of future.

Again, many alternative energy advocates accuse oil companies of
dragging their feet to protect the trillions of dollars they have invested in

conventional energy assets and systems — charges that are no doubt partly true. It is also true, though, that oil companies are genuinely daunted by the costs of refitting existing service stations to sell hydrogen — costs that, according to internal industry estimates, could run to thirty billion dollars in the United States alone, simply to offer hydrogen at 33 percent of all stations.[24] Then there is the matter of fuel price. Oil companies say they could make hydrogen as cheaply as gasoline — if they could sell it to consumers at the hydrogen refinery. Unfortunately, handling and transporting this strange fuel is extremely expensive. On-site reforming is easier and cheaper, though not by much: these tiny hydrogen factories manufacture such relatively small volumes that, according to one oil company, the most optimistic estimate of the cost of a "gallon-equivalent" of hydrogen — that is, a quantity of hydrogen containing the same energy as a gallon of gasoline — would be around three dollars — or roughly three times the pretax cost of gasoline. Even if one takes into account hydrogen's greater energy density and the greater efficiency of the fuel cell vehicle, consumers would be paying at least 50 percent more for hydrogen than for gasoline. As one oil company's fuels expert puts it, "That might fly in Tokyo or Italy, where gas taxes are high, but in America, hydrogen is going to need help." Perhaps most worrisome, though, is the high cost of producing a fuel cell small enough to fit into a car. Although larger "stationary" fuel cells are getting cheap enough that they might provide cost-effective backup power for hospitals or even serve as engines for freight locomotives or cargo ships (where weight and size are less an issue), automotive fuel cells remain vastly more expensive than internal-combustion engines, despite great success in bringing down the expense of the component materials. Whereas the average ICE costs around $50 per kW of power — or around $4,000 for a standard 90-horsepower engine — the most cost-competitive fuel cells are still well over ten times that amount for the same power output.

Many fuel cell advocates contend the high cost stems mainly from the fact that the devices are still "handmade by Ph.D.'s," to use the common phrase. Just as mass production brought down the cost of the ICE a century ago, once economies of scale kick in for the fuel cell, its costs will also fall dramatically, putting the fuel cell car well within reach of consumers. In other words, the barriers now are manufacturing barriers, not engineering barriers; the basic technology is there — it just needs to be brought to the masses. According to this view, all is still mainly a question of timing: if fuel cell companies like Ballard can simply hold on for a few more years, as volumes build and costs come down, the tipping point will be reached.

Yet this view still may be too optimistic. Many analysts believe that fuel cell developers still face major engineering hurdles — in particular, finding something other than high-cost platinum to make catalysts from — before the cars are ready for large-scale manufacturing. "If a certain level of mass production can be achieved, the cost should be dropped drastically," agreed Toyota's top fuel cell expert, Norihiko Nakamura, during a fuel cell conference in mid-2002. "But a great amount of effort is needed to bring down the cost to even two to three times that of a standard vehicle."[25] Consumers, meanwhile, do not appear to be holding their breath.

Consumer doubts, ultimately, may be the biggest hurdle for the fuel cell. For years, advocates have touted this brave new technology as the natural successor to gasoline and internal combustion. Yet as far as consumers, or the market, can tell, there is nothing wrong with either gasoline or internal combustion. Gasoline is cheap and plentiful, and the internal-combustion engine is probably the best-designed device in the history of the world. Nearly a century of continual refinement has created a staggeringly efficient machine — not to mention a design, engineering, and manufacturing process dedicated solely to discovering new ways to make the machine even better. "Automotive companies spend two to three billion dollars a year refining existing engines, making them more efficient," says Bob Shaw, a venture capitalist who invests in alternative energy companies.[26]

In other words, an automotive fuel cell revolution is not going to happen by itself. At this stage, it won't matter how much engineers can reduce the costs of the platinum catalyst or how many cents they can shave off a kilogram of hydrogen. At this stage, fuel cell cars are primarily a political question. Do we really want fuel cell cars? How much are the benefits worth to us? If we decide to go ahead with autos powered by fuel cells, revamping the automobile industry will require a massive global political initiative — extensive funding for fuel cell research, prodigious investments in a fueling infrastructure, and considerable political maneuvering to design incentives and regulations that give fuel cells the advantage they need to compete against an entrenched ICE technology. In short, we need a radical shift in the current automotive paradigm — a shift that, in today's political environment, is almost impossible to imagine.

In the meantime, the fuel cell seems to have stalled out, and worse, its image has become tarnished in the public mind. Investors, having been burned repeatedly, are understandably reluctant to put more money into hydrogen companies and have been investing their money elsewhere. Government initiatives have been similarly discredited. In January 2002, when the Bush administration launched its much-vaunted hydrogen initiative, FreedomCAR, the program was widely seen as yet another cynical ploy to avoid the politically dangerous task of raising fuel efficiency standards for existing car models. Worse, hydrogen and fuel cells seem to have squandered the public interest they attracted only a few years ago. Observes Frank Lynch, a former automotive engineer who now designs hydrogen generators, "I'm afraid that when we finally get people to stop associating hydrogen with bombs and the Hindenburg explosion, the next word they'll think of will be 'scam.'"[27]

4

ENERGY
IS POWER

ON THE TOP FLOOR of a stylish office building in downtown Riyadh, in an elegant office with gorgeous carpets, burgundy woodwork, and more floor space than the average American home, Ali Bin-Ibrahim al-Naimi, the Saudi minister of oil and, on some days, the most powerful man in the world, sits on an oxblood leather couch, glaring at a reporter. We had been talking about the future of energy and the long-term role that oil might play, and al-Naimi, who is sixty-eight and tiny, with a gray mustache, a cherubic face, and a famously sharp tongue, had been pleasantly insisting that oil would be around for decades. "There are alternatives," said al-Naimi, resplendent in the traditional white Arab *thobe* and headscarf. "But for the next twenty years, oil and gas will be it." Then I'd asked about a more current issue: a confrontation between Saudi Arabia, the world's largest oil producer and boss of the oil markets, and Russia, the Saudis' closest oil rival. Al-Naimi paused, his round face clouding. He glanced at his media man, who sat nearby, then smiled. "We don't like to talk about 'confrontation,'" al-Naimi gently lectured. "Rather, we seek a spirit of cooperation. We have a short-term situation here where the large producers need to work together to stabilize the market."

This was, of course, a load of diplomatic bull. Nine months before, in the aftermath of September 11, fears of global oil disruption had sent prices up nearly 30 percent to over thirty dollars a barrel. Although al-Naimi knew that no real shortage existed, that the price spike was driven by speculation, and that the markets truly did not need more oil, a gold-rush dynamic had set in. Other "large producers," such as Kuwait and Nigeria, had begun pumping at maximum capacity, in hopes of cashing in on the high prices. The Saudis had pleaded for moderation, for al-Naimi knew that such undisciplined production would glut the market and send prices tumbling. For a time, Russia, appearing to side with the Saudis, had promised to cut its own production. A few months later, however, Moscow not only reneged and began pumping at maximum capacity but started exporting its oil to the United States, the world's largest oil market — and the Saudis' most important customer.

Only a few years before, Saudi Arabia, as leader of OPEC and ruler of the global oil markets, would have dealt ruthlessly with such a challenge. At a nod from the crown prince, engineers at Shayba and Ghawar and other Saudi fields would have opened the oil taps, drowning Russia and any other rivals in a flood of cheap oil, then stepped in and taken their market share. But times have changed. It is the spring of 2002. U.S. planes are pounding al-Qaeda positions in Afghanistan, and reports are circulating that American military and energy strategists are already planning a second invasion of Iraq. Rumor also has it that the neoconservatives who now rule Washington are eyeing the entire oil-rich Middle East. Energy alliances have shifted, too. Whereas once the Saudis were the favored oil supplier to the United States — George Bush senior is said to have had a special relationship with Saudi King Fahd — now American analysts describe the kingdom as politically unstable, anti-Western, and undependable as a supplier. The United States has been actively courting new oil suppliers — in West Africa, the Caspian, and especially Russia, which American hawks regard as friendlier and more reliable. The Saudis, once confident masters of the oil universe, must today tread carefully.

Al-Naimi excuses himself to take a call from the Russian oil minister, and when he returns moments later, he is wearing a diplomat's smile. "You see?" he says of his phone conversation with Moscow. "We are friends, not competitors." I remain unconvinced. Haven't the Russians in fact just turned down a Saudi request to cut production? Isn't this a price war, even if an undeclared one? "The press is making a lot of hay out of this," al-Naimi says soothingly. "But in the end, cooler heads will prevail." What if they don't? Finally, al-Naimi seems to lose patience. "Saudi Arabia is *not* in competition with Russia," he tells me, his tone indicating that he wishes he were talking not to a journalist, but to the entire oil market. "If we were in competition, we would pull out all the plugs and put ten million barrels a day on the market and knock everyone out of business for two to three years." He looks at me. "We are not even in the same *league.*"

For anyone interested in the future of energy, the rivalry between the Kremlin and the House of Saud offered a dramatic window into the practical realities and the true priorities of the global energy order. Granted, such a rivalry may seem rather insignificant, especially by comparison with such issues as the depletion of world oil or the uncertain future of fuel cells. Nevertheless, in the here and now of energy geopolitics — that high, thin stratum where the business and politics of energy merge into a single, swiftly moving current — the Russian-Saudi fracas was the *only* story that mattered. On the floors of commodity exchanges in Tokyo, Moscow, London, and New York, in the boardrooms of international banks, currency traders, and energy conglomerates, within the chambers of every government on the planet, analysts scrambled to assess the near- and long-term impact of a Russian-Arabic schism. Had Saudi Arabia lost its authority over world markets? Could Russia maintain its production spurt? Were the Americans truly trying to wean themselves away from Middle Eastern oil? More pressing yet was the question how high, or low, speculators would drive the price of oil, and what that would mean for the world economy.

The obsessive focus on oil is hardly surprising, given the stakes. In the fast-moving world of oil politics, oil is not simply a source of world power, but a medium for that power as well, a substance whose huge importance enmeshes companies, communities, and entire nations in a taut global web that is sensitive to the smallest of vibrations. A single oil "event" — a pipeline explosion in Iraq, political unrest in Venezuela, a bellicose exchange between the Russian and Saudi oil ministers — sends shockwaves through the world energy order, pushes prices up or down, and sets off tectonic shifts in global wealth and power. Each day that the Saudi-Russian spat kept oil supplies high and prices low, the big oil exporters were losing hundreds of millions of dollars and, perhaps, moving closer to financial and political disaster — while the big consuming nations enjoyed what amounted to a massive tax break. Yet in the volatile world of oil, the tide could quickly turn. A few months later, as anxieties over a second Iraq war drove prices up to forty dollars, the oil tide abruptly changed directions, transferring tens of billions of dollars from the economies of the United States, Japan, and Europe to the national banks in Riyadh, Caracas, Kuwait City, and Baghdad, and threatening to strangle whatever was left of the global economic recovery.

So embedded has oil become in today's political and economic spheres that the big industrial governments now watch the oil markets as closely as they once watched the spread of communism — and with good reason: six of the last seven global recessions have been preceded by spikes in the price of oil, and fear is growing among economists and policymakers that, in today's growth-dependent, energy-intensive global economy, oil price volatility itself may eventually pose more risk to prosperity and stability and simple survival than terrorism or even war.

In this bleak context, it becomes easier to understand why nations as powerful and technologically advanced as Japan, Britain, and the United States have such abysmal records when it comes to long-term energy planning or alternative energy. Indeed, when the major nations speak of energy policy today, about energy for the future, or about the much-touted "energy security," they are not talking about depletion curves, or fuel cells, or a hydrogen economy. They are not talking about fuel efficiency, or solar power, or any of the potentially significant but speculative sources of energy. Rather, when nations discuss energy security today, what they are really talking about is the geopolitics of energy — and specifically, the actions, money, and alliances necessary to keep oil flowing steadily and cheaply through the next fiscal quarter.

The geopolitics of oil are vast, complex, and ever-changing, but three elements are of absolute importance. The first is the preponderant role of the United States. Since the earliest days of the oil industry, the country has been the dominant figure, first as the world's largest producer of oil and other energy and now as its largest consumer. Today, one out of every four barrels of oil produced in the world is burned in America, and this enormous, apparently limitless appetite exerts a ceaseless pull on the rest of the world's oil players and on the shape of the world political order.

American oil lust is a mixed blessing: on the one hand, such heavy dependence on foreign oil makes the United States vulnerable to disruptions in supply and to energy "blackmail" and has, in addition, fostered a long tradition of doing whatever is necessary, covertly or overtly, to ensure that the United States — and U.S. oil companies — have access to world oil supplies.

At the same time, however, the sheer extent of American demand, coupled with the country's own booming production (the United States is still the number-three oil producer), gives Uncle Sam a degree of influence over world oil markets and world oil politics that goes well beyond anything the U.S. might achieve militarily. America is not only the biggest oil market in the world, but the fastest-growing: in the 1990s, American oil imports grew by 3.5 million barrels a day, more than the total oil consumption of any country except China and Japan, and that trend has continued in the first decade of the new millennium. After the United States, no other market offers exporters like Russia or Saudi Arabia the same opportunities for both growth and volume of sales, and no oil producer, whether country or company, can afford to miss out. Today, a producer's share of the U.S. market is a critical measure of that producer's political standing and future prospects. Saudi Arabia, for example, is so desperate to maintain its share of the U.S. market that it sells oil to Americans at a discount. Even oil states with profoundly anti-American sentiments — Venezuela, Libya, and until recently Iraq — are exceedingly cordial when it comes to selling or trying to sell oil to Americans.

Within the oil world, no decision of any significance is made without reference to the U.S. market, nor is anything left to chance. Indeed, the world's oil players watch the American oil market as attentively as palace physicians once attended the royal bowels: every hour of every day, every oil state and company in the world keeps an unblinking watch on the United States and strains to find a sign of anything — from a shift in energy policy to a trend toward smaller cars to an unusually mild winter — that might affect the colossal U.S. consumption. For this reason, the most important day of the week for oil traders anywhere in the world is Wednesday, when the U.S. Department of Energy releases its weekly figures on American oil use, and when, as one analyst puts it, "the market makes up its mind whether to be bearish or bullish."

And woe to the markets should American consumers actually get excited about the issue. In late December of 1999, for example, as the world braced for blackouts, riots, and other fallout from a Y2K global computer meltdown, oil analysts were far more worried that American motorists might try to stock up on gasoline before the New Year. "Americans usually drive around with half-full tanks," one Saudi oil official told me. "If [they had all] decided to fill up their tanks that last week in December, the sudden demand would have completely disrupted world oil markets."

The second factor in the geopolitics of oil is, not surprisingly, the oil in the Middle East. If the depletion debate has focused mainly on non-OPEC

oil, oil geopolitics are concerned just as much with OPEC oil, and mainly
with what lies beneath the red sands of Saudi Arabia. The desert kingdom
possesses somewhere on the order of 265 billion barrels of oil, more than a
quarter of the world's known reserves, and some of the most sought-after.
The bulk of Saudi oil is known as Arab light — a thin, bluish crude that is
easily refined into almost any oil product and can be used by most refineries worldwide. More remarkable still is how easily Arab oil comes out
of the ground. Like the vast fields in east Texas back in the early days, the
enormous limestone reservoirs beneath the Arab desert and its offshore
waters are under great pressure: a field, once pierced by the drill, gushes like
a fountain. "Drive out into the desert and look at the wells," says one oil
company executive who has worked for years with the Saudis. "There are
no pumps. It just comes out of the ground."

This "easy" oil is extremely cheap to produce. Whereas crude from the
Gulf of Mexico or Siberia may cost $15 per barrel — and even more — to
find, drill, and pump, Saudi "lifting" costs are around $1.50 a barrel, among
the lowest in the world. (Only Iraqi oil is cheaper.) Given that world prices
have averaged $20 a barrel for the last two decades, this low-cost oil has
made the Saudis very, very wealthy. But the implications of cheap oil go beyond Saudi bank accounts. Low costs, coupled with a nearly unlimited reserve base, have allowed Saudi Arabia to develop steadily into the world's
top producer and exporter, pumping anywhere from 7.7 million barrels to
10 million barrels a day, or as much as a seventh of the global demand.

Low-cost oil has given the Saudis great flexibility. They could, for example, afford to maintain huge spare production capacity — that is, a vast
network of wells, pipelines, and loading facilities that essentially stand idle
until needed. Thus, on top of whatever they happened to be producing at
the moment, the Saudis could begin pumping and exporting half a million additional barrels of oil almost overnight. Within ninety days, they
could boost production by nearly two million barrels, and within eighteen
months by three million barrels. With so much spare capacity, Riyadh has
been the undisputed boss of OPEC: at a moment's notice, the Saudis could
flood the market — they call it capacity cleansing — thereby pushing
down world prices, driving out their high-priced competitors, and punishing any fellow OPEC members that cheat on the cartel's production quotas.

So much spare capacity has also made Saudi Arabia the bosom buddy
of the industrial world. It was extra Saudi oil that saved world markets

when Saddam Hussein invaded Kuwait in 1990, and it was Saudi spare capacity that calmed markets after September 11 and in the run-up to the second Gulf War. "With that much spare capacity," says one U.S. government energy analyst, "any number of producers could drop off the face of the earth and you wouldn't even notice: the Saudis would just open the taps."[1] Like the United States early in the twentieth century, Saudi Arabia is the "swing" producer, the big player who can cover shortfalls, impose order and discipline, and generally keep markets stable.

The third and final factor in the geopolitics of oil is far more prosaic: price. If the United States and its huge market determine who is in, and who is not, in oil geopolitics, and if the Saudis are the market enforcers, the price of oil is the impulse, the electrical charge that sets the entire geopolitical machine in motion. Price determines the direction and rate of flow of international money and political influence. Price dictates how fast or slow economies grow, and whether recoveries take or falter. Price also controls how much energy we use, and thus whether we consume or conserve, stay with current energy sources or develop new ones.

Because price is so critical, players are forever seeking to manipulate it. Big importers like the United States and Europe, whose economies are built on cheap oil, do everything they can to keep prices on the low side and will routinely bring diplomatic pressure to bear on OPEC when prices get too high. (The United States will also pressure OPEC when oil prices are too low, because low prices hurt U.S. oil companies and destabilize oil-dependent allies like Mexico.) Oil companies, too, try to manipulate the market, exploiting everything from rumors to artificial disruptions in supply to move prices and make money. In a tactic known as "squeezing the market," for example, oil companies will buy up twenty or thirty tanker loads of a particular grade of oil, such as Arab Extra-Light, or West Texas Intermediate. Such a move can temporarily drive up prices for that particular grade by as much as five dollars per barrel — and allow oil companies to make a tidy profit when the "squeezed oil" is sold.

Of course, oil states have tried to use price as a weapon, by withholding supplies in order to drive prices up — or, alternatively, flooding the market to bring prices down — although these tactics almost always backfire. Pushing prices too high or too low invariably sets off a destructive chain of events that has, on several occasions, started wars and come disturbingly close to wiping out the world economy. This is why, after fifty years of painful experimentation and catastrophe, price stability has become the overriding goal for countries as politically divergent as Saudi Arabia, Russia, and the United States. As a Middle Eastern oil executive once told me, "after price, everything else is secondary."

In the iconography of oil geopolitics, some of the most potent images are a series of grainy, black-and-white newsreels shot in 1939 on the Persian Gulf as the Saudis load oil aboard a tanker for their first export shipment. When I saw the film, it had been incorporated into a promotional bit shown to visitors at the headquarters for Saudi Aramco, the state-owned oil monopoly. The jittery images were now accompanied by jolly background music and the booming voice of an English narrator, and the effect was like that of a high school civics film. Here is the ailing King Abdul Aziz, the great military strategist and unifier of the Arab peninsula (and one of the more cash-poor monarchs of the early twentieth century), opening the spigot and officially allowing the oil to flow to the tanker. Here are the old pier and the tanker, a ramshackle old vessel a fraction of the size of today's supertankers. Here are the American oil engineers and executives, shaking hands with their Arab counterparts. The Americans smile easily and strike poses to suggest that this is indeed a Great Moment in Modern History. The Arabs seem vaguely bewildered. They stare into the camera with expressions of uncertainty, as if they have no idea what it all means.

Modern Saudis are very sensitive about the origins of the U.S.-Saudi oil relationship and are annoyed by suggestions that Aziz traded his country's oil wealth for a promise of protection from President Roosevelt. "The Saudis *welcomed* the Americans, at a time when the region was only too happy to get rid of the British and the French," one of Saudi Aramco's executives reminded me during one of many lectures on U.S.-Saudi relations. "The talk between Aziz and FDR was about principles, not just 'give me oil and I'll protect you.'"

Whatever the original intent, the effect of the oil venture was to draw Saudis into a modern-day deal with the devil. In exchange for steady oil revenues and a shot at modernity, the Saudis were forced to play the dual role of supplying cheap oil to a consortium of American oil companies (charitably titled the Arab-American Oil Company, or Aramco), while simultaneously buffering those companies against the price swings that make oil such a risky business.

For as the international oil companies had learned long before, oil is inherently volatile. Production is continually running ahead of demand or behind it, causing the market to shuttle back and forth between shortage and glut and creating huge swings in price. This volatility leads to additional costs: either for consumers, who must pay more for oil in a tight market, or for producers, who earn less for their oil in a glutted market. These costs are known as the burden of adjustment, and much of the recent history of oil has been characterized by a ceaseless battle between producers and consumers, each group trying to avoid the burden of adjustment by pushing it onto the other.

With the emergence of the modern oil state, the international oil companies had found a new beast of burden. When world oil prices fell — for example, when Moscow tried to destroy capitalism by flooding the market with Soviet oil or when the United States banned oil imports to protect its failing domestic oil producers or when the oil companies themselves simply screwed up and pumped too much oil — it was the oil states that paid the price. The international oil companies, instead of allowing their own profits to fall, would unilaterally drop their "posted price" for oil, thereby cutting the royalties they paid to the oil states. This tactic ensured huge profits for oil companies but forced the Saudis, the Iraqis, the Venezuelans, the Kuwaitis, the Libyans, and other oil-state vassals to swallow the cost of price swings.

The oil states complained bitterly, but they had few options. Although technically they owned the oil, the international oil companies had the technology, the expertise, the capital, and, above all, the markets, necessary actually to produce and sell the crude. Worse, while the early oil states were essentially acting on their own, the international oil companies enjoyed great fraternity, and could collectively set prices to keep the oil states in line. Thus, if a Saudi Arabia or a Venezuela got uppity and demanded fairer treatment, the international oil companies would jointly threaten to take their business to some other, more compliant oil state.

Oil colonialism was an arrangement that was hugely profitable for oil companies yet which inevitably created unbearable political and economic tensions within the host countries, until at last something gave. On September 14, 1960, Venezuela persuaded Iran, Iraq, Kuwait, and Saudi Arabia to form the Organization of Petroleum Exporting Countries, or OPEC, a political entity that would bring new meaning to the idea of oil as a political commodity — and in the process utterly reshape the world political order. No longer would the major oil companies dictate the price of oil: over the course of the 1960s, OPEC — which eventually included Algeria, Indonesia, Libya, Nigeria, Qatar, and the tiny United Arab Emirates[2] — began collectively raising the price of oil and forcing the majors, and oil-importing countries like the United States, to take back some of the burden of adjustment they had thrust upon oil exporters.

At first, the transfer of power was slow and almost cordial. Oil states gradually raised their royalty rates, and the majors promptly passed the new costs on to consumers. But in 1969 Libya escalated the divorce by brazenly cutting its production and causing the worldwide oil market to tighten. As prices rose, Saudi Arabia and the rest of OPEC, seeing their opportunity, raised their own prices to take advantage of the tight market. The majors managed to negotiate a price ceasefire with OPEC, but it was clearly a holding action. In 1971, Venezuela dramatically raised the royalty it charged oil companies for each barrel they pumped out to 70 percent and announced plans to nationalize its oil industry. As one OPEC nation after another followed suit, the majors, the big importing nations, and the rest of the old oil order confronted a world in which more than 53 percent of the world's most vital resource lay under OPEC control.[3]

Then, in 1973, all pretense of an oil "partnership" between exporters and importers vanished. First, OPEC unilaterally raised oil prices by another 70 percent, to $5.11 a barrel. Next, in response to the 1973 Arab-Israeli war, OPEC's Arab members embargoed oil shipments to the United States and the Netherlands. The effects were staggering. As OPEC members cut production, the United States was unable to cover the shortfall: the former long-term "swing" producer was hamstrung by its own falling production. Washington briefly considered taking Middle Eastern fields by military force but held back, in part because of threats from the Soviet Union.[4] Powerless, the West could only sit and watch as oil prices more than quadrupled, to well over $20 a barrel. The oil crisis had begun.

The 1973 oil embargo utterly transformed the map of world power. The OPEC cartel had emerged as the new kid on the geopolitical block, an international bogeyman that controlled more than half the world's oil and was capable of laying low the once-invincible Western powers. In a matter of months, the global flow of revenues and power essentially reversed course, as the United States, Europe, and Japan began exporting enormous sums of cash to OPEC. By 1979, as the Iranian hostage crisis drove oil prices to thirty-four dollars a barrel, OPEC's annual earnings soared to the modern equivalent of nearly three-quarters of a trillion dollars. It was the largest, most sudden redistribution of wealth in history — an economic revolution on a scale never before imagined. For a time, OPEC members were earning more money than they could spend, a bizarre situation that caused a temporary shortage of cash in the world's financial markets.

On the main drag in downtown Riyadh, you can see where much of that money has gone. The city boasts department stores, convenience shops, hotels, and even a few skyscrapers, including a bizarre forklike glass tower built by one of the many Saudi royal princes. True, the sidewalks are virtually empty of women, and the few I see are accompanied by men and wrapped head to foot in the black purdah. Still, the streets are wide and smooth and crowded with taxis and large American cars and offer easy access to a network of freeways that look remarkably Western.

Indeed, from certain angles, and at certain times of day, downtown Riyadh is hard to distinguish from any major urban center, and this similarity, apparently, has not been lost on my hosts at the Ministry of Petroleum. Between interviews, my driver, a large and very friendly man named Hamdan, has been instructed to show me the glories of the Saudi economic miracle. We do the requisite Western journalist's tour of the gorgeous markets with their gold and incense; the old citadel where King Aziz took power nearly a century ago; and, of course, the infamous "chop-chop" square outside the governor's mansion, where public executions are held each Friday. Hamdan and presumably his employers also want me to see the new Saudi Arabia, which, in Riyadh, manifests itself in a veritable explosion of expensive modern infrastructure: huge government buildings, fantastic museums, sprawling universities, and a hospital bigger than any I have ever seen in the West — all paid for with oil money.

Of course, not all the petrodollars have been spent so prudently, in Saudi Arabia or elsewhere. Like paupers who have suddenly won the lottery, a generation of royal families, military dictators, and autocrats went crazy, spending lavishly on everything from statues and mansions to racehorses and yachts — and, in some cases, sophisticated American and European weaponry — and in the process recycling billions of petrodollars. In Venezuela, for example, oil wealth encouraged not only huge public works projects but artificially high wages and luxury imports, not to mention one of the highest per capita consumption rates of Scotch whiskey in the world. Even when oil revenues dipped, Caracas continued pouring billions into dubious state-owned enterprises and, when deficits appeared, began borrowing heavily from international lenders in the expectation of future oil earnings.[5] "There must be examples of worse fiscal management than that of Venezuela in the last eight or nine years," noted one retired Venezuelan diplomat in 1983, "but I am not aware of them."[6]

Oil states quickly found political uses, as well, for their wealth. Venezuela lent billions of oil dollars to its poorer neighbors and, with Mexico, created a pact to fend off U.S. "economic hegemony."[7] Iran funded terrorists in Lebanon and Palestine. The Saudis were particularly shrewd. Desperate to curb fundamentalist opposition at home, Saudi royals gave huge bribes to radical mosques, then financed the Islamic "revolution" in places like Afghanistan and Pakistan, essentially exporting a generation of young Saudi radicals and sowing the seeds of today's militant Islam.

Predictably, so much new wealth and power were difficult to manage. Within OPEC, jealousy and rivalries arose among fellow oil states. Iran jousted with Saudi Arabia for domination over OPEC affairs. The Saudis fought with Venezuela over pricing policy. Iraq accused Kuwait of stealing oil from a shared field on their border and then, in 1980, invaded Iran with the goal, among others, of capturing an enormous Iranian oil field.

But OPEC's biggest weakness was its profound misapprehension of the mechanics of oil power, particularly the setting of prices. As the owner of the world's cheapest oil, OPEC could easily have used its lower production costs to outsell its rivals, like Russia or Mexico — countries that needed to charge more per barrel to make a profit. Such a low-cost strategy would have let OPEC gain a majority share of the world oil market, while still earning a reasonable price for its oil. To succeed at such a strategy, however, OPEC couldn't get too greedy. If cartel members tried to push prices too far up by withholding their own production (and thus tightening world supply), the effects would be disastrous. Importing nations would either turn to non-OPEC suppliers (thus reducing OPEC's precious market share), or they would simply use less oil, either by switching to cheaper fuels, like coal or gas, or by becoming more energy-efficient.

So when oil prices skyrocketed during the 1970s and early 1980s, OPEC would have been wise to pump a little more oil and let prices fall slightly. That way, the cartel would have ensured a long-term market for oil by reassuring the big consumers, like the United States, Europe, and Japan, that oil was a reliable, economical, long-term energy source. True, OPEC's revenues would have fallen off a bit; but by protecting its market share and its customers, the cartel could have make up any losses later, when prices recovered, as they inevitably would have.

Instead, OPEC did the opposite. Addicted to the higher oil revenues of the 1970s, OPEC members refused to reduce their prices. The high prices acted as a brake on global economies accustomed to cheap energy, and the entirely predictable result was widespread recession. Energy demand fell, and importing nations tried to "wean" themselves from "foreign" oil. Utilities and other industrial users switched to coal, natural gas, and nuclear power, which were now cheaper. Homeowners began heating with natural gas instead of furnace oil. Governments in the United States, Japan, and Europe, embarking on a crusade for energy conservation, poured billions of dollars into alternative fuels and technologies and forced automakers to build fuel-efficient vehicles. For the first time in nearly a century, oil was losing its allure as the miracle energy source, and the impact was staggering. By 1986, world oil demand had fallen by five million barrels a day.

Worse, just as oil demand was falling, a wave of new oil production hit the market. Norway, the United Kingdom, the United States, the Soviet Union, and other non-OPEC countries, whose oil was normally too expensive to compete with OPEC's, now scrambled to take advantage of the high oil prices. Between 1978 and 1986, non-OPEC oil production jumped by fourteen million barrels a day — and most of this increase came at OPEC's expense. Between falling demand for its own oil and rising non-OPEC production, OPEC saw its share of a dwindling market shrink from more than 50 percent to just 29 percent.[8] In retrospect, says one former U.S. State Department official, it is clear that "OPEC [members] had no idea what they were doing. It was totally unrealistic of them to think they could keep prices that high for as long as they did and not have a huge impact on demand."[9]

Desperate to avoid further damage, Saudi Arabia, OPEC's most powerful member, tried enforcing a production limit, or *quota,* on each member, to reduce supply and shore up prices. But other OPEC members refused. While most saw that cutting production could bring higher prices eventually, in the short term, it would mean an immediate loss of oil income — something no formerly free-spending petrostate could withstand. In Nigeria, desperate oil officials actually cut their prices in an attempt to boost sales and grab back some market share from non-OPEC countries. Mexico, too, lowered its prices.

The Saudis now found themselves in the classic cartel bind: the only way to keep prices high was to cut their own production, as they reluctantly did, letting it fall from 10 million barrels a day in 1980 to a mere 2.5 million by 1985. However, this remedy too proved disastrous. Although prices did rise, the Saudi market share was now so tiny that its overall oil revenues remained dangerously low. As the situation worsened, the Saudi royal family felt it had little choice but to turn the "oil weapon" on OPEC itself. Opening its taps, the Saudis flooded the world market with cheap oil.

This first use of "capacity cleansing" was brutal but effective. As prices plunged below ten dollars a barrel, Venezuela and other OPEC quota-busters capitulated and cut their production. Saudi Arabia regained its lost market share. Better still, from OPEC's point of view, rival oil operations in high-cost areas like the North Sea and Alaska suddenly became uneconomical, and many were scaled back or even put on hold. These developments hit the Soviet Union, until then the world's largest producer of oil, particularly hard. As falling oil prices cut Moscow's hard-currency income in half, the Soviet oil industry — and the Saudis' biggest oil rival — was knocked out for years.

For political leaders in the West, however, OPEC's floundering seemed further proof that the cartel would only continue to destabilize the world's most important commodity. The price collapse was undermining oil-exporting U.S. allies like Norway and England. Just as important, it was hammering the oil industry. Western oil companies, including the five United States–based majors, lost billions of dollars in revenues. Many American independent oil companies were ruined; others were snapped up in a merger-and-acquisition craze that would reshape the business. Houston, the once-regal center of the oil universe, was a ghost town. In 1987, George Bush senior, then vice president under Ronald Reagan (and, not co-incidentally, a Texas oilman), was sent to Riyadh to persuade the Saudis to stanch the flood of oil and bring prices up to eighteen dollars a barrel — a price everyone agreed would be fair to both consumers and producers.

This was hardly the first time the United States had intervened in the oil markets, nor would it be the last, for in fact, by the end of the 1980s, the politics of oil were about to reach an entirely new level of intensity. Again, the issue would be price, and again the major players would be Saudi Arabia and the United States — along with a relative newcomer, Saddam Hussein, the Iraqi strongman who would be the pivotal figure in an oil saga that blended the drama of international politics with the pettiness of a family feud.

In 1989, having just finished a long and costly war with Iran, Saddam was desperate to sell as much of his oil as he could to replenish his depleted treasury. His neighbors, however, had no interest in seeing Saddam get any richer or stronger. Kuwait in particular feared Saddam and, in an effort to deprive the Iraqi leader of oil revenues, stepped up its own production, intentionally flooding the market and as a result depressing prices. Saddam was not amused. He regarded the Kuwaitis' tactics as tantamount to economic war — he could claim that Kuwait was "stealing" Iraqi oil revenues — and made it clear he would take military action. Too late, the Saudis saw the danger: if Saddam invaded Kuwait, he would probably press on into Saudi Arabia. Desperate to placate the well-armed Iraqi dictator, the Saudis cut their own production and begged Kuwait and other OPEC states to do the same, to push prices back up to twenty-one dollars — high enough, it was hoped, to mollify Saddam and dissuade him from attacking anyone.

The tactic might have worked. Now, however, Venezuela refused to play along. Still reeling from the price collapse of the 1980s — and never terribly interested in Middle Eastern politics — the Venezuelans opened the taps. That move, coupled with similar cheating by United Arab Emirates, effectively destroyed any hope of price appeasement. By 1990, Saddam had massed troops on the Kuwaiti border and, believing the United States to be unwilling to risk a war just for oil, launched his invasion.

The first Gulf War was the first military conflict in world history that was entirely about oil. Interest in oil had not only prompted Saddam to invade but had largely defined the world's reaction. While government spokesmen made much of Kuwaiti suffering, behind closed doors, diplomats were focused almost entirely on the loss of Kuwaiti oil and, more to the point, whether the attendant price spike would tip the world into recession. Indeed, Washington's first decisive act was to obtain assurances from Riyadh that Saudi Arabia would pump enough extra oil to cover the loss of combined Kuwaiti and Iraqi production, thus preventing what everyone agreed would be the worst aspect of Saddam's aggression — a long-term disruption in oil supply.

Oil was also central to the speed and scale of the United States–led military response. Despite much talk in Washington, London, and other Western capitals about "protecting the sovereignty of Kuwait," the only reason the United States so quickly won international support for military action was that no industrialized nation could countenance having so much of the world's oil supply under Saddam's control. Saddam's imperial ambitions were well known; after Kuwait, he was sure to turn south toward the militarily weak Saudi Arabia, which offered him potential control over nearly a fifth of the world's oil production and nearly a third of its oil reserves. (Indeed, the merest thought of Saudi oil fields ablaze was sufficient to send prices through the roof.) Oil is also the key to the relative lack of opposition, even among Arab states, to Washington's plans to attack Saddam. Not only did Saddam's neighbors wish him ill, but most realized that, with Iraqi oil temporarily off the market, Saddam's three-million-barrel-a-day market share would be up for grabs.

Oil was even the basis for a grotesquely fitting final act of war: as Saddam's elite Republican Guard fled before the advancing U.S. tanks, Iraqi soldiers torched Kuwait's oil fields, in the process creating a continent-sized black cloud that would linger in the atmosphere for years.

With the end of the Gulf War, the geopolitics of oil shifted yet again, moving the world a step closer, or so it seemed, to an era of energy stability. Riyadh emerged as the undisputed leader of OPEC. More important, the United States was now the dominant power in the Middle East — and, for better or for worse, responsible not only for regional stability but for the security of two-thirds of the world's oil supplies.

In a real sense, Washington had regained its old role as the world's oil power — even if this time most of the oil no longer belonged to the United States. Beyond acting as the Gulf's regional policeman, the United States had assumed the role of protector of the world oil market and guarantor of oil price stability. If a disruption threatened to send up oil prices, the United States had shown a new willingness to restore stability, and market confidence, by releasing its strategic oil reserves and, if necessary, by using military force. Moreover, the United States had persuaded a newly grateful Saudi Arabia to use its huge surplus capacity — its ability to bring to market, virtually overnight, a flood of oil — not as an oil weapon, but as a supply cushion, a hedge against the disruptions and price spikes that had proved so disastrous to economic growth. (So grateful were the Saudis, in fact, that during the postwar recession, they willingly overproduced, to keep oil prices down and thus encourage a U.S. economic recovery.)

Together, the United States and Saudi Arabia had emerged as the managers of the global energy order. The Saudis would supply the oil; the Americans would supply the protection, via an expanded military presence in the Gulf — including a billion-dollar military command center near Riyadh — as well as a growing network of military bases and stepped-up diplomatic missions in and around the entire region, from Africa north to the Caspian.

The oil markets, too, appeared to be achieving the stability that had eluded oil players for so long. Decades of price volatility had left painful scars on both producers and consumers of oil, and although oil would continue to be among the most powerful factors in international relations, the oil weapon itself was seen more and more as a suicidal option. Oil policy for both exporters and importers, whether they stated it or not, shifted toward the general goal of stabilizing prices at a level that satisfied all the dominant players within the oil regime: not so low as to harm oil companies and oil states, but not high enough to harm the economies of the major oil consumers — or worse, encourage conservation or alternative energy technologies.

On nearly all levels, the 1990s seemed to be the golden years for the oil order. Oil demand had returned in a big way. The world economy was booming again, especially in the United States and Asia. Also, earlier trends toward energy conservation and efficiency seemed to have diminished under the corrosive influence of low energy prices, and conspicuous energy consumption was experiencing a resurgence. Nowhere was this more manifest than in the United States. Convinced that the energy crisis had been laid to rest by the defeat of Saddam Hussein — this was, after all, a war over oil — and inspired, perhaps, by the display of off-road vehicles during the Gulf War, America became a nation of truck- and SUV-drivers, in the process helping to spark a new boom in the oil markets. Oil companies thrived, and OPEC regained much of its old power. With Russian oil still off the

market — the Russian industry still hadn't recovered from Riyadh's mar-
ket-cleansing move in the 1980s — nearly all the new demand was met by
OPEC producers, who saw their market share rise to 40 percent.

Yet for many in the West, the Gulf War had simply reemphasized the
fundamental flaws in the oil order. Even if OPEC had declared an era of
price stability, Western observers, particularly in the United States, continued to argue that as long as oil remained under the political control of
states like Saudi Arabia and Venezuela, volatility would pose an enormous
risk to the fast-growing global economy. Research showed that after each of
the six major oil price spikes since the Second World War, global economic
activity had begun to fall within six months; typically, every five-dollar increase in oil prices brought a .5 percent decline in economic growth. Worse,
the effects of price hikes were "asymmetrical." When prices came back
down, economies usually regained only about a tenth of what they had lost
in the preceding spike. Cumulatively, according to energy economist Philip
Verleger, price spikes had cost the economy 15 percent in growth, and more
than a $1.2 trillion in direct losses, "as well as uncountable costs in personal
dislocations."[10]

On top of these concerns about volatility, a new, related worry was
emerging: political instability. Although the first Gulf War was supposed to
have increased the security of the world's largest oil reserves, world oil supplies actually seemed less secure. Members of OPEC were still fighting
among themselves, cheating on their quotas, and making it impossible for
Saudi Arabia to enforce discipline and keep prices stable. The secrecy within
OPEC — many members refused to publicly state how much oil they were
shipping on any particular day — left markets in a permanent state of anxiety, as traders could never be sure whether supply would actually meet
demand. In Venezuela and Nigeria, civil unrest and strikes, spurred in part
by popular dissatisfaction over management of oil revenues, had nearly
caused a civil war and had repeatedly shut down oil exports.

The biggest source of instability, though, seemed to be Saudi Arabia
itself. The keystone of the global oil industry showed increasing signs of an
imminent meltdown. The country's finances were in shambles. The fabled
Saudi oil fortune from the 1970s was gone, depleted by the lavish lifestyles
of an estimated fifteen thousand royal princes, as well as by a welfare state
that had once rivaled northern Europe's but which was now collapsing
under a population explosion. Other costs came into play: the fifty billion

dollars owed to the United States for defending the kingdom against Saddam; plus hundreds of millions of dollars the royal family was still spending to placate Islamic fundamentalists. By some estimates, to meet its various obligations, the Saudis needed to bring in at least twenty-five dollars for each barrel of oil. In other words, the low production costs of Saudi oil were increasingly irrelevant. With such financial obligations, Saudis had no choice but to keep oil prices high. And that pressure, industry analysts and Western policymakers now believed, all but ensured market chaos: as Saudi-led OPEC tried to control supply to keep prices up, the cartel was bound to make mistakes, thus causing even greater price volatility.

These fears were quickly confirmed. In 1997, Saudi Arabia again tried to punish an overproducing Venezuela by initiating another round of capacity cleansing. But Riyadh, focused on Venezuela, had failed to notice that the overheated Asian economy was slipping into recession and cutting worldwide oil demand — just as the Saudis were flooding the market. This one-two punch sent prices down to ten dollars a barrel — a boon for big consumers, but a costly hit for international oil companies and a potentially devastating blow for Saudi Arabia and other oil exporters. Desperate to raise prices, Saudi Arabia and OPEC then made a series of deep production cuts in 1998 and 1999 but, not surprisingly, went too far. The markets tightened, sending the price of oil up well past thirty dollars a barrel, a hike that hammered the world economy and sparked a highly politicized "energy crisis" in the United States just in time for the election of U.S. president George W. Bush, another former oilman who, within less than a year, would embark on a controversial campaign to reassert U.S. control over the world energy order.

From the moment Bush took office in 2001, critics derided his energy policies as representing little more than the agenda of an international oil industry — an agenda that included maximizing oil production, and, above all, regaining access to the big, fat Middle Eastern oil fields. Exhibit A, critics say, is the Iraqi war, which, although repeatedly justified by the administration as part of the "war on terror," was actually a ploy to regain control of the world's second-largest oil reserves and revive American oil imperialism. These complaints have some merit, not least of all because Bush's strongest political support (and the bulk of his 2000 campaign contribu-

tions) comes from the oil industry. In the context of oil geopolitics, however, it is more useful to see the president's foreign oil policies as part of a
larger, older campaign by U.S. neoconservatives against what they saw as
one of the biggest threats to American power: oil price volatility.

According to this so-called neocon view, in the twenty-first century
the United States no longer has conventional rivals for global dominance.
In the post–Cold War era, the only real risks to American primacy are the
threats posed by energy disruption and, to a lesser degree, world terrorism.
And in the minds of many neoconservatives, these two threats neatly in-
tersect in OPEC's continuing control over Middle Eastern oil. The artificially high prices OPEC imposes have insulated oil-state autocrats from
the winds of political change, while allowing them to fund their increasingly anti-American paramilitary agendas. At the same time, and perhaps
even more significantly, OPEC's self-serving and shortsighted efforts at
"price management" have brought decades of high prices and volatility that
have eroded economic growth and, therefore, American power.

As far back as 1975, as the Arab oil embargo slowly strangled American
economic might, conservative economists and policymakers were searching for ways to defeat OPEC. Although the Nixon administration's plans to
take OPEC's Middle Eastern oil fields physically were shelved, the dream of
a post-OPEC oil order was kept alive by a cadre of neoconservative American analysts and policymakers — among them, Paul Wolfowitz, now deputy defense secretary, Richard Perle, a top adviser to Defense Secretary
Donald Rumsfeld, and, of course, Rumsfeld himself.

In the 1980s, the neocons had supported sanctions against oil sales
from Libya and Iran, in hopes of depleting their terrorist budgets — a
move that earned them the scorn of big oil companies. A few years later,
some neocons began arguing that even Saudi Arabia, that stalwart oil ally,
was looking less and less loyal: not only were members of the Saudi royal
family reported to have spent five hundred million dollars to export radical
Islam, but Riyadh was the ringleader of a pricing regime that was hurting
American interests. "For a lot of conservatives, the Middle East, or a significant part of the Middle East, has effectively been at war with the United
States ever since the 1970s," says a policy analyst with close ties to the Bush
administration. September 11 "was just one final argument that these elements need to be taken care of."[11]

And the key to "taking care" of those elements was Iraq, a country that had at least 150 billion barrels of crude and, except for Saudi Arabia, the cheapest production costs in the world.[12] Months before the September 11 attacks, when Vice President Cheney (another former oilman) was drawing up a new national energy policy, he and other White House energy strategists had pored over maps of Iraqi oil fields to estimate how much Iraqi oil might be dumped quickly on the market. Before the war, Iraq had been producing 3.5 million barrels a day, and many in the industry and the administration believed that the volume could easily be increased to seven million by 2010. If so — and if Iraq could be convinced to ignore its OPEC quota and start producing at maximum capacity — the flood of new oil would effectively end OPEC's ability to control prices. As supply expanded, prices would fall dramatically, and not even the Saudis with their crying revenue needs would be able to cut production deeply enough to stop the slide. Caught between falling revenues and escalating debts, the Saudis, too, would be forced to open their oil fields to Western oil companies, as would other OPEC countries. The oil markets, free at last from decades of manipulation, would seek a more natural level, which, according to some analysts, would be around fourteen dollars a barrel, or even lower — a price much more conducive to long-term economic growth.

Toppling OPEC wouldn't be easy. Reviving Iraq's moribund oil industry would take massive infusions of capital. By some estimates, it will cost five billion dollars just to resume prewar production levels, and at least forty billion over the long haul. That kind of money could come from only one source — the international oil companies — which would invest in Iraq only if a) Saddam were gone and b) they received some assurance that they would have a share in production revenues and that the market, and not OPEC, would determine production levels.

To be sure, after the September 11 attacks, the question of Iraqi oil became vastly more complicated. Bush officials now insisted that Iraq possessed weapons of mass destruction and had links to al-Qaeda. After a few ill-advised comments by Cheney about the threat that Saddam posed to regional oil supplies, White House officials ceased talking about oil supplies as a rationale for war — and indeed began strenuously objecting to suggestions that the war was "about oil."

Such denials were patently absurd. As the war unfolded, even casual observers could see the priority that U.S. forces gave to securing the massive oil fields in Kirkuk and cordoning off the Ministry of Petroleum in

Baghdad (while the rest of the city collapsed into anarchy): this war obviously had at least *something* to do with oil. Nonetheless, though the war was "about oil," that was true in a way that most of Bush's critics failed to grasp. It wasn't simply that an Iraq without Saddam would enrich Bush's energy industry allies (although it would). Nor was the connection merely that war in Iraq would bolster America's military and economic presence in the region — or keep Iraqi oil from falling into the hands of Chinese, Russian, and French oil companies — although this too was an intended effect. Rather, it was that liberating Iraq, and its oil, was key to the neoconservatives' vision for the future of American power — and for the new geopolitics of oil.

It is a radical vision. At a stroke, the administration hopes to depoliticize what has for nearly a century been the quintessential political commodity and, in the process, remove the last real obstacle to American power. As Michael Klare, professor of world security studies at Hampshire College, told the *Toronto Star* last year, in the eyes of the Bush administration, unlocking OPEC oil, "combined with being a decade ahead of everybody else in military technology, will guarantee American supremacy for the next fifty to one hundred years."[13] Cheney and Rumsfeld "see control of oil as merely part of a much bigger geostrategic vision," argues Chris Toensing, an analyst who works on the Middle East Research and Information Project. "By controlling the Gulf and the Middle East, the United States gains leverage over countries that are more dependent on the Gulf for oil, like China and Europe."[14]

In this context, it is hardly surprising that the Bush administration's energy policy has been so lopsidedly slanted toward oil. Whereas many energy experts, particularly those in the left-of-center advocacy community, saw 9/11 as a prime opportunity to renew the efforts to move away from oil altogether, the Bush administration drew the opposite lesson. For Bush, the lesson to be learned about energy insecurity was not that the West should use less energy, as it did in the early 1980s, but that the West should be willing to make energy more secure and less unpredictable, as America had tried to do during the first Gulf War. During that war, rather than simply retreating into a defensive energy policy, the West had taken a bolder, more muscular internationalist approach and had simply removed the threat to price stability.

A decade later, U.S. officials saw no reason this Gulf War policy should not continue, even be enlarged. Indeed, given the increasing importance of energy to economic growth, given our greater understanding of the risks of disruption and volatility in energy supply, and given the growing shakiness of OPEC, any policy that did not seek to stabilize oil permanently would merely postpone disaster. For the United States and its partners in Europe and Japan, retreat was no longer possible: energy security meant oil stability. "It was as if there were two separate debates," recalls a former State Department energy official. "Outside the Beltway, it was 'Oh my God, this is what we get for importing Saudi oil.' Inside the Beltway it was 'We can take care of the oil problem by redrawing the map of the Middle East.'"[15]

Not surprisingly, the rest of the world, and especially Middle Eastern OPEC countries, are not quite ready to have that map redrawn. Although OPEC has been working hard since the September 11 attacks to deflect neoconservative criticism by keeping prices stable, the cartel has also made clear that it has no intention of relinquishing its control over prices anytime soon. Saudi Arabia, for example, has been mending fences with the Russians, and last year, Moscow and Riyadh announced a new oil pact that, if successful, could give these two top producers even greater control over prices. So bold is OPEC feeling that last fall, as rumors swirled that the White House was pressing Iraq to leave OPEC, the cartel boldly announced a production cut of nearly a million barrels a day, pushing prices up by a dollar per barrel and, in the minds of many observers, sending a clear signal to Washington that Iraq was not up for grabs. President Bush, firing back his own diplomatically worded warning, seemed to take the point: "I would hope our friends in OPEC don't do things that would hurt our economy," Bush told reporters.[16]

At a time when OPEC is supposed to be on the ropes, such boldness reflects a wider confidence within the cartel that American plans for a new, freer oil market may be somewhat premature. In the first place, in OPEC's view, world oil demand is set to resume its rapid growth, especially as the Asian markets heat up, and no amount of new oil from Russia or the Caspian will cover the coming shortfall. According to this calculation, in the short term, demand for OPEC oil will remain high, and the cartel will retain its pricing power. In the longer term, OPEC sees an even brighter future. Like analysts elsewhere, OPEC officials believe that the current surge in non-OPEC oil production is only temporary; it could peak as early as 2015. At that point, the world will be forced to turn to OPEC, and in particular to the Middle East, for its oil, effectively ending any American dream of a "free" market and fourteen-dollar oil prices. "The highest-cost producers come in first," one Saudi oil official reminded me, "and when they run out of the last barrel that needs twenty dollars to come in, then our oil comes in."

Admittedly, the ultimate question in oil geopolitics is whether Saudi Arabia and its unstable and corrupt OPEC brethren can survive until the high-cost producers run out. But the cartel has made numerous comebacks and repeatedly confounded predictions of its demise. "What is amazing about OPEC [countries] is just how well they've managed to manage the market," says one former U.S. State Department official who has watched OPEC since the 1970s. "Everyone keeps predicting 'the end,' but look where prices have been for the last few years. I'd say they're doing pretty well."

In Riyadh, meanwhile, on the spacious upper floor of the Ministry of Petroleum, al-Naimi had resumed his calm and controlled demeanor. When I asked whether OPEC could withstand a sudden deluge of non-OPEC oil, he nodded with assurance. "If there were a need for four million barrels of new production, we would make room," he promised. More oil was actually better than less oil, he said. Surplus keeps prices moderate and discourages new oil production in Russia or the Caspian or Africa. "It doesn't help the cause of Arabs to leave the oil in the ground," he said.

And besides, al-Naimi told me, OPEC was not under any particular threat. Earlier rumors about countries defecting and threatening to flood the market with oil had always proved unfounded. At one point, the United States had reportedly been trying to pry Nigeria loose from the oil club. And only a few months before, as the Saudis and the Russians moved closer to a price war, President Bush was said to have asked the Russians *not* to cooperate when OPEC ministers begged Moscow for help in cutting production and halting a price slide.

Al-Naimi then offered what sounded to me like a cautionary tale for any country that might try to take on OPEC at its own game. In March 2002, after Venezuelan president Hugo Chavez had been temporarily driven out by a coup, the markets were awash with rumors that the new, pro-business junta meant to pull Venezuela out of OPEC. According to the theories, the new regime, at the behest of Washington, would begin pumping oil at full throttle, thereby flooding the market with an additional four hundred thousand barrels a day and driving down prices faster than OPEC could prop them up. Confident that the cartel had finally been outfoxed, many oil traders bought up "short" futures contracts in the belief that oil prices were soon to plummet. "That was the analysis that was driving the market, and already the price was dropping," said Al-Naimi. But two days later, oil prices reversed and began climbing again. And, said Al-Naimi, smiling faintly at the recollection, all the oil traders who had bet against OPEC "lost their shirts."

5

TOO HOT

THOUSANDS OF MILES from the red Saudi sands, in the frigid waters off the northern coast of Siberia, the cause and effect of the world's most complex energy problem circle each other in a bizarre dance. Since the 1970s, average temperatures here have risen five degrees Celsius, enough to cause the vast Arctic ice sheet to recede by about 3 percent and open a channel of ice-free water along the coast. Rising temperatures are also melting ancient ice fields farther inland and exposing dirt and rock — along with the occasional woolly mammoth — that have lain hidden for thousands of years. Because the newly bared soil is darker than the ice that once covered it, the dirt absorbs heat more readily, raises surrounding temperatures faster, melts even more ice, and exposes still more soil — a treacherous feedback loop that helps explain why Siberia is warming so much faster than other parts of the globe and why the Russian Arctic has become the poster child for those who want strong policies to reduce catastrophic climate change.

Not everyone is appalled at the situation. Whereas global warming is already bringing drought, crop failures, famine, flooding, and other calamities to parts of Africa, Asia, and southern Europe, higher temperatures may actually mean a net gain for northern countries like Russia. In Siberia and elsewhere, milder winters and longer growing seasons may act like a growth hormone on certain farm and timber yields; potato crops, for example, are expected to jump by one-third. Russian shipping companies are already dreaming of an ice-free northern sea route along the Siberian coast that would allow oil tankers and other vessels to sail from Europe to Japan two weeks faster than they can via the Suez Canal.

Yet perhaps the oddest beneficiary may be the Russian oil companies and their Western partners. Although warming temperatures will turn the

Siberian tundra into a swamp and bog down drilling operations there, warming could make it easier and cheaper to reach some of the billions of barrels of oil and trillions of cubic feet of gas currently locked away offshore, beneath the Russian Arctic. Thinning ice will make offshore drilling less complicated and costly. Shorter, less severe winters will mean fewer disruptions in exploration and drilling, while a decline in the number and size of icebergs will reduce "downtime" for offshore oil and gas rigs.

To put it another way, while many observers and scientists see climate change as catastrophic, a warmer planet may be a boon for the cash-strapped Russian government, which today depends on oil exports for a third of all its revenues and which, as existing reserves decline, will have to tap its Arctic wealth if it wants to maintain exports, fulfill its new role as preferred oil supplier to the United States, and bring economic development to its own inhospitable north. "Some people believe Russia's north has no chance of surviving," says Vyacheslav Popov, a Russian politician from the port city of Murmansk, which is being rapidly developed as a loading port for oil exports to America. "I think this is totally wrong." Arctic oil, says Popov, "will boost Russia's economic security and help to restore our previous glory."

Given such hopes, one begins to see why Russia isn't aggressively championing climate protection and, in particular, wants little to do with the 1997 Kyoto Protocol, an ambitious if controversial international effort to reduce emissions of carbon dioxide, or CO_2. Russian officials do not deny that the earth's climate is warming, or that the primary cause is man-made CO_2. Nor do they disavow Russia's complicity: the Russian Federation currently produces 17 percent of all CO_2 emissions (in part because Russian factories, cars, and power plants are so obsolete and polluting), a figure that gives Russia the dubious honor of being the world's third-largest CO_2 exporter, right behind the United States and China.

Where Russians and Western climate specialists do part ways is on the question of how much can be done, or should be done, to mitigate Russia's contribution to the climate mess. The Russian economy is only just emerging from a near-death experience in the early 1990s, when the GNP shrank by nearly half and the government was on the brink of collapse. Even now, Moscow can barely afford to feed its veterans and pensioners, pay its soldiers, and guard its nuclear arsenal from privateers — much less save a climate that the entire world shares. In the eyes of Russian policymakers, reducing CO_2 emissions means either spending hundreds of billions of rubles replacing industrial infrastructure or cutting energy consumption and therefore economic output — neither of which fits in with Moscow's policy of maximum economic growth. Factor in the attitude, prevalent among some ordinary Russian citizens, businesspeople, and even scientists, that a warming climate might actually be beneficial, and it becomes clear why climate policy is not a priority for Russian politicians — and why optimism is not a strong suit among climatologists. As one American energy economist told me, "as far as the Russians are concerned, a little warming would definitely *not* be a bad thing."

Climate change is the latest and possibly greatest confirmation that our great mastery of energy may be more accurately described as a series of accounting errors. Though cheap, plentiful fossil fuels have clearly been key to our industrial success and continued economic vitality, we are discovering that our rosy picture of energy as the Key to Prosperity has omitted a number of serious costs, from geopolitical instability and oil price volatility to, now, rising global temperatures due to centuries of carbon dioxide emissions.

Just what climate change will end up costing us is unclear, but the early numbers hardly give cause for cheer. Estimates for the cumulative economic impact of rising sea levels, more frequent hurricanes and droughts, higher rates of infectious diseases, and other climate-related calamities range up to tens of *trillions* of dollars over the course of this century. Nor are the costs of halting climate change any less frightening. Because 90 percent of man-made CO_2 comes from the burning of gas, oil, and especially coal, and because gas, oil, and coal provide more than 85 percent of the world's energy,[1] we cannot "fix" our climate problem without making substantial changes to our energy economy — changes that go beyond privatizing OPEC or finding ways to drill through the Siberian ice. According to one analysis, making all the changes to our energy economy that would be necessary to slow CO_2 emissions could cost the United States alone a full percentage point of its GNP every year for the next century. As a consequence, Russia is not the only country to voice serious misgivings about climate policy, or to question whether the climate change is even worth stopping.

Although the causes of climate change are complex, most evidence points to a buildup in the atmosphere of "anthropogenic," or man-made, industrial pollutants, especially carbon, in the form of carbon dioxide, or CO_2.[2] Any activity that burns fossil fuel produces carbon, in surprising quantities. Burning a single gallon of gasoline, for example, releases five pounds of carbon — the equivalent of a small bag of charcoal briquettes. This means that most Americans generate a ton of carbon a year, simply by driving their cars. Burning a ton of coal yields nearly a ton of carbon, because coal is nearly pure carbon.[3]

As the carbon, in the form of CO_2, rises into the atmosphere, it disrupts the natural cooling mechanisms of the atmosphere. Like a one-way mirror, the CO_2 allows sunlight to pass through the air and warm the earth but then prevents much of the resulting heat from radiating back from earth into space. This is the infamous "greenhouse effect." Over the past hundred years, it has boosted average global temperatures by between one and three degrees Fahrenheit, depending on where the temperatures are measured.

To be sure, not everyone believes that temperatures have climbed so high or that the greenhouse effect exists or that anthropogenic CO_2 is playing any significant role in global warming. For years, skeptical scientists — some of them financed by skeptical energy companies — have claimed that the greenhouse effect is overblown and that the current warming trend is simply the latest in a progression of natural warming trends that have occurred throughout history. Scientists like Fred Singer, a prominent atmospheric physicist, former U.S. Environmental Protection Agency (EPA) official, and sometime energy industry spokesman, have argued that although anthropogenic CO_2 emissions might contribute to warming, the effects would be negligible. "Even if we do notice it, it will be extremely small and actually inconsequential," Singer told an interviewer several years ago. "It will be an interesting scientific curiosity, but it won't be of any practical importance."

In recent years, however, as new temperature data continue to confirm climate forecasts (as even some energy companies have acknowledged), skepticism about global warming now comes chiefly from a tiny minority of scientists, including Singer, plus a chorus of conservative policymakers, many of them American. These skeptics believe climate change to be a vast environmentalist conspiracy, launched by Eurosocialists and Luddites, to further regulate business and cripple the great industrialized economies of the world, foremost among them the United States. To cite but one of the more colorful examples: James Inhofe, a Republican U.S. senator from Oklahoma and chair of the powerful Environment and Public Works Committee, dismisses climate change as "the greatest hoax ever perpetrated on the American people."[4]

Among the majority of climate scientists, however, as well as a great many policymakers in Europe and even the United States, the consensus is that man-made greenhouse gases have pushed temperatures up by as much as three degrees Fahrenheit over the past century. Such an increase may seem fairly innocuous — until you realize that the end of the last Ice Age was triggered by an increase of only three degrees. Just as important, whereas the post–Ice Age planet required five thousand years to warm up by three degrees, our own small warming trend has taken less than a century and has already been sufficient to trigger some fairly frightening alterations, including a 15 percent shrinking of polar ice caps, a four-inch rise in sea level, and widespread retreat of glaciers, in addition to longer, more severe droughts, warmer winters, more floods and hurricanes, the spread of tropical diseases, and a string of record-breaking hot years. Of the sixteen warmest years since records were first kept in 1860, fifteen have occurred since 1980. The seven warmest years occurred during the 1990s;[5] one year — 1998 — appears to have been the warmest since the Norman Conquest.

Yet these changes, research suggests, give only the palest intimation of what is to come. Although the precise impact and timetable of climate change are still under debate, most climate researchers, including the respected United Nations Intergovernmental Panel on Climate Change (IPCC), contend that unless CO_2 emissions can be dramatically lowered in the next several decades, global temperatures will climb by as much as seven degrees Fahrenheit by 2050 and by as much as ten degrees by 2100. At these temperatures, we could expect a kind of Endless Summer, in which icecaps melt away completely, seas rise by twenty inches (and keep rising for centuries), island nations drown, entire tropical landmasses turn into deserts, species go extinct, and storms become more frequent and deadlier.

Much of what we call normal life would change. In the temperate zones, such the United States and much of Europe, prairies and farmlands would quickly become barren dustbowls. Forest fires would be more frequent and far more devastating. Summers would scorch, whereas winters would be much wetter. Wildlife populations would shift. Colder-weather birds and animals would migrate away or die off. Insects, bacteria, and viruses now confined to the tropics would move in, bringing a host of diseases not seen in temperate climes. According to a study by Belgian and British researchers, a five-degree (Fahrenheit) increase in temperature could give rise to eighty million new cases of malaria a year and allow the disease to spread to Australia, the United States, and Southern Europe.[6]

Cost estimates for such calamities are speculative at this point, but even the conservative figures are stunning. One study by British Energy found that a temperate increase of around four degrees Fahrenheit could lead to agricultural losses from crop failure, soil erosion, desertification, and flooding in excess of $265 billion per year worldwide. Potential climatic impact on drinking water supplies — ranging from the silting up of rivers and reservoirs to salt-water intrusion into drinking wells as the sea level rises — could exceed $300 billion a year. Human health costs would also climb. The spread of disease would drive up medical costs. More frequent heat waves, like the kind that killed thousands in Europe in 2003, would cause an even greater number of deaths. In one scenario that assumes a nineteen-inch rise in sea level, a 25 percent increase in hurricane activity, and a 10 percent increase in winter rain and snow, many nations would suffer a massive jump in mortality rates and billions of dollars in lost earning power.[7]

Worse, climate change is not an equal-opportunity disaster. Whereas the northern, and richer, countries might suffer relatively minor detriment or might even benefit in certain ways from global warming, the severest effects — rising sea levels, floods, and crop failures — will be felt disproportionately in Africa, in parts of Asia, and among some of the tiny island nations. These places are already battling drought, disease, and civil unrest and are far too poor to have even a hope of protecting themselves from a new onslaught. Famine is a critical concern. According to an Oxford University study, even a half-degree change in temperature will alter the monsoon patterns that now provide much of Asia with critical rainfall and will as a consequence reduce crop yields and displace tens of millions. With even a small temperature change, more than twenty-six million Bangladeshis would become refugees. As many as twelve million people would flee Egypt, while more than twenty million Indians would be forced to migrate.[8] As Rajendra Pachauri, chairman of the IPCC, told a reporter last year, for many of these countries, climate change "will represent the last straw on the camel's back."[9]

Researchers are particularly nervous about China, where the climate-related agricultural impact is expected to be the highest in the world and, according to some projections, could produce a hundred million famine refugees and generate political instability on an almost unimaginable scale. "As China's domestic pressures escalate," writes journalist Ross Gelbspan, "national security advisors all over the world will report periodically to their presidents on their country's military readiness to engage China if the need should arise."[10]

Gelbspan may be going too far. But it is not hard to understand why climate change is regarded as the ultimate energy problem, and perhaps the greatest challenge facing the architects and users of the next energy economy. Because hydrocarbons provide 85 percent of the world's energy today and, given current trends, will play a dominant role for decades to come, substantially reducing CO_2 emissions would entail more or less chucking the existing energy system and finding something new. Not surprisingly, this hasn't been an especially easy package to sell. Despite widespread understanding of the links between energy and climate, and despite broad consensus on the need to move *eventually* to a carbon-free energy economy, most governments, companies, and individuals are not yet ready to commit to such a radical program.

Developing countries have essentially refused to curb CO_2 emissions on their own. China, for example, is busy building its next energy infrastructure around coal, the most polluting, least climate-friendly fuel, and by 2020 will have surpassed the United States as the leading producer of CO_2 emissions. For their part, industrialized countries, though they emit the most CO_2 today (and though they have benefited most from the hydrocarbon economy), are no more eager to slow their consumption of fossil fuels or spend trillions of dollars replacing their existing energy systems. Indeed, in some cases, industrialized nations — most notably, Russia and the United States, which together account for two-fifths of all CO_2 emissions — have actually worked against efforts to launch a coherent global climate strategy.

Thus, whereas one might hope to see, after some twenty years of scientific and political debate, at least the outlines of an action plan, climate policy has instead stalled out, in a mind-numbing blame game in which governments and corporations and advocacy groups argue over who is most at fault for past emissions, who should cut future emissions, and who should pay for it all — while the collective gaze of the global public glazes over in boredom and incomprehension.

And yet, despite its near absence from public discussion, climate change remains not only one of the most pressing long-term issues but one of the most critical influences on the development of the next energy economy. How we choose to deal (or not to deal) with climate change will determine, among other things, what kind of energy systems we build, how quickly we build them, and whether life in the future will be fundamentally better or worse because of it. As one former Clinton climate negotiator puts it, "climate could be that huge-scale force for change in patterns of energy use, in the ways that you and I use energy."

In a sense, climate change is emerging as the *only* real driver for an entirely new energy economy. Although other energy concerns, such as oil depletion and price volatility, are also reshaping the energy economy, their impact is not fundamental, because they do not challenge our basic reliance on hydrocarbons. We "solve" depletion simply by switching to natural gas — or coal or the superabundant heavy oil or tar sands. We "solve" volatility of oil prices by changing regimes in Iraq. In other words, if the criterion for the next energy economy is simply that we find another fuel to burn or that we stabilize our sources of existing fuels, we can do that without altering the current energy paradigm. But if the requirement is that we somehow find a way to make and use energy without emitting carbon — which is where any sound climate policy must take us — then this is a whole new ball game. If we want to see how our energy economy needs to change over the long haul and where it needs to go, we need to take a closer look at the strange and ancient dance between energy and climate.

Climate change is probably best understood as a gigantic accident, an unintended interruption in the billion-year-old process by which earth transformed itself from a seething, poisonous hell into a lush and hospitable cradle of life. In primordial times, our atmosphere was much more like that of Venus, composed almost exclusively of carbon dioxide, which, because of the greenhouse effect, kept temperatures far too high to support any kind of terrestrial animal life.

As the eons passed, however, most of this troublesome CO_2 was soaked up by a system of carbon reservoirs, or "sinks." Seawater, for example, naturally absorbs CO_2, and our oceans today are chock full of primeval carbon — about thirty-five trillion tons of the stuff. More to the point, CO_2 is also captured and stored in green plants. Over the millennia, earth's primordial fens and forests have collectively inhaled trillions of tons of CO_2, then converted it, via photosynthesis, into life-giving oxygen and carbohydrates.[11] This so-called plant capture is sometimes only temporary: when a green plant dies, burns, or gets eaten, its stored carbon may return to the atmosphere as CO_2, where it can float for centuries before being recaptured — in a great loop we now call the carbon cycle. But over billions of years, most of the CO_2 captured by plants was stored, or "sequestered," in more permanent, geological forms — including limestone, shale, and of course hydro*carbons:* coal, oil, and gas.

Until around five hundred years ago, this massive carbon-capturing process was going swimmingly. Plants were sucking up so much CO_2 that atmospheric concentrations of carbon dioxide — that is, the actual number of individual CO_2 molecules floating around among other molecules in the air — had fallen to a minuscule level: around 270 parts per million (ppm), thus dramatically reducing the greenhouse effect and lowering temperatures substantially. The carbon cycle had achieved a rough equilibrium: for every molecule of CO_2 emitted by a decomposing plant or by a fire, another would be reabsorbed by forests or oceans.[12] In fact, the carbon cycle was actually running slightly ahead of the game: every year, the earth's natural processes released some 210 billion tons of CO_2 into the atmosphere; yet every year, the earth's forests, prairies, jungles, and vast algae farms soaked up around 213 billion tons — leaving a "safety" margin of around 3 billion tons a year,[13] enough to sponge up any extra CO_2 emissions produced by, say, a forest fire, a volcano, or even a pre–industrial era chimney.

Around 1500 or so, this favorable trend began to slow. As expanding agriculture and demand for firewood had depleted forests, the planet's natural capacity to reabsorb carbon had declined. More important, the Industrial Revolution with its insatiable demand for energy sparked a dramatic increase in the burning of hydrocarbons — unintentionally reversing the hundred-million-year-old carbon storage process.

Since the late 1700s, emissions of anthropogenic CO_2 have climbed from a paltry hundred million tons of carbon per year to around 6.3 *bil-*

lion tons a year — about twice what the biosphere can easily absorb.[14] Because more carbon is now entering the atmosphere than can be captured — by about 3.2 billion tons every year[15] — atmospheric concentrations have begun climbing again, and are currently up to around 370ppm.

Worse, because we have so dramatically exceeded the capacity of the natural system to reabsorb carbon, and because natural reabsorption can take centuries, atmospheric concentrations will keep rising for centuries, regardless of what we do. Even if we were somehow, for example, to halt CO_2 emissions at current levels — that is, around 6.3 billion tons a year — concentrations would still rise at a low but steady rate of 1.5ppm a year, reaching 520ppm by 2100 and leading to significant warming along the way. Thus, there is a huge lag between the time we take action and the time it begins to produce a beneficial effect. Notes Ben Preston, a researcher at the Pew Center on Global Climate Change, one of the top environmental non-profit organizations that works on climate policy, "Even if human beings stopped emitting all carbon today, we're still looking at about two to three centuries for the natural sinks to remove the excess CO_2 that is already in the atmosphere and return CO_2 to its preindustrial level."

At this point, ceasing all emissions all CO_2 emissions, or even freezing them at current levels, is simply not an option, given forecasts for population growth and economic activity, and given the momentum of an expanding energy economy that is dominated almost entirely by coal, oil, and gas. Emissions are currently increasing at around 3 percent a year, and are on track to hit twelve billion tons a year by 2030 and more than twenty billons tons by the end of the century.[16] At this rate, atmospheric concentrations will reach 1,100ppm by 2100 — a level at which even skeptical climate scientists concede that all hell (so to say) will break loose.[17]

Precisely how much higher concentrations can rise before we fry is still being debated. But most climate models indicate that once concentrations exceed 550ppm, we will start to witness "dangerous" levels of warming and damage, especially in vulnerable areas, such as low-lying countries or those already suffering drought. Most climate scientists would much rather see concentrations stabilized at 450ppm, which is about 20 percent higher than current levels. At 450ppm, models suggest, we might avoid most long-term effects and instead suffer a kind of "warming light," with moderate loss of shorefront land, moderate loss of species, moderate desertification, and only a moderate increase in hurricanes, winter floods, summertime droughts, forest fires, and other climate-related weather conditions. Adhering to a limit of 450ppm might also be sufficient to save at least some of the low-lying countries and other vulnerable regions.

But here's the rub. In order to stabilize at even 550ppm, researchers say, some extraordinary things would need to happen. First, CO_2 emissions would need to peak at around 11 billion tons a year somewhere between 2030 and 2035, before falling off quickly and then tapering toward the end of the century.[18] By 2100, according to Gerry Stokes, director of the U.S. Joint Global Change Research Institute, "we'd need to be back down to 6 billion tons of CO_2 from a population that is not only much larger than it is now, but richer." Such a drastic reduction, given current technological trends, is hard to envision. In other words, in order to peak at 11 billion by 2035, each person on the planet must be emitting no more than 1.2 tons of carbon a year — which is around one-third the per capita emissions for industrialized countries today — and just one-*sixth* the per capita emissions in the United States! In fact, to meet that goal, the United States alone would, by the end of century, need to cut its total emissions by 70 percent[19] — despite the fact that it will have more people and a much larger economy and, by current forecasts, will be using dramatically more fossil fuels.

Can we hit that target? True, we're becoming much less "carbon-intensive" than we used to be: we emit far less carbon, per watt of energy produced, now than, say, a hundred years ago, because we've gradually shifted to a mix of fuels that is lower in carbon (more gas and oil, and proportionately less coal and wood). Also, our existing energy technology — ranging from internal-combustion engines and oil furnaces to gas-fired power plants — is becoming, on the whole, more efficient and cleaner. However, this trend toward lower carbon intensity will be more than offset by the projected massive increases in population and economic growth, the twin catalysts of energy consumption and, therefore, the main factors in CO_2 emissions.

Given anticipated growth in worldwide population and economic activity, most climatologists believe that even 550ppm is a flatly impossible ceiling to impose on an energy economy that is still even partly reliant on fossil fuels. That's not good news in a world that, according to current trends, will still be getting more than half its primary energy from coal, oil, and gas in 2050. "Given a snapshot of where we are now, we will hit 550ppm by the middle of this century," says the Pew Center's Preston. "If we're to have any chance at stabilizing at that level, it means divesting ourselves of fossil fuel sometime during the next four decades." In other words, Stokes told me, not only do we need a new generation of energy technologies, but we must begin developing them now, and then deploying them on a massive scale "in the next twenty years."

Nowhere is the sense of urgency and anxiety more clearly reflected than in the 1997 Kyoto Protocol. There, under intense pressure from environmentalists and progressive politicians — including U.S. vice president Al Gore — and in the face of heavy lobbying by energy companies and petrostates — most nations of the world solemnly promised one another that by 2012 global carbon emissions would actually be lower than they had been back in 1990. It was an audacious commitment: in light of the rate at which emissions are growing and the inability of the developing world to help out economically, the burden of Kyoto was to fall almost entirely on the shoulders of the industrialized world. Germany, Britain, and France, and other members of the European Union (E.U.), for example, were ready to cut their emissions by 8 percent below 1990 levels. Japan agreed to 6 percent. In the most impressive concession, the United States, the world's biggest CO_2 emitter — and until then the industrialized nation that was least cooperative on climate policy — had agreed to a 7 percent cut.

At the time, many climate activists hailed Kyoto as a breakthrough, and its subsequent collapse has been popularly ascribed to the self-serving politics of various corporations and industrialized nations, most notably the United States. In this view, the deal unraveled after the Clinton administration suddenly realized that fulfilling its Kyoto cuts would require the White House to take on the big U.S. emitters, such as car companies and coal-fired utilities — a move that would have been political suicide for Gore's intended 2000 election campaign. Despite Gore's clear interest in signing a climate treaty, "in the end, it came down to raw politics," recalls one U.S. climate policy analyst who, like many, would speak only off the record. "Clinton and especially Gore regarded climate as important, but they didn't want to do anything that would offend car companies, the utilities, or the coal states."

Thus, even as U.S. negotiators were pushing for emissions cuts in Kyoto, administration officials back home did little to prepare a domestic program to carry out the cuts. Instead, according to observers, the United

States secretly planned to meet its treaty obligations as cheaply and pain-lessly as possible. First, Washington planned to fulfill its 7 percent goal by demanding "credit" for its vast forests, which, theoretically, can sequester carbon (as energy experts put it) and thus should count as a credit against U.S. emissions. Second, the United States intended to buy emission credits from other countries; for example, Russia, whose economy had collapsed after 1990, and whose CO_2 emissions had fallen by a third, was allowed un-der Kyoto sell those "unused" emissions as credits to countries like the United States that did not want to make cuts of their own.

These two maneuvers infuriated environmentalists, as well as many government officials in Europe. "No one is discounting that carbon se-questration or buying emission credits from abroad can help," says Eileen Clausen, a former U.S. State Department assistant secretary for environ-mental affairs who now directs the Pew Center's climate policy. "But the notion that the world's biggest economy and biggest emitter could do it all that way, so that you wouldn't need to do anything domestically with your industry or your energy sector, is a real stretch. Despite the rhetoric of Clinton and Gore, they negotiated a treaty that no one had any intention of implementing." In the end, Clausen told me, the Clinton administra-tion's main aim at Kyoto was "looking good to constituencies that were im-portant to Clinton — environmentalists and swing voters."

Yet while politics of national self-interest, especially those of the United States and Europe, have certainly obstructed Kyoto — and much of the de-bate since — the real problem was that Kyoto failed rather dramatically to reflect the nature of the problem it was intended to solve. Paradoxically, al-though specific targets and timetables for emissions cuts may have given the comforting impression that *something* was being done, sticking to the Kyoto targets — and making the required changes in the energy infrastruc-ture — would probably have done more harm than good. "Everyone got all worked up in setting these targets and timetables for emission reductions," says David Victor, director of Stanford University's Program on Energy and Sustainable Development and one of the top experts on climate economics, "but no one looked very carefully at whether the targets were achievable. Industry wanted no targets. Environmentalists wanted targets and timeta-bles that were not achievable, and then you had Gore who really wanted it to succeed and so agreed to a policy that, in the end, wasn't achievable."

What Kyoto failed to reflect is that climate change is a long-term, cumulative problem. The amount of warming is determined not by how much CO_2 we emit in a particular year, but by the *accumulation* of CO_2 in the atmosphere over the course of centuries. In short, we humans have a "carbon budget" — or a total amount of carbon we can pump into the atmosphere over the long term before we get into trouble. According to the Intergovernmental Panel on Climate Change and many other climate experts, if we emit no more than 1.225 *trillion* tons of carbon between now and the year 2300, atmospheric concentrations will remain below 450ppm and allow us to avoid most of the worst effects of warming. Spread over three hundred years, 1.225 trillion tons works out to an average of 4.1 billion tons a year. What that means is that even if current emissions are at 6.3 billion tons and rising, it won't matter in the long run — so long as we eventually bring annual emissions down far enough to keep our average yearly emissions at 4.1 billion tons.

In short, we have a little room to breathe, which is handy, because at present we have very little sense of what the ultimate climate solution will be. Not only are we depending on energy technologies yet to be invented, but our understanding of climate change and its costs — and therefore our certainty about the "best" solution — will surely continue to evolve.

Whatever course of action we choose today will need to be modified, perhaps substantially, within the decade, or even sooner. In this sense, climate policy is a kind of gamble: ten years from now, the climate crisis will be considerably more dire, but we will also know vastly more about the severity of the problem and our options for solving it.

"These conditions pose a dilemma for policy makers," notes Bill Lahneman, head of the Washington-based National Intelligence Council, which advises intelligence agencies on energy and other U.S. security threats. "Some sort of strategy is required to combat climate change, but any comprehensive strategy would be quite expensive to implement. What if such strategies turned out to be inappropriate because global warming phenomena were misunderstood at the time the strategies were devised and implemented? The world would have devoted large amounts of scarce capital [to] the wrong fixes, leaving [fewer] resources available to combat global warming effectively once the process became better understood." Climate policy, says Lahneman, "must be a hedging strategy because, if the world's nations devote too much wealth to large-scale, multinational efforts to reverse climate change and it turns out these steps have been ineffective, then it will be difficult to recover and begin to take the proper measures."[20]

Thus, while many hard-core climate activists, such as Greenpeace, still insist on strict, Kyoto-style limits, the emerging consensus among climate economists is that climate policy must remain flexible. Fixed targets for atmospheric concentrations of carbon dioxide are essential in the long run, but we should allow ourselves considerable flexibility regarding how, and when, those targets are met. For example, because trees and other green plants truly do remove CO_2 from the atmosphere, countries should be allowed to cut their emissions by planting new forests. Differences between industrial sectors must be considered: although transportation generates a third of all CO_2 emissions, it may actually be more cost-effective to tackle the power sector, because one power plant is easier to replace than forty thousand cars and trucks.

Regional differences should be taken into account as well: for example, China and India will soon be the biggest emitters of CO_2; yet because their respective energy economies are so different, their approaches to CO_2 reduction should be different as well. Whereas China, which is likely to build an energy economy based on coal-fired power, could get the most benefits from a policy emphasizing efficiency and "clean-coal" technologies, India, with its huge agricultural base, could most effectively cut CO_2 emissions by switching from oil and coal to the cleaner biofuels made from crops, crop wastes, and other biomass. "The most important thing we've learned over the last decade is the need for flexibility," says Richard Richels, a Stanford economist who helped draft the IPCC reports and who advises the coal industry on hedging strategies — "flexibility as to where carbon is cut, and by whom, as well as who pays for the reductions."

Indeed, whereas simple fairness might dictate that all countries cut emissions equally, it is actually much more effective to go after the cheapest reductions first, regardless of where they are. In China, for example, the country's energy sector is so inefficient and polluting that a relatively small investment in energy efficiency would cut more emissions much faster than the same dollar investment in Europe or the United States, whose power sectors are already more efficient and less polluting and so would yield small emissions cuts. To put it another way, per ton it costs less to avoid CO_2 emissions in China than it does in Europe or America, and since the climate doesn't really care where the CO_2 comes from, the cheaper route is almost always better.

This least-cost approach has a great many implications for climate policy. Countries where it is cheap to reduce carbon emissions, like China, can essentially sell their emission-reduction services to countries where cutting emissions is costly, like the United States or Europe. A German utility faced with high-carbon reduction costs at home could simply pay a Chinese power plant to cut emissions in Beijing. The Chinese get much-needed funds to improve efficiency; the German utility avoids high costs — and a ton of carbon has been withheld from the atmosphere.

Many climate experts envision a global carbon-trading system as the only feasible climate policy, because it harnesses the power of the marketplace to cut emissions most efficiently and economically. Studies by the Stanford University's Energy Modeling Forum show that if industrialized countries have the flexibility to buy their emission cuts anywhere in the world, the total long-term costs of climate policy would drop by nearly half. Such savings are critical, because currently the costs of controlling carbon through traditional methods are expected to be colossal — enough to curb economic growth in countries like America and Japan by as much as 2 percent a year. "From a perspective of economic efficiency, it's more important that we reduce emissions wherever it is cheapest, regardless of who pays for it," Richels told me. "So if there are opportunities in China — so-called low-hanging fruit — we need policies that take advantage of that, even if the Chinese can't, or won't, pay for them."

So far, this notion of flexibility doesn't seem too radical, but advocates like Richels take it a step further, proposing flexibility in timing as well. Although many environmentalists insist on making deep emission cuts immediately, à la Kyoto, this may not make the most sense economically. For example, although we're making great progress on new, carbon-free technologies — such as wind power or solar energy or hydrogen generated from renewable energy — these technologies may still be years or even decades from being cost-competitive. Trying to deploy them prematurely, on a global scale, would be so expensive that even the richest countries would go bankrupt — that is to say, there would be no deployment.

Even if carbon-free energy technologies were available today on a cost-competitive basis, we would still be stuck: the existing fossil fuel infrastructure is worth around ten trillion dollars, and its components — everything from power plants and supertankers to oil furnaces and SUVs — must be operated for ten to fifty years before their capital costs can be paid off. Any "premature retirement" of this hydrocarbon infrastructure would force power companies, oil firms, and other owners of these assets simply to write off hundreds of billions of dollars in lost value — an economic blow that few owners and investors would accept without a nasty political fight.

The only practical solution, experts say, is to pace ourselves — time our strategy for emission reductions to match the natural "turnover" rate of the capital stock more closely. We would still replace coal-fired power plants and diesel buses and inefficient air conditioners — but we would do so toward the end of their economic life. This strategy not only avoids huge premature retirements but gives the replacement technologies, whatever they turn out to be, more time to mature and, equally important, more time to come down in cost. Further, when we take into account the fact that a dollar spent today always buys less than one spent in the future (going by the "time value of money" principle), it really pays to delay at least some emissions cuts.

Granted, such a strategy is risky: it means letting emissions rise in the meantime — perhaps so much that atmospheric concentrations exceed 450ppm or even 550ppm, at least temporarily. Still, by lowering replacement costs and allowing alternative technologies to mature, Richels says, we dramatically improve our chances for making even greater reductions later on. "People think in terms of an absolute ceiling in concentrations, but what if you allowed for an overshoot of 450ppm, due to capital stock turnover, but then reduced emissions more quickly later on?" Richels asks. "Right now, we just don't have energy alternatives that are sufficiently low-cost; but we probably will in twenty to thirty years."

Not surprisingly, many environmentalists and energy experts are deeply skeptical of this "delayed action" policy (they note that many of its proponents, like Richels, just happen to be associated with the industries which would dearly love to delay any kind of expensive action on emissions). Regardless of the biases of experts like Richels, though, the case they make against immediate deep emissions cuts simply for the sake of taking action cannot be overlooked. The United States may be profoundly guilty of foot-dragging on climate policy. But it is also clear that the country probably could not have hit its Kyoto targets without large economic disruptions, even if it had started cutting emissions in 1997, and it certainly couldn't hit them now: between 1990 and 2000, U.S. carbon emissions climbed by more than 10 percent. Whether climate activists like it or not, the United States will be far more effective as a player in climate policy if it moves into emission reduction gradually.

Of course, in order for this back-loaded approach to work, and not simply turn into a rationale for permanently deferring action, the United States and other industrial nations will need to take a number of immediate steps — and so far, they have not.

First, if we are going to develop and deploy low-cost, carbon-free energy technologies within twenty years, we need to commit huge amounts of capital — on the order of hundreds of billions of dollars — toward research and development right now. Today, nearly every major alternative energy technology — whether solar, wind, or hydrogen — along with technologies to capture and sequester CO_2 — has a potential for substantially improved performance and lower cost that could be exploited with additional R & D dollars. Instead, since 1985, government funding for energy-related R & D has fallen in every industrial country except Japan (an island nation wholly dependent on energy imports, and far more committed than the United States to the Kyoto process). "We have been able to document less than $15 billion annually invested in the development of energy technologies by the world's governments and private firms," lament the authors of a recent study of climate and technology. "Although the U.S. commitment [to R & D funding] is one of the world's largest, it represents less than 0.05 percent of the U.S. gross domestic product and less than 2 percent of all R & D conducted in the U.S."[21]

Researchers must also look more closely at economic questions — in light of both the effects of global warming and the price tag for cutting emissions. Estimates of the total economic impact on the United States, for example, range from $37 billion to $351 billion, depending on which economic models are used and how various factors, such as storm frequency, are evaluated.[22] Such wild variations make it impossible to debate the issue meaningfully. Likewise, the cost estimates for *avoiding* climate change vary considerably: one survey, by the World Resources Institute, found that estimates of the cost to the United States of meeting its Kyoto obligations for CO_2 reductions ranged from $20 per ton of carbon to $400 per ton of carbon. Without a clearer sense of the costs they may face, governments simply cannot make smart decisions about where to focus policy efforts, spend climate dollars, or push other countries in climate negotiations.

Second, even as we acknowledge that it makes sense to wait for new energy technologies to develop, we cannot gamble that these technologies alone will be sufficient to cut emissions to the necessary levels — particularly if nothing has been done in the meantime to reduce carbon emissions or overall energy consumption. Bill Chandler, an economic analyst with the U.S. Joint Global Change Research Institute, suggests an immediate effort to improve energy efficiency and to encourage a gradual shift to low-carbon fuels, such as natural gas or biofuels, thereby reducing the burden on the new energy technologies when they are eventually deployed. "It would be incredibly risky to say, 'Let's just develop new technologies that we can start introducing in 2030,'" Chandler told me. "It's far wiser and much less risky to try to reduce our growth rates in energy consumption and greenhouse gas emissions through incremental measures, like increased fuel economy standards and better appliance standards. That way, when the new energy technologies become available, there is some hope that they can support this new, lower level of energy demand."

Third, and perhaps most important, if we are going to delay making cuts in the near term, we need a system to ensure that the cuts do eventually get made. We need a political framework that locks in emission targets, lays out schedules for meeting those targets, prescribes how much each country will cut, describes such market mechanisms as carbon trading, and contains provisions for enforcing cuts and penalizing countries that don't fulfill their obligations. We need, above all, to have the will to remake our economic system in a way that accounts for climate change and the costs associated with it by recognizing the "costs" of carbon. In short, we need an international system for addressing climate change — and, given that the politics of climate have actually deteriorated since the collapse of the Kyoto Protocol, this requirement may prove to be the most elusive.

The politics of climate today are driven primarily by a rivalry between America and Europe — a transatlantic feud that is both economic and, to a surprising degree, ideological, and which has intensified with the election of George Bush and the war on terror. Within months of taking office, Bush made clear that he had little interest in climate policy — and little fear of international opinion on the subject. He reneged on a campaign pledge to regulate CO_2 as a pollutant, then rejected the Kyoto treaty as "fatally flawed" because the international agreement did not require developing nations to cut their share of emissions. The White House also attacked the basic science of climate change and demanded that the prestigious National Academy of Sciences (NAS) independently review the conclusions of the IPCC report. When that tactic backfired (the NAS essentially confirmed the IPCC findings), the administration blithely claimed that the United States could not cut emissions because the nation was deep in an "energy crisis." When the IPCC issued a report suggesting that in fact a wide range of technologies and policies were available that could dramatically reduce the costs of emissions reductions, the Bush administration quietly pressured the IPCC to dump its chairman, Robert Watson, a longtime critic of U.S. recalcitrance.

Why is Bush so avowedly opposed to climate policy? The charitable answer is that the Bush administration, with its intensive background in the energy industry, understands what steep emission cuts would mean for the world's most energy-intensive economy. Per capita, Americans use more coal, oil, gas, and other energy than any other nation. And though we use that energy magnificently, generating more wealth per capita than any other nation, we also produce more carbon per person in the process. Cutting CO_2 emissions would therefore cost this country more than it would others — more to replace our cumbersome fossil fuel infrastructure, but also more in terms of lost growth for our energy-intensive economy. According to some analyses, implementing Kyoto could cost America as much as 2 percent of its gross national product, every year, for centuries. By contrast, Japan would lose just 1.2 percent, while Europe would lose 1.5 percent.

In fact, in the eyes of the Bush administration, and of many other climate analysts, this disparity in costs explains a great deal about Europe's great enthusiasm for rapid, deep emissions cuts. As it turns out, many of the emissions cuts that Europeans agreed to under Kyoto had already been made.[23] Since 1990, emissions in Europe had risen more slowly than in the United States — not because Europe was any greener or more ethical, but because its economy was growing slowly. Emissions have fallen further because the English coal industry was all but shut down a decade ago, as were polluting, Soviet-era power plants in the newly liberated East Germany. In short, the Europeans knew it would be easier for them to meet their reductions than it would be for the Americans to meet theirs; the U.S economy is booming and America still relies heavily on coal, especially in its power sector. As one former White House analyst told me, "some in Europe regarded Kyoto as an economic weapon, something to slow the American economy down."

By most accounts, the White House fears Kyoto's potential to do not only economic damage, but political damage as well— especially to the administration's allies. Like Clinton before him, Bush understands that energy policy is intimately tied to three politically powerful American industries: cars, coal, and coal-fired electric utilities. Bush has particular reason to be protective of American coal: many observers contend that he owed his victory in 2000 largely to winning electoral votes in the "coal belt" — Virginia, West Virginia, Kentucky, and Tennessee. Because coal produces more CO_2 than oil and gas combined in the United States, nearly all climate policies are decidedly anticoal, making them nonstarters as far as the Bush administration is concerned.

The cynicism of that view carried over to the way Bush gauged public sentiment on the issue. Observers say the White House was genuinely convinced that although certain key business constituencies cared deeply about climate, most of the public at large was too frightened by the energy crisis of 2000, and too blasé about climate issues in general, to object to, or even notice, what Bush did about Kyoto. "They had been told that dumping Kyoto would be one small story in the newspaper, below the fold," says a climate expert who writes policy for an environmental group; "but in fact, the polling had been fairly consistently the other way: Middle America doesn't understand the science of climate or know exactly what the Bush administration has done, but it does have a strong, visceral response to the issue itself."

Predictably, the Bush climate policy provoked fierce criticism — from environmentalists and liberals in the United States and from many European policymakers.[24] Even some Republicans found the White House's moves to be clumsy, arrogant, and incredibly poorly timed: stories were already coming out about how the Bush national energy policy had been written largely by the president's allies in the energy industry, including Enron, the largest contributor to Bush's 2000 campaign. "The criticism the Bush administration was getting was that he was out to kill Kyoto but had no alternative to propose," recalls one observer, "and they were getting hammered from all sides."

Slowly, the White House realized that simply ignoring the climate problem was not an option, and that while opposing Kyoto might be justifiable, the United States needed some kind of an alternative plan to cut emissions. "It's one thing to protect your core constituency, but if you do that by losing the center, there is no way you can govern — or get re-elected," Stanford's Victor told me. "The administration finally understood that, and it was finally prepared to do something more serious." During the summer of 2001, the administration reportedly held half a dozen high-level cabinet meetings outlining several climate strategies, including a modest proposal to have industry begin reporting its CO_2 emissions — the first step toward a carbon-trading scheme, and one that many climate economists favor. Indeed, by early September 2001, according to one participant, the White House had scheduled a major cabinet meeting to "get to some final policy on global warming" in preparation for an upcoming climate summit in Delhi that fall.

Whatever momentum existed for a substantive U.S. climate policy vanished with the attacks on September 11. When the issue reemerged several months later, the administration's earlier interest in even a modest climate policy had vanished. Although internally many administration officials maintain a fairly sensible take on climate — one top-level staffer at the Energy Department I spoke with lectured me with great intelligence and verve on the logic of phasing in emissions cuts — the public message continues to reflect the administration's earlier intransigence. With national and international attention firmly focused on Iraq and terrorism, the administration has clearly felt less pressure to offer any serious climate proposals. In fact, although White House rhetoric on climate is much more polished than it was, the focus remains primarily on creating the appearance of action without actually committing the United States to any expensive steps — steps that might alienate any key electoral blocks.

This concern is reflected in the Bush climate plan of 2002. Though the White House did push for modest additional funding for solar and wind power and other noncarbon energy, and although Bush personally talked up the idea of a hydrogen economy, the administration's basic position hadn't changed substantially. No mention was made of any kind of carbon-trading system, nor any commitment to reducing emissions.

At the same time, the plan was certainly framed to give the appearance of action. In some rhetorical sleight of hand, the White House issued a statement promising to cut "energy intensity" by 18 percent over ten years — as if this were some substantial sacrifice. True, energy intensity — or the amount of energy required to produce a dollar's worth of wealth — is a critical factor in reducing emissions. But the White House proposal was seriously disingenuous. In the first place, the press release itself was intentionally unclear, making it seem as if Bush were actually proposing an 18 percent reduction in CO_2 *emissions* — a reduction that, if true, would have outdone Kyoto.

In the second place, the kind of energy intensity reductions the White House was calling for were pathetic. According to Lawrence Goulder, an analyst with the Stanford Institute for Economic Policy Research, given economic projections, an 18 percent reduction in energy intensity would have actually allowed emissions to *rise* by 10 percent by 2012 (whereas Kyoto would have required a 19 percent reduction in emissions). More to the point, an 18 percent reduction in energy intensity over ten years works out to 1.8 percent a year — an absurdly unambitious goal, given that the United States had between 1996 and 2000 reduced energy intensity by 2 percent without any kind of federal program. In terms of actual reductions, Goulder says, the Bush plan differs little from what economists call a business-as-usual scenario — that is, it would be the same as doing nothing. As Goulder noted in a critique of the Bush climate plan, "various forecasts (including the administration's own estimates) indicate that the [Bush] plan allows emissions in 2012 to be over 95 percent of what they would have been with no policy."[25]

The more the United States resists a coherent climate policy, the more it becomes clear that the one country that could make the biggest difference — in reducing emissions but also, and perhaps more important, in using its wealth and technology to lead the way to a postcarbon energy order — has become the biggest obstacle to any meaningful progress. Europeans have grown tired of waiting for Washington to join in and have begun implementing Kyoto without the United States. Countries like Germany and

England have carbon budgets and are implementing carbon "caps," or limits for various industrial sectors, such as utilities and manufacturers. Yet everyone understands that the programs are of limited value without U.S. participation. America is not only the biggest CO_2 emitter but probably is the only party capable of bankrolling the programs — or persuading China and India, or holdouts like Russia, to join the process.

As a consequence, most climate experts now believe that earlier hopes of stabilizing atmospheric CO_2 at 450ppm are simply untenable, given the current rate of growth in emissions and the lack of a concerted international climate policy. "To stabilize at 450ppm, global emissions would have to peak by 2010," says Robert Watson, the former head of the IPCC. "And they're not going to."

Instead, the emerging consensus is that stabilization at 550ppm is the best that can be hoped for — and even that will be a stretch. In fact, many climate experts have begun to argue that a more realistic approach might focus less on trying to mitigate climate change and more on simply *adapting* to it. Some adaptation is inevitable, given that warming is already occurring and won't stop for decades, no matter what we do. Low-lying and coastal areas *will* suffer some degree of flooding. Disease *will* spread, crops *will* fail, forests *will* burn, and to pretend otherwise is foolish. "Whatever we do today, we are committed to a certain level of climate change," Rajendra Pachauri, chairman of the IPCC, told a reporter last year.[26] "In the short term, we have no choice but to adapt."

The question that arises now is whether adaptation will become the de facto climate policy, the default mechanism, simply because no one is prepared to lead the more difficult and complicated effort required for us to mitigate climate change. As our skyrocketing emissions make clear, whatever progress we have achieved in making our economy less carbon-intensive has been more than matched by growth in population and economic activity, and this pattern is certain to continue. We may be producing less carbon per person or per dollar of wealth, but we will shortly have a lot more dollars and a lot more people.

In short, our energy technology is being outpaced by the very economic success it engendered, and the ramifications are alarming. As we've seen, according to IPCC forecasts, even with enormous improvements to existing fossil fuel energy technology — including steady improvements in energy efficiency and a gradual shift toward nonfossil fuels in the power

sector — the resulting emission reductions will still not be sufficient to stabilize atmospheric CO_2 at 550ppm by midcentury. Instead, says Stokes, the
somewhat pessimistic director of the U.S. Joint Global Change Research
Institute, "we're going to need a set of energy-related technologies that basically emit nothing into the atmosphere" — technologies that, Stokes and
his colleagues readily admit, are nowhere near becoming feasible and to all
intents and purposes haven't even been imagined.

Stokes continues: "Most of the people who worry about the climate
problem have seriously underestimated how hard it will be. We are going
down a path that, if we stay on [it], will see us triple the amount of CO_2
that we emit by 2100. If we're going to avoid that, we need to have a set of
energy technologies in place, and these are technologies that are not going
to just magically appear. We need to start working on them now." To put
this in perspective, by 2050, assuming that we have managed to keep concentrations of CO_2 below 550ppm, more than half our emission reductions
will be coming from energy technologies that do not yet exist.

PART II

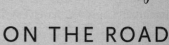

ON THE ROAD
TO NOWHERE

6

GIVE THE PEOPLE
WHAT THEY WANT

DURING THE LONG, hot summer of 2002, while most of the world focused on the prospects of war in the oil-rich Middle East, police in eastern China found themselves in a surprisingly brutal oil war of their own. For months, the country's two biggest petroleum companies, PetroChina and Sinopec, had been battling for domination of the burgeoning Chinese gasoline market. Within three years, the government planned to open the retail gasoline market to foreign oil companies; and PetroChina and Sinopec, both commercial spinoffs of China's state-owned oil company, were working feverishly to build market share before then. The fight was getting expensive and nasty. Both companies were spending billions of dollars, buying up prime real estate along busy highways, building new stations and renovating old ones, and pressuring hundreds of independent stations to sell out — often through a mixture of price cuts and intimidation tactics that would have made John Rockefeller proud. By midyear, workers at more than 90 percent of China's gas stations were wearing either the red vests of PetroChina or the blue vests of Sinopec.

The oil wars had turned particularly fierce along the new highway in Henan Province, five hundred miles south of Beijing. Although Henan had traditionally been blue-vest Sinopec territory, PetroChina had entered the region aggressively, by buying up independent stations and persuading the local highway construction company to put in dozens of new stations at key truck stops. All told, the fifty PetroChina stations along the highway would bring in an estimated twenty-four million dollars annually — a modest sum by Western standards but not one Sinopec planned to give up without a struggle. After touring the province, angry Sinopec officials bragged to one local newspaper that they intended to "resolutely conquer the Henan highway."

Local government officials quickly banned all news coverage about the rivalry, but the battle was already under way. In July, in a scene reminiscent of America's nineteenth-century oil-field wars, gangs of men armed with iron rods swarmed over dozens of PetroChina construction sites along the highway. "It was the most violent scene I have ever seen in my life," a Henan police officer told reporters. The gangs "smashed windows, doors, took away equipment and also drove the construction workers out of their rooms with fire extinguishers and beat them with iron rods."[1]

China's gasoline wars have since cooled down somewhat, and the market remains split between the two Chinese majors. Yet what is significant about this battle is not so much who controls China's expanding gasoline market as the fact that anyone would want to control that market in the first place. As recently as 1990, China was a nation with little use for gasoline. Most of the vehicles on China's roads were diesel trucks; the few passenger cars in the country were mainly Soviet-designed limousines built for Communist party officials. In the past decade, though, China has transformed itself from a preindustrial nation of bicycles and badly run mass transit to a country in the thrall of car culture. While most Chinese are still far too poor to afford private automobiles, China's gradual economic revival has spawned a burgeoning middle class, a small but substantial cohort of businessmen and entrepreneurs who have discretionary income and want to spend it on wheels.

China's automakers are responding. Teaming up with General Motors, Volkswagen, and other Western automakers, Chinese carmakers are rolling out a wave of new models in all sizes and price ranges, from the full-size Buick sedans still favored by party bigwigs to a compact model based on the Ford Fiesta. In 2002, Chinese automakers for the first time ever produced and sold a million new cars — up more than 50 percent from the year before. While such growth isn't sustainable for long, analysts do expect the industry to expand at an enviable clip of between 15 and 20 percent over the next decade, making China the hottest car market in the world.[2] In 2003, an ecstatic General Motors predicted that China alone would account for nearly a fifth of all new car sales between 2002 and 2012 — nearly twice as many as the United States.[3]

China's embrace of car culture is bringing other, less desirable recompense. Traffic congestion in Beijing and other crowded cities grows worse by the month, and China can now claim seven of the world's ten smoggiest cities. Worse, the millions of eager new Chinese motorists have substantially increased the demand for oil and forced China, for the first time in its history, to import oil and wade into the unstable waters of oil geopolitics. As a consequence, the Chinese have developed a welter of relationships with oil producers like Russia, Nigeria, and Venezuela and are playing an increasingly prominent role in the politics of the Middle Eastern oil states — a cause of considerable anxiety to many in the West. This "tightening embrace of necessity between [China] and the Middle East . . . over the next generation, could fundamentally challenge the Western-dominated global order," notes one nervous commentator.[4] Another sees China's new oil appetites as leading to "the West's worst nightmare: an 'Islamic-Confucian' coalition."[5]

Moreover, transportation is only one sector where Chinese energy use is soaring. Demand for electricity and industrial fuels in China is outpacing growth even in industrial countries, and this dynamic makes it more and more likely that the world's second-largest consumer of energy — outstripped only by the United States — will be in first place within fifteen years. By 2020, China will be responsible for two-fifths of all coal burned, one-tenth of all oil consumed, and one-seventh of all electricity used — as well as for nearly one-fifth of all energy-based emissions of carbon dioxide.[6] It is no surprise when energy experts contend that the "center" of the energy world is shifting eastward, away from America and Europe, and toward China and the rest of Asia — a shift that will be among the most significant changes in the global energy economy.

⌘

China offers a dramatic and disturbing hint of the size and shape of the emerging energy economy, and the tensions that will keep that economy in a constant state of flux. Already, our energy system is straining to produce adequate energy supplies. Oil is becoming physically and politically harder to find. The environmental costs of burning it, and other fossils, are becoming more and more apparent. On top of these concerns, we must now add another: a worldwide appetite for energy that is growing faster than previously thought possible — and which will soon test the very core of our energy system.

Over the next two decades, global energy use is expected to climb by between 1.5 and 2.5 percent each year, depending on the strength of the economy and long-term energy prices. Described in percentage points, such a growth rate may seem harmless, but taken year by year, a rate of 2.5 percent means that by 2032, demand for oil and other energy will double. Much of this staggering growth will take place in developing economies in Africa, the former Soviet Union, South and Central America, and especially Asia — countries that for decades have lagged far behind the West in energy use but are eager to close that gap as quickly as possible. By 2020, these developing countries will account for 60 percent of the world's total energy demand, up from around 45 percent today.

Wherever this new demand occurs, it will play an enormous role in shaping everything about the next energy economy, from the amount of energy we require to the kinds of fuels we will use to how quickly we will need them. Demand will determine which countries provide the energy and which countries consume it. Demand will set the price of those energy sources, which will dictate how easily we change from one fuel to the next and whether the shift will be managed by existing energy players, like oil companies, or by some new entrants to the market. Above all, this huge demand will vastly complicate the task of building an improved energy economy: to put it simply, we will be too busy scrambling for existing energy sources to devote attention or resources to finding something new.

Demand is, in short, a veritable force of nature, almost like gravity, that pulls the energy economy forward through time. Powerful multinational corporations may control the machinery of the energy economy; national governments may control the military forces that defend the world's energy supplies; but at the end of the day, the driving force behind the energy economy, the reason those supplies have value, is the billions of energy transactions that take place every day in every nation on earth and which together embody the world's insatiable thirst for energy. In other words, we cannot plan for or even imagine the next energy economy until we grasp the realities of energy consumption.

In a very basic way, the next energy economy is already being constructed, one sale at a time. Every act of economic activity is also an act of energy consumption. The hamburger I order from the takeout window is the ultimate, concrete form of demand in a cascade of individual energy decisions and transactions and uses: from the diesel in the tractor that tilled the feed grain to the electricity that powered the lights in the slaughterhouse to the gas that fired the restaurant grill (and this calculation, of course, does not include French fries).

Energy and economic activity are in fact two forms of the same substance: the one cannot occur without the other. Historically, the more economically active we humans have been, the more wealth we have created, and the more energy we have used to create it with. It is an endless cycle: more wealth leads to more purchases; more purchases increase demand for products, which in turn calls for more factories, more raw materials, and more trips by truck and train from factory to warehouse and from warehouse to the Wal-Mart and the Pottery Barn. The entire global economy is like a huge machine, steadily converting energy into wealth.

One can trace a country's material advancement by its growing appetite for energy, and by its success at feeding that appetite. The richest nations use great quantities of energy and do so with stunning sophistication and startling obliviousness: beyond occasional complaints about gasoline prices or the electric bill, the vast majority of Americans and Europeans are no more aware of using energy than they are of breathing air.

In the poorest nations, by contrast, energy use is scanty, rudimentary, primitive, and wholly conscious: for the poor, every act of energy consumption is calculated. In fact, when we talk about poverty and the conditions of poverty — lack of access to clean water or education, for example, or an inability to produce sufficient crops — what we are really talking about is lack of access to energy: electricity to run a pump for water or illuminate a classroom; diesel to fuel a tractor. Not surprisingly, when governments and international aid agencies propose development programs for the poorest nations, access to energy is a key component.

In this context, it becomes easier to see why the Chinese are so obsessed with acquiring gasoline filling stations, automobiles, and other accoutrements of the modern energy economy. Like the rest of the world, China has seen the future and knows that it depends on vast quantities of energy, and China's energy ambitions have serious consequences for the evolution of the entire energy economy. Over the next two decades, the countries that drive world energy demand will increasingly be those, like China, which are determined to have the same energy-based industrial prosperity that exists in the developed world, yet are so poor and technically backward that they cannot be counted on to make the most enlightened energy choices. Instead, the world's largest users of energy are likely to take the most expedient route: choosing existing fuels and technologies and forms of energy use — thereby adding tremendously to the inertia of an obsolete hydrocarbon energy system and effectively putting the brakes on development of something new.

One of the most peculiar symptoms of China's energy poverty is the spectacle of public spitting. From the largest city to the smallest agricultural hamlet, men and women of all ages and in all walks of life can be seen hawking gobs of spittle as frequently and nonchalantly as a Westerner might light a cigarette. Why? Despite China's incessant efforts to modernize, most Chinese either are still peasants themselves or come from peasant families. For generations, peasants have suffered near-constant respiratory infections — and thus, have acquired the hawking habit — in no small part because they cannot afford to heat their homes in winter. Coal is too expensive. China has mountains of it, but traditionally most coal has been used to fuel the country's great industrial campaigns and power stations. In some northern provinces, peasants' homes get so cold that the inside walls turn white with frost. "In my village," recalled one peasant, "when a girl was preparing to marry, the first thing the parents checked was, will the back wall of the would-be son-in-law be white or not? If not white, they approved the marriage, because this meant his family was wealthy enough to keep the house warm."[7]

In many respects, China epitomizes the effects of energy poverty, and its potential and pitfalls as the developing world fights its way into the modern energy economy. Despite mass campaigns to exploit their huge coal reserves, most Chinese are still mired in an energy economy that has not evolved significantly since the nineteenth century and is still powered by wood, dung, and other biomass.

China's energy poverty comes across most starkly in the transportation sector. The decade-old emphasis on industrial development instead of consumerism created a transportation sector devoted almost entirely to cargo — trucks and trains. As far as the Communist party was concerned, the private car was a capitalist luxury. China's Great Leap Forward wasn't about consumers; it was about producers — factories and freight. The only worthy mode of personal transportation was the bicycle (unless you happened to be a party bigwig, in which case, you rode around in a Soviet-made limousine). For most Chinese, transportation still meant walking, bi-cycling, and, in the city, using a public transit system of buses and trains so broken down, dirty, and overcrowded that passengers routinely urinated on the seats and exited through the nearest window. In fact they still do.[8]

In the 1980s, when the Chinese sought to drag their nation and its sickly centrally planned economy into the twentieth century, Beijing was forced to rethink its ideological disdain for the private automobile. Not only was an emerging class of entrepreneurs creating a demand for private cars and taxis, but Chinese economists now recognized that an auto in-dustry itself could be an engine for economic growth. Specifically, China wanted to do for its economy what Detroit had done for the United States, where carmakers generate 4 percent of the American GDP and millions of jobs. By 1994, the Chinese auto industry had been rechristened one of five "pillar industries" (the others being housing, petrochemicals, machinery, and electronics) that would underpin China's great leap into modernity. According to Beijing's ambitious plan, China would be producing a million cars by 2000, and 3.5 million by 2010 — more than 90 percent of them to be sold domestically. China was going to become an auto nation.

The creation of a Chinese auto industry could not happen overnight. Chinese manufacturers hadn't the first idea how to make a car. The few ve-hicles they had were trucks, imported from the Soviet Union. No sooner was Beijing's car policy launched than dozens of enterprising Chinese auto companies sprang up (at least 118 separate entities by 2003), but most have been so poorly managed and have such antiquated technology that they produce only a few dozen cars a year, if that.

Sensing disaster, the Chinese invited Western automakers to set up joint ventures with Chinese companies to roll out the first generation of Chinese cars; but this, too, has hardly been smooth sailing. Because the Chinese market is so risky — the much-touted Chinese middle-class con-sumers may be desperate for cars but still earn, on average, less than three thousand dollars a year — Western automakers have been reluctant to in-vest much money designing vehicles specifically for the Chinese market. In fact, in some cases, such Western automakers as General Motors, Ford, and Volkswagen have simply recycled car models from the 1980s and early 1990s. This formula has allowed automakers to reuse design and engineer-ing processes and even assembly-line equipment that have already been paid for — but that result in a vehicle that is obsolete before it rolls off the production line.

This is not the fault of Western automakers alone: so desperate are the Chinese for economic growth and a thriving auto industry that they have refused to impose fuel-efficiency standards or put into place the kind of stringent air-quality protections that most industrialized nations do. Even if they did, Chinese gasoline and diesel are of such low quality that they would ruin some of the more advanced engine technologies, especially those geared toward pollution control, like catalytic converters.

The result has been a mixed bag. Joint ventures are thriving; companies like Honda and DaimlerChrysler plan to spend billions of dollars building new cars for the Chinese market. Auto sales are growing at 20 percent a year (the world's fastest), producing 1.5 million jobs and adding twelve billion dollars to the national economy — or the equivalent of 5 percent of the manufacturing base. At the same time, however, the "new" Chinese auto fleet is made up of vehicles that, in many cases, are far more polluting and far less fuel-efficient than anything being sold in Berlin, Tokyo, or Los Angeles. "The Chinese auto industry is still very primitive," William Moomaw, a professor of international environmental policy at Tufts University, told me. "They started out forty years behind, and none of their partners has given them much that isn't already ten or fifteen years old. The Americans didn't offer much, but the Chinese didn't ask for much, either. They have got to get their act together and realize that now is the time."

One way to see where these energy trends may take China, India, and other would-be industrial powers is to look at their economic idols — the United States, Japan, Germany, and South Korea. It's no coincidence that four of the richest nations on the earth also have the greatest per capita energy usage: as we have seen, the more economically powerful a country is, the more energy it uses. More factories running at higher capacity will make for a greater energy demand. In addition, faster economic growth yields higher personal income, spurring consumer spending on such energy-intensive goods and services as dishwashers, big-screen televisions, and, of course, cars. This is one reason that, in an advanced energy economy like the United States, the average person burns through the equivalent of 7,500 gallons of oil a year, while the average Chinese burns through just 800 gallons.[9]

Happily, the curve of energy use and economic growth is not fixed. As capitalist societies advance materially, competitive market pressures force improvements in technologies and processes that almost always result in better energy efficiency. Over time, modernizing economies become less energy-intensive — that is, they require less energy to generate the same amount of wealth. Equally important, in advancing economies, energy use becomes cleaner. Cars and factories emit less pollution, thanks to better exhaust technologies. More efficient engines and boilers need fewer gallons of gasoline or tons of coal to produce the same output. As a consequence, factories are emitting lower levels of carbon and other pollutants. In the United States, for example, energy intensity actually peaked around 1920 and has declined ever since; meanwhile, carbon intensity is also declining. Globally, energy intensity peaked around 1955.

Specific events, like wars or price shocks, also lead to improvements. Following the Arab oil embargo, businesses and consumers dramatically reduced the energy they used. Factories and gadgets and toys all became more efficient, capable of producing more power, more value, more wealth — more bang for the same buck. Between 1970 and 1986, the amount of energy the United States required to generate a dollar of wealth fell 30 percent. In other words, the U.S. economy was able to grow substantially without any increase in energy consumption. Europe and Japan achieved even higher levels of efficiency, and many analysts believed that the historic link between economic growth and energy consumption had snapped.

The efficiency evolution is not seamless, however. A cleaner, more efficient energy economy requires substantial investments, so historically, before a country could develop a clean, lower-carbon energy economy, it had to pass through a maturation phase in which energy was a dirty, unsubtle business. All industrial countries go through this — think of the pollution-choked American skies and rivers in the 1960s, before the upsurge in concern over energy and the environment, or of England in the early 1950s, when coal smog killed thousands of Londoners.[10] If history is any guide, all developing countries will have to pass though this phase as well, and that is a big reason energy experts are so fearful of the future: the idea that a megastate like China or India might pass through a full-blown industrial revolution simply boggles the mind.

What is more, once a country moves into a mature, efficient energy economy, progress quickly stalls: even as energy becomes, in effect, cheaper, we simply use more of it. If a widget maker's energy costs go down thanks to improved efficiency, he simply ups his energy use in order to make more widgets and generate more wealth. In short, he invests his efficiency "dividend" not in lower overall energy use but in greater production, in order to maintain his competitive advantage.

The same thing happens on the consumer level. Today, the average new home in the United States, Europe, Japan, and other developed nations is many times more efficient than even twenty years ago. More efficient furnaces and better-insulated windows mean that it now takes less money to heat a square foot of house space. Yet most homeowners have "spent" this efficiency dividend in ways that completely negate this trend toward energy savings. In the United States, houses are now substantially larger (although no increase in household size has occurred), and at least twice as energy-intensive as European and Japanese households.[11]

This "supersize" trend in housing is most visible in the phenomenon of the "minimansions": expensive, spacious tract homes built to the very limit of local zoning allowances — anywhere from three thousand square feet on up — and packed with a surplus of rooms and every amenity imaginable, from Jacuzzis to track lighting — all of which require energy. "No one who buys one of these homes is explicitly saying, 'Hey, I want to use more energy,' " John DeCicco, an energy analyst with the environmental lobbying group Environmental Defense, said to me; "but that's implicit in their choice. They like the Jacuzzi. They like the bathroom that is as big as the bedrooms most of us grew up in."

Perhaps the most discouraging example of how developed nations misspend their efficiency dividend is transportation — and nowhere more so than in the United States. Before the 1975 Arab oil embargo, the American transportation model rested on the assumption of an inexhaustible supply of cheap gasoline. Because energy costs were trivial, carmakers made no effort to build cars that were fuel-efficient. After the 1975 Arab oil embargoes, this changed. Not only did consumers pay attention to gasoline prices, but lawmakers looked for ways to cut U.S. dependence on "foreign" oil. In 1975, Congress passed the Corporate Average Fuel Economy (CAFE) standards, which forced the auto industry to design cars that got more miles per gallon. Carmakers complained bitterly, claiming that the new laws would put them out of business. Amazingly, Detroit not only survived but, after a sorry period of experimentation (recall the AMC Pacer, for example), began rolling out machines of stunning efficiency and even elegance. By 1985, new American cars were averaging twenty-five miles to the gallon — up from fifteen a decade before. Better still, automotive engineers had only begun to squeeze more efficiency from the internal-combustion engine, the transmission, and body aerodynamics. By some estimates, average fuel efficiency would, according to then-current rates, easily reach forty miles per gallon by the end of the century.

Beginning in the mid-1980s, however, the efficiency incentive vanished. As the Saudis flooded the market to regain their market share and as world oil prices declined, American politicians saw no need to continue the campaign for energy efficiency and froze CAFE standards at their 1985 levels. Detroit, now desperate to regain the customers it had lost to Japanese automakers, quickly dropped the economy car approach and began making much larger, more powerful, more "American" vehicles. It wasn't that car technology had stopped improving; compared with the internal-combustion engines of 1970, the new engines generated much more power for the same gallon of gasoline. Yet instead of using this "efficiency dividend" to save more fuel — that is, instead of keeping power constant and cutting fuel consumption — Detroit, and eventually its rivals in Europe and Japan, went the other way, making larger, heavier, more powerful cars and trucks that could carry bigger loads, accelerate more quickly, and offer more features but that used more fuel in the process.

A quick look at the numbers shows how dramatic the change has been. In 1975, the average new American car got around fifteen miles to the gallon and had enough power to accelerate from zero to sixty miles per hour in around fourteen seconds. By 1985, after ten years of oil shocks and government fuel-efficiency mandates, U.S. cars averaged twenty-five miles per gallon, but acceleration had improved only marginally. In other words, Detroit had invested the efficiency dividend in better fuel economy; power and performance were secondary.

But from the mid-1980s on, automakers began using their more efficient engines and transmissions to achieve greater acceleration, to drive heavier vehicles, and to provide power for more onboard features. By 2002, the average American "car" not only was heavier but could go from zero to sixty in less than 10.5 seconds — a huge increase in power. At the same time, though, fuel efficiency had slumped to about half what it could have been had Detroit kept its focus on miles per gallon.

It's tempting to blame automakers for this missed opportunity. But

American consumers are just as culpable. We Americans are driving more than we used to — around twelve thousand miles a year, nearly a third more than in 1980 — in part because we live farther from work. We are taking more trips, with fewer passengers in the car. And to be honest, we haven't really cared about fuel economy since the early 1980s. Despite the occasional spikes, gasoline prices in America, when adjusted for inflation, are as low as they have ever been — and certainly lower than anywhere else in the world. For most car buyers, fuel efficiency simply isn't a factor in their decision to buy a car.

What is a factor, however, is size and power and "features," for as it turns out, Americans were never content with the trend toward smaller, fuel-sipping cars. Many felt the smaller cars unsafe; others simply pined for the Camaros, Mustangs, and other muscular chariots from the automobile's glory days. As oil prices fell, America's passion for large, powerful, vehicles roared back with a vengeance — only this time the "muscle car" wasn't a Camaro or a Mustang. Pickup trucks, which for decades had been marketed mainly to farmers, contractors, and other real working types, suddenly became a hot ticket for a burgeoning class of urban cowpokes — city slickers and suburbanites anxious to look tough.

Even more popular than the pickup was its cousin, a large, powerful, four-wheel-drive rig dubbed the sport-utility vehicle. Originally designed for work crews, residents of the snow country, and other folks who might actually have need for a vehicle that could travel off-road, SUVs have since become the car of choice for executives, sports stars, and gangster rappers, as well as house husbands, soccer moms, and tens of millions of others who will never intentionally leave the paved roads. In fact, fewer than one in twenty SUV owners ever goes off-road, and only one in ten pickup drivers ever actually carries anything in the back of the truck. But that hasn't stopped Americans from buying huge vehicles.

The SUV represents the height of conspicuous energy consumption. The extra size, weight, and power of the vehicles are rarely justified by the way their owners drive them. Even though owners and carmakers counter that the SUV's greater size, weight, and capabilities provide an extra margin of safety, studies indicate that SUVs not only are more likely to kill people in cars they hit but, because they roll over more easily, are actually more dangerous to their occupants as well.

Whatever their actual utility, SUVs and pickups are exceedingly popular among American drivers of all ages and incomes. In fact, the "light truck" category, which includes pickups and SUVs, is the largest-selling category in the United States, accounting for 48 percent of all new vehicle sales in 2003, and it may reach 60 percent by 2015. This, more than anything else, explains why the fuel economy of the average new vehicle sold in the States is now less than twenty-one miles per gallon[12] — the lowest level since 1988, the peak year for fuel efficiency. To put it another way, of the nearly twenty million barrels of oil that America uses every day, more than a sixth represents a direct consequence of the decision by automakers to invest the efficiency dividend in power, not fuel economy. Or as the U.S. Environmental Protection Agency concluded recently, if the 2003 vehicle fleet had the same average performance and weight distribution as vehicles made in 1981, the average fuel economy would be a third higher.

One consequence of these trends was that even as most of the rest of the U.S. economy was in the doldrums in 2003, oil demand was growing at nearly 3 percent, a reminder to the rest of world why America is the most important oil market. More generally, the trend toward larger cars and trucks, coupled with the expected growth in number of vehicles and in miles traveled, helps us understand how oil consumption has increased in the United States from seventeen million barrels per day in 1990 to twenty million today, and may rise as high as thirty-two million by 2020.[13]

Yet what is most disturbing about our desire for ever-larger cars and houses, more gadgets, and ever-greater demand is that it is difficult to see where it all ends. Where are the natural limits? Barring some massive disruption in energy supply, it is hard to see why consumers or companies would willingly use less energy — or for that matter, why any political leader would suggest that they use less energy, or even that they slow the growth in their energy demand. For all our astonishing improvements in technology and energy efficiency, an expanding economy is still seen as inseparably linked to constant increases in energy use. And the rest of the world, especially the developing world, has noticed.

Is this the shape of the energy future? Are the energy trends and tendencies of the United States and the rest of the postindustrial West indicative of what is to come in developing countries over the next two or three decades? Should we expect to see two-car garages, big-screen televisions, and three-quarter-ton pickup trucks in places like Rio, New Delhi, or Beijing? Will people in China and India not be satisfied until they consume as much oil, gas, and electricity as Americans do?

One could argue that America is a special case. One could insist that American consumers are unique in their energy obliviousness, or that the status of the United States as a superpower and world policeman somehow entitles Americans to worry less about energy. It may be that other countries won't follow this path, won't trade away their efficiency dividend for an energy-lavish lifestyle or yield to the escalating spiral of consumption.

History suggests otherwise, though. Even if many factors, such as population growth or economic development, will ultimately determine how closely the third world parallels the first in energy consumption, it's also clear that the historic link between economic development and rising energy use still holds true. Indeed, most developing nations have explicitly acknowledged that their economic goals are tied to the ability to gain access to more energy.

Without question, individual consumers in the developing world will demand more energy goods and services. As wealth rises, people in developing nations do precisely what their richer counterparts do — they build larger homes and fill them with more energy-consuming appliances. Look at China. Before 1985, only 7 percent of all Chinese had refrigerators; today, the figure is more than 75 percent. The number who own TVs has climbed from 17 percent to 86 percent.[14] Air conditioners have multiplied by a factor of fifty. All of this explains how demand for residential electric power in China more than quadrupled since 1984 and 1996, and why power generation is one of the fastest-growing sectors in the developing world.[15] In China alone, the government says it must build as many as sixty electric power plants every year for the next decade, simply to keep up with demand.

Even more dramatic is the trend in transportation energy in developing countries. Despite low incomes, poor roads, and lack of access to quality fuels, car ownership is an increasingly important personal goal — even in "middle-income" countries, where personal income is only just above the poverty level — between five thousand and fifteen thousand dollars per annum — and the cost of a car can be equal to a year's earnings. In booming Thailand and other Southeast Asian economies, for example, the private car fleet grows 30 percent a year. In South Korea, a rapidly industrializing Asian country that may well indicate the energy future for developing nations, the number of passenger cars quadrupled between 1987 and 1997.[16] Gasoline consumption tripled during the same period, and will double again by 2020. "Multiply this phenomenon of an emerging middle class, already in the case of China, roughly 100 million, by the perhaps 200 million more in India and Southeast Asia, and the proportions of the consumer revolution in developing Asian states begin to register," writes Robert Manning, an expert on Asian energy issues.[17]

Of course, not all developing countries will grow as fast as South Korea did, especially large rural nations like China and India, where car ownership and per capita oil consumption remain minuscule. Yet as Manning suggests, the fact that Chinese and Indian citizens now consume only a tiny percentage of global transportation energy provides no comfort, since it hints at the massive changes in store for the region and for the world energy economy as these nations proceed with their economic awakening and strive for Western-style energy usage.

In India, where a middle class of some hundred million people is flexing its economic muscles, car ownership has tripled in the last decade and is expected to more than triple again by 2020. This will bring about a fourfold rise in use of fuel for transportation, not to mention a daily import requirement of some 3.5 million barrels of oil. All told, demand for oil in the developing world is expected to increase by more than 250 percent, from around twenty-five million barrels a day in 2003 to as much as sixty-seven million barrels in 2020. In less than two decades, nearly half the oil produced worldwide will be consumed by developing nations.

Even these figures, however, understate what is happening already, and how dramatically countries like China are altering the balance of supply and demand. As recently as the summer of 2003, several forecasting agencies were predicting that worldwide oil demand would grow by just 1.3 percent a year in 2003 and 2004. At this rate — roughly the same as during the "booming" 1990s — the world's daily energy use would grow by a million barrels each year. Yet so rapidly has China's economy, especially its automotive sector, grown in the past eighteen months that forecasts must be revised upward. New estimates suggest oil demand will climb by 2.7 to 3.3 percent a year, meaning consumers will instead use another 2 to 2.6 million barrels a day each year. These new forecasts point to ever tighter markets and even more stresses on an overtaxed energy economy.

Such concerns are not confined to oil. By 2020, natural gas consumption in the developing world will have nearly tripled. Demand for electrical power will have increased by a factor of two and a half, as will the demand for coal, the primary fuel for power plants.

From these projections, several disturbing realizations emerge. First, without an ever-expanding supply of energy, the economies of the most populous nations in the world — and the aspirations of billions of people — will come crashing to a halt. And second, such continued growth in energy production and consumption, necessary though it may be for economic growth, will become harder and more dangerous to maintain without some kind of disruption. Thus, when we say that the developing world is driving most of the growth in world demand for energy, what we mean is that most of the growth is taking place in the areas that are least equipped, politically or economically, to manage the related issues of pollution and energy security.

Nothing illustrates the risks of such growth more dramatically than what is happening in Shanghai, a thriving, densely populated port city on the Yangtze River Delta. Visitors to the "Chinese Detroit" — both General Motors and Volkswagen are headquartered here — are confronted by an explosion in new construction: office towers, houses, and, above all, roads and bridges. In the past decade, Shanghai has spent more than ten billion dollars — an enormous sum for China — to modernize and expand its transportation infrastructure and has begun construction of two new bridges, a giant tunnel, a new "ring" highway, and hundreds of miles of new roadway. As one analyst has noted, the pace of new construction has been "something like building the Brooklyn and Manhattan Bridges in New York and the Lincoln and Holland tunnels between New York and New Jersey — all in five years."[18]

Shanghai is, in fact, primed to become China's car capital. Its thirteen million residents enjoy one of the highest per capita incomes in any Chinese city — four thousand dollars a year, nearly twice China's average — and can afford cars. A nascent car culture is taking hold, as an emerging middle class has come to see the private automobile as both a personal status symbol and a sign of national success. Cars are even becoming a necessity: in an attempt to ease inner-city housing shortages, planners have created a constellation of suburbs around the city center — thereby spawning the first Chinese commuter culture.

The results are hardly surprising. In 1995, the three most popular forms of personal transportation among Shanghai's millions were bicycle (33 percent), foot (31 percent), and bus (25 percent). Cars accounted for less than 5 percent — a smaller share, even, than scooters. By 2000, car use had tripled to 15 percent, and by 2020, cars will account for more than *half* of all personal transportation. But this motorization will come at a high price: most forecasts envision heavy congestion, chronic respiratory illness from rising air pollution, and a sevenfold increase in emissions of carbon dioxide and other climate-altering gases. Even today, the amount of soot and other suspended particulates in Shanghai's air exceeds maximum international standards by a factor of nearly four.

This is a price that all of urban China may soon be paying. The Chinese campaign to develop a domestic auto industry, coupled with declining car prices and the growing prevalence of consumer credit, truly is creating a nation of would-be motorists. Although car ownership is still negligible — fewer than eight in every thousand Chinese have a private car — it is growing at a phenomenal rate. Among urban Chinese, for example, nearly three-quarters say they plan to buy a car within the next five years, and a third already have their driver's licenses. More to the point, say auto industry executives, per capita incomes are reaching a critical threshold. As Nissan chief executive officer Carlos Ghosn told a Chinese reporter several years ago, "an increasing number of people in China earn annual salaries equivalent to the price of a new passenger car. As has happened in other markets, this is exactly the point in time when domestic car sales begin to take off."[19]

By some estimates, China could have anywhere from a hundred million to two hundred million cars by 2020 — far fewer than the United States, on a per capita basis, yet enough to create nationwide motor-related problems. Urban air pollution now accounts for around four million deaths a year in China, and most experts expect this figure to rise, despite ambitious government efforts to reduce tailpipe emissions. Cars will also add significantly to China's greenhouse gas emissions. Today, Chinese motorists produce only a tiny fraction — less than 3 percent — of the greenhouse gases churned out by American motorists. Within twenty years, however, China could be producing more than a sixth of the global total, with much coming from its growing population of cars.

Equally worrying is the impact that China's growing energy appetite will have on oil geopolitics. Since China became a net importer of oil, the

Chinese government has scrambled to boost oil production by developing onshore fields and moving aggressively into the offshore, and by building huge, expensive pipelines from its oil-rich western provinces to its heavily populated east. Since 1990, the country has raised domestic production from 2.8 million barrels a day to 3.2 million barrels a day, but levels may be approaching their apex: China has more than one-fifth of the world's population, yet less than 2 percent of the world's oil reserves. China's oil boom is largely over: today, more than half China's oil output comes from two supergiant fields — Daqing and Shengli — that have been overdrilled and are already in decline.[20] Nationally, total production is increasing at less than 2 percent a year, while demand for oil is growing by 7 percent.

At best, according to U.S. analysts, China will never produce more than 3.2 million barrels a day, which means that by 2020 the country will be importing up to eight million barrels a day. The implications aren't pretty. Beyond draining China's hard-currency reserves — money Beijing would much rather spend buying "clean" Western energy technology — rising imports will make China increasingly vulnerable to the vicissitudes of the oil market and oil politics generally and will have a corresponding effect on world markets. Just as American imports are hugely important to oil markets today, so, too, will China's rising oil demand be: in the not-so-distant future, any shift in Chinese domestic economic policies or cooling off of China's red-hot economic growth could send oil prices reeling. Chinese demand is already pushing Beijing into a headlong rush to find oil and gas suppliers and setting up a showdown with other regional consumers, like Japan and South Korea, over access to Middle Eastern oil.

How inevitable are such scenarios? Clearly, the rise in energy consumption in China and elsewhere throughout the developed and developing world depends on a host of factors. If world economic growth slows to just 2 percent a year, instead of the current 2.6 percent, daily world oil consumption would reach just 101 million barrels by 2020, instead of 120 million barrels — and that difference would dramatically lessen the pressure on world oil markets.[21] Likewise, a gradual rise in the world price of oil would also keep demand down. According to one estimate, if the price of oil were to climb from its historic average of twenty dollars a barrel to thirty dollars a barrel and remain there, in real terms, for the next two decades, demand would be pushed down to around 106 million barrels by 2020.[22]

Fate and world events are not the only determinants of oil demand. Similar reductions might be possible through active campaigns to cut back on energy use through new technologies. (Many oil analysts, for example, believe world demand forecasts are essentially meaningless until it becomes clear whether China's auto industry is going to eschew or embrace energy efficiency.) Moreover, many developing countries are pushing hard to move from coal and oil to natural gas, and, as we'll see in the next chapter, gas systems are among the most common new energy projects in China, India, and other developing countries. China has also launched an ambitious fuel cell technology program, and many cities have embarked on comprehensive programs that, while aimed at cutting traffic congestion and air pollution, will have the indirect effect of reducing energy demand.

In Shanghai, for example, far-sighted city officials have adopted the most stringent clean-air standards in all of China, which may well persuade automakers to develop more fuel-efficient and therefore less polluting cars and trucks. The city's huge taxi fleet is being refitted to burn cleaner natural gas, and electric scooters are being promoted. A cap has been set on new car purchases, and would-be owners must pay a steep fee to register their new cars. Automakers are being encouraged to build so-called China cars — ultracompact vehicles suitable mainly for quick urban trips. Shanghai has also invested heavily in a passenger rail system and a new bus system and, in January 2003, inaugurated the world's first commercial magnetic-levitation, or "mag-lev," train, a sleek, energy-efficient train (capable of reaching speeds of 260 miles per hour) that links the city's bustling financial district with its airport, nineteen miles away.

Yet as we have seen with the United States and other developed nations, such mitigating factors run up against a powerful array of economic and political forces — countervailing influences that steadily push up energy demand and favor expediency at the expense of fuel efficiency. Oil prices, for example, could just as easily fall, at least in the short term, especially if countries with enormous reserves but little current production, such as Iraq and Iran, obtain the investment they need and start adding supplies to the world market. As we have seen, low prices discourage conservation and fuel efficiency, as well as reliance on alternatives like natural gas or hydrogen, or renewable energy, such as solar or wind. By one estimate, if oil prices fall to fifteen dollars a barrel and stay there until 2020 (a scenario fervently desired by the Bush administration), world oil demand will surge to 124 million barrels a day by 2020[23] — around 20 million barrels more than in average, or "business-as-usual," forecasts. Such an increase would put an enormous strain on oil producers, not to mention add significantly to pollution and other oil-related problems — among them, more cars, greater suburban sprawl, and a far slower emergence of even such conventional alternative technologies as gasoline-electric hybrids. According to one study, a scenario in which prices averaged twenty-three dollars a barrel would encourage so much additional energy use that U.S. CO_2 emissions would jump 50 percent by 2035, effectively destroying any chance at meeting a carbon target.

Even if prices remained high, it's hard to see how alternative energy technologies could succeed in developing countries. Alternative technologies are tremendously expensive and uncertain, even in developed countries — think of Ballard Power's struggles to survive in the richest, most technically advanced market in the world. It is very hard to imagine a technically backward country such as China or India successfully embracing solar technology or rolling out a production-ready fuel cell model before Detroit does. "R & D *is* a huge priority among Chinese automakers, and they are really hoping to leap-frog existing technology and go right to more advanced technologies," says Kelly Sims-Gallagher, an expert on the Chinese auto industry and researcher at Harvard University's Belfer Center for Science and International Affairs. "But the fact is, Chinese engineers couldn't design a complete car worth anything to save their lives. That's how little they have learned in the last twenty years."[24]

The larger truth is that developing countries like Chile, China, and South Korea don't really *want* to use less energy. Although developing nations fear energy dependence and the related environmental problems, they are in many cases more than willing to accept these costs — for the time being, at least — if higher energy consumption can bring them greater economic growth. Many countries regard a clean environment or energy independence as luxuries that are not possible, and therefore not worth worrying about, until solid economic growth can be established. For decades, "Grow first, clean up later," was the unofficial policy in Chile,[25] and it remains a standard refrain throughout the developing world, where politicians worry not simply about GNP but about providing the most basic services — housing, food, education, and medical treatment — for rapidly growing populations whose members are primarily interested in basic survival.

In this context, it is hardly surprising that leaders in Nigeria or Chile or China are not terribly enthusiastic about reducing energy use or curbing greenhouse gas emissions — especially when such proposals almost invariably come from developed nations whose economies are already strong and which, in many cases, seem reluctant to adopt such policies themselves.

Ultimately, the problem goes beyond national policy. Energy use may be tied intimately to the wealth of nations, but it is driven mainly by individual demand — demand from manufacturers and other industrial users and, above all, demand from *consumers* — both in their purchases of direct energy and in purchases of goods and services that require energy. Not all consumer demand is equal. For some two billion people — a quarter of the world's population — demand for energy goods and services remains at a sustenance level. Past a certain level of economic growth, though, energy consumption is much more a matter of individual choice and even desire. People want things, and the benefits and pleasures that come from things — ovens, water heaters, and big-screen TVs. They want mobility, and the freedom and status it provides. They want convenience, the luxury of spare time, and entertainment and communication to fill those free hours. All of it requires energy. Few consumers ask about the energy implicit in each of these desires — how much energy they require, where it comes from, and what its economic, political, and social costs are. Instead, consumers mainly want the goods and services the energy provides. As a Chinese researcher told environmental writer Mark Hertsgaard a few years ago, "if you talk to Chinese people, many of them will tell you, 'To have a car is my dream.' The car represents affluence to the Chinese, and until they have had a chance to own one, it will be difficult to convince them not to use a car because of its environmental effects."[26]

If any doubt existed about the power of China's incipient car culture, it would be dispelled by the Seventh Beijing International Auto Show. At the most recent biannual event, nearly half a million Chinese thronged the massive International Exhibition Center for a glimpse of hundreds of new foreign and domestic cars. The crowd was mainly young, urban, and professional; as in the West, people seemed to enjoy the spectacle as much as the cars themselves. At any number of displays, clusters of amused Chinese — many of them men — gaped at beautiful young models, clad in scanty bikinis and draped over muscle-bound performance vehicles and luxury cars that cost more than the average Chinese middle-class family would earn in a century.

Yet this was hardly a fantasy fair. Many of the model cars on display here had been designed specifically for the nascent Chinese market: the compact Elysée, from Dongfeng Citroen Automobile Company, the tiny Bora from First Automotive Works Volkswagen, and the sporty Buick Sail from Shanghai GM. There were the minivans, which the Chinese call "bread loaves"; there were large luxury sedans for the new business class; and for the up-and-coming Chinese executive who dreams of an SUV, there was the sleek Volvo CrossCountry, as well as several models from Beijing Jeep. "It was completely jam-packed," Sims-Gallagher, the researcher, told me. "You had no elbow room — you were just being carried along on this wave of people. I've been in China many times, but I had never experienced crowding on this scale, and that was when it hit home how many people there are who are absolutely serious about buying a car."

Indeed, by the end of show, automakers had received orders for more than ten thousand vehicles — including more than eleven hundred sedans from Shanghai GM — and were optimistic about future sales. "When I squeezed into the crowds, I could imagine how the market will thrive," said Dieter Laxy, senior vice president of the Volvo Car Corporation. Other auto executives were even more confident. "You could see consumers' fever," crowed Qie Xiaogang, an official from the Beijing Asian Games Village Automobile Exchange. "Many of them will buy cars soon."[27]

7

BIG OIL
GETS ANXIOUS

FOURTEEN MILES NORTH of Ensenada, Mexico, along the Pacific coast of the Baja Peninsula, the future of the global energy order is ready to rise up out of the coastal desert. High atop the Costa Azul plateau, just minutes away from vacation villas and a golf resort, a California company called Sempra Energy wants to build a four-hundred-million-dollar, state-of-the art factory that will transform ice-cold liquid methane imported from across the Pacific into natural gas vapor for the energy-hungry north.

Beginning in 2008, if all goes according to Sempra's rather daring plans, the turquoise waters off northern Baja will host a steady parade of gargantuan refrigerated tanker ships, each filled with superchilled liquefied natural gas, or LNG. Fresh from the giant gas fields in South America, Russia, Indonesia, and even Australia, the vessels will moor at a long concrete pier, then carefully discharge their liquid cargo into huge storage tanks on the plateau. From there, the LNG will go to special "regasification" units where it will be slowly warmed and allowed to expand to its natural vaporous state. A forty-mile pipeline, nearly three feet in diameter, will carry the gas to several sites near the U.S.-Mexican border, where gas-fired power plants will turn most of it into electricity for the local market or send it north to power-starved Southern California. "We see California and Baja California as one region," Michael Clark, the Sempra spokesman, told *Power & Gas Marketing*, "and our goal is to make sure that the region has adequate energy infrastructure to meet its future needs."

Not everyone shares Sempra's enthusiasm. Critics, attacking the project as "energy colonialism," say that the only reason Sempra has come to Mexico is that the rich residents of Southern California, where most of the electricity is headed, won't allow Sempra to locate the enormous LNG facilities there. Yet as legitimate as such complaints may be, it is hard to see how they will matter, for the Costa Azul project is merely the leading edge of a giant wave of LNG that may soon wash over the Baja Peninsula. Drawn by the insatiable North American energy market, companies like Shell and Conoco are already planning at least five other "regas" terminals for Baja, and another dozen projects are under consideration. Over the next decade, proponents hope, Baja will become an energy hub for the American Southwest and help ease a U.S. gas crunch that seems to grow tighter with every winter heating season.

The United States is hardly alone in this "dash for gas." In the past five years, the global energy industry has been positioning itself to profit in an energy economy fueled increasingly by natural gas, an abundant, relatively clean hydrocarbon that many believe can close the worrisome gap between global energy supply and energy demand. In the summer of 2003, as the United States entered the "cooling season" with what appeared to be a whopping natural gas shortfall, no less a spokesman than Alan Greenspan, chairman of the U.S. Federal Reserve and one of the most influential voices in the global economy, urged lawmakers to pursue "a major expansion of LNG terminal import capacity" or risk the economic poison of continued price volatility. Greenspan's warning, echoed by many in the energy industry, touched off a storm of media interest and almost overnight transformed LNG from an industry acronym to the Next Big Thing on Wall Street. Notes Ira Joseph, an LNG analyst at PIRA Energy in New York: "Ever since Greenspan came out and was actually talking about LNG, I've been getting about three calls a day from investors asking me, 'How can I make money in LNG? What is the LNG play in my business?'"

On many levels, Costa Azul and the rest of Baja's gas boom are the perfect metaphor for the latest twist in the evolution of the global energy economy. As worldwide demand for energy continues its relentless rise, and as the sources for oil become riskier, the energy economy is shifting inexorably toward fuels that will completely reconfigure the world energy mix. Just thirty years ago, oil dominated the energy economy utterly, accounting for nearly 50 percent of global demand, and leaving just 31 percent to coal and barely 20 percent to natural gas. Since then, natural gas — or simply "gas," in industry parlance — has emerged as the fuel of choice for everything from power generation and industrial heating to home furnaces and cooking. Gas's share of the world energy mix now matches that of coal, and by as early as 2025, gas, not oil, could be the world's dominant energy source. Already, in places as diverse as Nigeria, Qatar, Trinidad, India, Siberia, Iran, and South America, energy companies, utilities, investors, and entire governments are throwing hundreds of billions of dollars into a sprawling gas and LNG infrastructure that will change the energy world completely.

In many respects, the rise of gas parallels that of oil half a century ago. Gas is now the abundant fuel: by some estimates, the massive gas fields in Qatar, Iran, Turkmenistan, and Russia, which hold more than half the known global reserves, could fuel the world for more than a half century, and these are only a fraction of total gas assets. Gas is also incredibly versatile. It can be used in everything from power plants to gas-powered buses and taxis. It can be converted to liquid fuels — gasoline, for example — and compete directly with oil. Gas also contains less carbon and more hydrogen than either oil or coal does, so not only does it emit less pollution and climate-altering CO_2, but it is also easily refined into pure hydrogen to power fuel cells and other energy technologies of the future. For these reasons, gas is widely touted as a "bridge" fuel — the one existing fuel that can simultaneously power much of our current energy economy and drive the transition to a more ideal system in the future.

Still, gas hardly offers an unencumbered pathway to the next energy economy. Despite its lower carbon content, gas produces emissions. The promise of its great abundance is compromised, because, as with oil, the largest gas supplies are located so far from the biggest markets that in switching from oil to gas, we may simply be trading one insecure energy infrastructure for another. Even more significantly, gas is far more expensive to handle and transport than oil is. Gas facilities cost billions of dollars, take decades to pay off, and pose massive financial risks for the energy companies undertaking them: a single LNG operation can cost a company four billion dollars — so much money that most energy companies simply cannot handle gas deals by themselves, and those which can often do so only by forgoing investments in oil and other, more proven lines of business.

In fact, as energy companies like Shell and BP and ExxonMobil scramble to capture a piece of the burgeoning gas market, it's clear that this mad rush to gas is not occurring just because gas is a better fuel than oil or because gas might be a bridge to something else or because the United Nations says we need a cleaner, more climate-friendly energy economy. What is driving the move to gas is mainly that oil is no longer the sure bet it used to be. With all the questions about long-term worldwide supply, and specifically with the generally bleak prospects for non-OPEC oil production, energy consumers are having to rethink their sources of supply, while energy companies are being pushed into completely new lines of business. In short, the big players in the energy order are moving into gas because, in the not-too-distant future, it may be the only way some of them can make any money.

Until fairly recently, the notion that gas might be a moneymaker would have struck most oil executives as absurd. Gas was mainly a waste product, the nasty, potentially explosive stuff unlucky oil drillers found when they were looking for crude. True, gas burned hotter and cleaner than coal or oil. But the very diffusive nature of gas — a house-sized volume of gas contains less heat energy than a single barrel of oil — made dealing with gas a costly, complicated procedure. Whereas oil could be shipped by the crudest means (wooden pipelines, say, or wooden barrels in horse-drawn wagons, or converted sailing ships), gas is far less amenable to half-measures. The best method is to pipe it directly to your customers, but early pipelines leaked so badly that most of the precious vapor never reached its destination. (Interestingly, the gas era of late nineteenth-century Europe and the United States — symbolized by gas lights, lamplighters, and periodic, lethal explosions — was fueled not by natural gas, but by "town gas," a synthetic miasma cooked from coal at factories in the cities.)

As pipeline technologies improved, regional gas networks developed in the United States and Western Europe, especially after World War II. But the dismal economics — a pipeline more than a few hundred miles long cost more to build than the gas could be sold for — kept the gas industry from attaining oil's global reach. The geographic distance was simply too great between the largest sources of supply — such the vast gas fields in Siberia or Iran — and the biggest consumer and industrial markets, such as Europe, Japan, and the American East Coast. To borrow an industry phrase, the great majority of gas was a "stranded" asset — something whose great potential value was negated by its distance from paying customers. Unless you had a source within a few thousand miles of a market, gas was simply a pain. Pipelines are expensive to build (at around a million dollars per mile), and oil companies were still making far more money producing oil. Until very recently, a dollar invested in an oil field returned twice the profits of a dollar invested in a gas field. Given these disincentives, most gas never left the field at all, but was either reinjected into the reservoir to keep the oil pressurized or, more commonly, simply burned, or "flared off," at the wellhead.

Then, three things changed. First, demand for gas took a sudden leap in the 1970s. As the Arab oil embargo drove up oil prices, big industrial oil users — factories, and especially power generators — were forced to look for alternative fuels, including gas. At the same time, the rise of the environmental movement, and the emphasis on reducing air pollution, encouraged many big coal users — again, mainly electrical utilities — to switch to cleaner fuels, including gas. As demand for gas steadily climbed, price followed, encouraging energy companies to reexamine their "worthless" gas assets and figure out new ways of getting them to market. More pipelines were built, especially between the Soviet Union and Europe, and between North Africa and Southern Europe. To bridge the greater distances, oil companies like Mobil began developing processes to liquefy gas commercially and carry it aboard tankers, and a small LNG trade developed between Africa and North America.

As LNG technology improved, energy companies realized they could profitably sell not just bulk methane, but also some of the higher-value elements in the gas. In addition to low-value methane, natural gas contains small quantities of so-called natural gas liquids, or NGLs — ethane, butane, and propane, which can be separated and sold for good prices. (Ethane, which is made into plastics and synthetic rubber, is especially valuable.) Better still, some natural gas fields contained something called condensate, a gas-based liquid that behaves like an ultralight crude oil and is easily refined into gasoline and other high-value products. In fact, NGLs are the profitable part of the gas business (and, according to many analysts, the main reason oil companies can afford to bother with gas). By the mid-1970s, as NGL technology improved, the oil industry suddenly found itself with an incentive to develop its formerly worthless "stranded" gas fields.

Another, larger change was afoot in the oil industry, however, which would promote gas from a reasonably profitably sideline into something that could transform the oil business. After decades of high profitability, the major oil companies found themselves confronting the bizarre possibility that oil was no longer the cash cow that it had been for nearly a century.

For most of the twentieth century, the oil business was a relatively straightforward enterprise, at least by comparison with other sprawling global enterprises. The industry was dominated by a handful of large, integrated companies — the majors — that controlled the entire chain of production, from oil well to gasoline pump. The industry's considerable profits came mainly from refining and marketing heating oil, fuel oil, and especially gasoline. In fact, the oil industry's entire business model — from the kind of crude oil it sought to the kind of refineries it built to its intense focus on retail marketing — was built around the gasoline pump.[1] And what ensured its profitability was the control this integrated model gave the majors over the most critical variable: supply. When demand rose, the major oil companies simply stepped up production and quickly captured any potential profits.

This is no longer the case. The rise of OPEC in the 1970s effectively "disintegrated" the majors, severing them from most of their supply and cutting their upstream production to as little as a third of pre-OPEC levels. To cite one example, in 1972, before the OPEC nationalizations, Exxon and Mobil could boast a combined production of 7.3 million barrels a day; today, the merged ExxonMobil, largest of the majors, produces fewer than 4.2 million barrels of oil a day, or less than half the volume of the Saudi Aramco.

As a result of disintegration, the majors were forced to buy more of their crude on the open market, usually at higher prices, or go look for it in new places, a quest that generally raised production costs. That was fine for a while: oil companies simply passed on higher costs to consumers and continued to make fabulous profits — to such a degree that they began investing in such nonoil businesses as computers and financial services.

In the 1980s, however, two new trends became apparent. First, the oil "downstream" — as refining and marketing are called — was invaded by a slew of outside players. State-owned oil companies in Saudi Arabia and Venezuela and elsewhere began refining their own crude and selling it at their own gasoline stations and convenience stores in Europe and the United States, in effect bypassing the majors and stealing some of their lucrative retail market share. Second, the market gluts of the 1980s drove oil prices down, and that price dip not only hurt the majors' profits but pushed their stock prices down. The depressed stock prices ignited a mergers-and-acquisitions frenzy that forced the industry to change the way it did business.

Big oil companies, long accustomed to the luxury of fat margins and huge budgets, now found themselves selling off ancillary businesses, furloughing thousands of workers, and slashing costs, all in hopes of boosting share prices and fending off takeovers. Acquired companies were also subject to brutal cost cutting. The effects were far-reaching. To cut costs, for example, oil companies stopped carrying as much "excess" oil inventory in storage as they had — in effect, they removed much of the slack that had historically buffered the international markets from small disruptions and price swings.

The larger impact of the merger mania was to create a new breed of oil company that simply needed more oil to survive. Today, these "supermajors" — ExxonMobil, ChevronTexaco, TotalFinaElf of France, and BP (which swallowed up both Amoco and Arco) — are so outsized that the task of maintaining their reserves — that is, of replacing every barrel sold with a freshly discovered barrel — has become an epic struggle. Not only must these companies discover lots of oil each year, but because they are so large, with such high operating costs, each discovery must be huge in order to be profitable. Exploration and production costs are now so high, for instance, that no large company can afford to search out and drill a great many smaller fields. Instead, they need the efficiencies and economies of scale of a single massive score — a billion barrels or more — to operate profitably. To use a baseball analogy, says Fadel Gheit, an energy analyst at Fahnestock & Company in New York, "these large companies don't strive to hit singles. They really need a home run . . . something in the hundreds of millions of barrels, to make a dent."[2]

The paradox, of course, is that the same trend that encouraged those mergers has also made it harder for the huge corporations to find the big oil they need. As we have seen, outside the OPEC countries and the former Soviet Union, large oil discoveries come less frequently every year, despite increasingly sophisticated and expensive exploration technologies — and despite a powerful price incentive that has encouraged companies to search more diligently. Instead, companies must settle increasingly for a diet of smaller fields. The few big fields that have been discovered, like Kazakhstan's massive Kashagan, are so expensive to work that only a consortium of companies can operate them. "Name me an international oil company that on its own is developing an oil field with production of more than a million barrels a day," challenges PIRA's Joseph. "You can't."

The result is that the majors are laboring to maintain their barrels-per-day production rates. ExxonMobil's production has been flat since 1999.[3] In 2002, BP was forced to actually downgrade its target for production growth three times — in part, some analysts say, because the super-sized company was hard-pressed to find enough oil in its existing fields.[4] Shell, too, has missed its growth targets. "Shell has struggled with production," oil analyst Jon Wright told *Wealth Manager Magazine*. "To replace a billion barrels every year is a real challenge."[5]

Shell's woes are particularly illuminating. Despite record profits, the venerable company seems to be depleting its oil reserves faster than it can find new oil. In 2000, Shell admitted that it had replaced only three-quarters of its annual production — that is, for every four barrels pumped, Shell was able to find or buy only three new barrels. In 2001, replacement fell to half. Replacement has improved recently, but in January 2004, Shell announced that it was reclassifying 20 percent of its "proven" reserves as "unproven" — a move that stunned analysts, angered shareholders, and revived concerns over the decline of non-OPEC oil.[6]

From the standpoint of an oil company's long-term profitability, this inability to hit targets or replace reserves is akin to a diagnosis of cancer — and the industry knows it. In today's global economy, companies that cannot grow will not survive, and companies that cannot even maintain their current size are in even more serious trouble. The market now watches company production numbers and so-called reserves-to-production ratios — or how many years a company's reserves will last — as closely as it used to watch profits. The slim geological pickings help explain why international oil companies have increasingly begun searching for oil on other companies' balance books. In yet another industrial-sized irony, many of the recent oil company mergers — especially the moves by Western companies to partner with Russian oil firms — have been driven at least in part by the buyers' interest in gaining the "booked" discoveries of the purchased company without the expense or risk of actual exploration. Nevertheless, oil companies, like oil, are finite in quantity, and with most of the marriageable companies already acquired, this strategy has only a limited lifespan.

Shell, BP, and other majors have vowed to reverse the trend of declining discovery, by investing in riskier, more expensive oil fields — such as those in deep water, the Arctic, or politically unstable terrain — in the hopes of striking it rich and restocking their inventories. Others are betting heavily on "unconventional" oil projects, such as the tar sands in Alberta or the heavy-oil fields in Venezuela. But most observers doubt that the trend can be reversed in the long run, short of oil companies' being allowed back into the Middle East. "Unconventional" oil faces high political hurdles, largely because refining the stuff is so polluting.[7] And as we have seen, venturing into ever-riskier oil regions offers an ever-lower probability of paying off. In fact, in an era of declining discoveries and worsening geological prospects, such desperate moves serve only to increase production costs and raise the price of failure. A single ultra-deep-water well can easily run a hundred million dollars; and according to some recent accounts, four of every five wells come up dry. Indeed, in a tacit admission of their own pessimism, BP and other majors are giving up on the traditional practice of publicly stating production goals. Goals are becoming too difficult to hit, and failure to hit them hurts company share price, which increasingly is the industry's new "production target." Despite Herculean efforts, the majors "are fighting an uphill battle to maintain reserve growth," says Herman Franssen, an oil industry analyst based in Washington. "Senior executives of several major oil companies have told me at recent seminars that they *must* have a stake in oil developments in Russia and the Middle East to increase their future reserve base."

It may be hard to feel much sympathy for the majors when they seem to be making money hand over fist. With the sustained high oil prices of the past two years, the industry as a whole has realized extraordinary profits. In late 2002, for example, ExxonMobil reported fourth-quarter profits of $4.1 *billion*, while Shell's $9 billion take for the entire year was the fattest in all of corporate Europe. For the majors themselves, though, such fantastic profits are bittersweet. Oil prices are high because markets are tight, and markets are tight in large measure because non-OPEC production — the hunting grounds for the majors — faces a long-term decline. Oil companies may be making more money per barrel, but over time they will be selling fewer barrels.

What the majors are looking at, then, is a long-term trend that means they will have a smaller number of fields to work, and thus lower production and shrinking market share. Over time, the majors face a grim

choice: continue to lose market share to those who *do* have large reserves
— namely, Mexico, the OPEC countries, and the fast-growing Russian oil
companies — or find something else to sell. This, as much as anything, explains the almost lemming-like rush by Western oil companies to get a
piece of the newly opened Russian oil business — and outside Russia, prospects look bleak for the kind of big oil strike that the majors need.

Shrinking reserves also explain the push by the majors into nontraditional markets, like hydrogen, solar power, and, above all, natural gas. For if
Western oil companies are short on oil, they're "long" on gas, in a big way,
and they have begun, in the last decade, to transform themselves into gas
giants, and to vie for the right to invest billions in complex LNG deals.
"The majors are getting shut out of most of the new major plays in the oil
market," says Joseph, "and the best alternative is gas."

Just off the Mexicali-Tijuana Highway, three miles south of the U.S. border
and some two hundred miles southeast of Los Angeles, sits a shiny new
piece of the gas economy — the Termoelectrica de Mexicali (TDM) power
plant. Completed by Sempra Energy in 2003 at a cost of $350 million, its
twin stacks rising high above the flat landscape of dun-colored sand and
brush, TDM runs off of natural gas piped in from the north — and potentially from the company's Costa Azul regas project. The gas is pumped into
what is known as a combined-cycle gas turbine, an ultraefficient, two-stage
generating system that burns the gas to turn a turbine, then captures the
superhot exhaust to make steam for additional power generation. When
running at maximum capacity, the TDM plant produces six hundred megawatts of electricity — enough power for six hundred thousand homes in
Mexico, or around a quarter that many across the border in the United
States.[8]

In fact, although the company says the plant's electricity is being made
available to customers on either side of the border, analysts say it is mainly
the U.S. power market, with its rapid rate of growth and yawning supply
problems, that Sempra and other owners of border-town power plants are
targeting. Indeed, one of the main reasons Sempra chose Mexicali was its
proximity to the Southwest Powerlink, a high-voltage transmission line
that connects Arizona and California.[9] "This is only about the Mexican
market in a very small way," says gas analyst Joseph. "There is no way most

people in Baja can pay the price that these power companies will have to charge. This is about Southern California and the very high prices that people there are willing to pay for electricity."[10]

Not surprisingly, it is the power sector that has emerged as the strongest impetus for the gas business, and the first piece in a gas "bridge" to the next energy economy. Electrical power, even more than transportation fuel, is the critical resource for modern economies that are increasingly based on technology and services. Power generation is thus *the* hot growth market, in developed places like Europe, Japan, and the United States, but especially in such developing countries as China and India, where the majority of people lack access to any electricity at all — but will demand it.

Like everything else in the energy world, the supply of power is not keeping up with demand, in large part because traditional methods of producing power are failing. Oil, used for decades to fuel power plants, has become too expensive. Most of the rest of the world's power comes from coal and nuclear energy, both of which are hugely problematic. Nuclear power, which today provides 18 percent of world electrical needs, faces so many hurdles that few governments and companies view it as a viable option. Not only is it politically difficult to select sites for nuclear plants (owing to fears about Chernobyl-style accidents), but the economics of nuclear power are dismal. The plants themselves cost nearly two billion dollars each, require years for the permitting process, and then take at least five years actually to build. In most cases, the costs up front are so high that plants can be built only with heavy government subsidies or with the promise of some kind of clean-air credit. (Nuclear is clean, after all.) Future technologies may overcome these and other problems, such as where to put all the spent nuclear waste and how to keep the fuel out of the hands of political malcontents. If so, nuclear power, with its near-total lack of emissions, would be the perfect carbon-free energy source for the next energy economy.[11] For now, even countries with large nuclear energy programs have actually begun phasing the plants out, and, as of 2001, only five countries — China, India, Japan, Russia, and South Korea — had plans for expanding their nuclear base.[12] Despite the Bush administration's calls for a "nuclear renaissance," energy forecasters believe that nuclear power's share of the world electricity will actually fall to as low as 10 percent by 2015.

The hurdles for coal-fired power are only slightly lower. Coal-fired power is the dirtiest of all power. The plants emit great gobs of sulfuric soot, which causes acid rain, and about twice as much carbon dioxide as a gas-fired power plant producing the same number of kilowatts. And coal-fired power is expensive. Although the coal itself is dirt cheap (about a dollar for the equivalent of a million British thermal units, or Btu's) — and plentiful (world reserves should last about two centuries), a new coal-fired power plant costs two billion dollars to build, faces all kinds of pollution rules, and takes thirty years to pay off. In other words, although the *operating* costs (the fuel) for a coal-fired power plant are low, the capital (construction) costs are substantial. This equation has discouraged utilities in industrialized countries from building new coal-fired power plants, or even replacing older, dirtier coal-fired plants with new ones.[13]

Gas, by contrast, is a breeze. It burns more cleanly than coal does, especially in the new ultraefficient combined-cycle gas turbine generators. More to the point, even though gas itself is more expensive than coal (prices have averaged around two dollars per million Btu's over the last several decades), a new gas-fired power plant costs half as much to build as does a new coal-fired power plant and has a shorter payoff period — sometimes as little as five or six years — which dramatically reduces financing costs. As John Browne, chief executive of BP, told a reporter, "one dollar invested today in gas-fired generation capacity produces three to four times the amount of electricity [as] the same dollar invested in coal-fired generation capacity."[14] And because gas turbines can be started and stopped more quickly than coal-fired systems can, utilities are firing them up as needed: during periods of peak power demand, for example, or in emergencies, when other power sources fail.

This versatility helps explain why gas-fired power has been the trend in Europe, Japan, and even in the developing world and why more than 90 percent of new power plants in the United States burn gas.[15] With the wave of deregulation sweeping power markets around the world and forcing governments to stop subsidizing the coal industry, gas is rapidly becoming the number-one fuel for producing electricity. Even though gas is still more expensive than coal, the huge demand for electricity, and the high prices that power companies have been able to charge (especially in the United States and Italy) have made the gas-fired power sector enormously profitable. Indeed, the so-called gas power market is so large, and has been so lucrative, that it is transforming both the gas industry and the power industry and, in some ways, merging the two. Gas companies not only are selling gas to power companies but, increasingly, are *becoming* power companies themselves, à la Enron: the Houston-based firm began in the gas pipeline business and morphed into an "energy merchant" with thousands of megawatts of generating capacity.

This move into the power business has allowed companies to exploit the so-called spark spread, or the difference between their costs for gas and the price for which they can sell the power. In theory, since demand for electricity is expected to grow, if energy companies can continue to bring down the cost of supplying themselves with gas, the "spark spread" offers the potential for generous profits and new markets — enough, perhaps, to replace the business some companies stood to lose in oil. Here, finally, is a way to "refine" natural gas — by turning it into electricity.

The spark spread may become one of the most important factors in the new energy economy, and it explains a host of developments in the energy business — from multibillion-dollar LNG deals in Baja to the rise, fall, and uncertain future of energy traders like Enron, which made most of its money, and many of its mistakes, trying to buy gas cheap and sell the generated power at as high a price as possible.

More to the point, the spark spread and the new interest in a "gas economy" explain why oil companies have been racing to attain a stronger position in gas, by shifting from a primary focus on oil to a more gas-centric model (or in the delightful parlance of the industry, to be less "oily" and more "gassy"). That is why companies like BP, Shell, and ExxonMobil have recently been on a shopping spree for new production rights in gas-rich areas or have been partnering up with gas-producing countries like Russia, Algeria, Qatar, or Indonesia.

It is also part of the driving force behind many of the megamergers in the oil industry during the 1990s; for example, although Exxon and Mobil glossed their 1999 merger as a joining together of complementary strengths, most in the industry saw the union as a tacit admission by Exxon that it had failed to exploit its stranded gas assets, whereas Mobil was one of the acknowledged leaders in LNG. Likewise, when BP bought Amoco in 1998, BP executives were primarily motivated by the recognition that only 20 percent of BP's own assets were in gas — it was the "oiliest" of all the majors — whereas Amoco was the big dog in U.S. gas production. With the acquisition, BP not only could brace itself for a future in which oil played a smaller role, but moreover could exploit a more progressive image, including the rather extravagant claim that BP now stood for "beyond petroleum."

The move into gas isn't limited simply to big Western oil companies. Many oil-rich states in the Middle East and elsewhere also find themselves looking beyond petroleum, although for slightly different reasons. Even if countries like Saudi Arabia or Algeria or Venezuela or Iran still have plenty of oil left in the ground, oil is not the growth business it used to be, at least, in the near term. Although global demand for oil remains strong, with the wash of Russian oil entering the market, other oil producers can't "grow" their exports as much as they would like; in fact, in September 2003, markets were temporarily so replete that OPEC actually threatened to cut production by nine hundred thousand barrels a day. To be sure, this problem is indeed temporary: when non-OPEC oil production peaks, the oil business will again be a growth market for those who can still pump it out of the ground. In the meantime, oil states whose bloated and corrupt governments require ever-larger infusions of revenue — which is to say, pretty much all of them — need to get in on the gas boom as well. "If the new reality is that you can't rely on oil to achieve economic growth, then the next major play has to be gas," says Rick Gordon, executive vice president at Connecticut-based John S. Herold. Because demand for energy is high, says Gordon, and no production limits have been imposed on gas to parallel OPEC's quotas on oil, "a country like Qatar or Iran can triple or quadruple production tomorrow," assuming they can muster the capital and build the necessary relationships. The real question, Gordon says, is who will get in first.

Now, with LNG, the missing link in a truly global gas market is finally in place. As improvements in technology have made it possible to liquefy, transport, and regasify great volumes of gas in economical fashion, the gap between the world's biggest gas reserves and the world's biggest markets in theory looks bridgeable; that is to say, gas could be a truly global market. LNG marks a turning point, not just for the gas industry but for the direction of the entire energy economy. With LNG tankers, suppliers and buyers have flexibility. They can commit to long-term contracts, as is currently done with gas pipelines, or they can sell their LNG on the "spot" market, one tanker-load at a time, wherever demand, and price, are highest. The lack of this kind of flexibility, analysts say, contributed to the power crisis in the western United States in 2000. "Just imagine," Gordon points out, "if it had been possible to deliver a few tankers of LNG to California."

With the emergence of LNG and the accelerated shift toward gas, the entire energy economy appears to be moving toward a system that is cleaner and more flexible, one that could provide a springboard to the next energy systems. Not only is gas the cheapest existing method for making hydrogen, but it offers numerous options for transitional power systems, as well as entirely novel ways to look at power and power distribution. Gas-fired turbines can now be made in nearly any size, from the six-hundred-megawatt power plant models to tiny thirty-kilowatt "microturbines" that can run six to seven homes or a small business. The full range of turbine sizes means that individual companies and even communities can become independent generators of power, buying gas instead of electricity from utilities and creating their own self-contained micro–power grids that would be far less prone to blackouts than the current national system is.

These so-called distributed power systems, which many experts believe will eventually replace our traditional centralized power systems, would also let consumers create their own mix of power sources. Owners of a microgrid in a city or state with strict laws on air quality, for example, could choose to emphasize wind or solar power as their main power source, then fill in any supply gaps with the quick-starting gas-powered microturbines. Because microgrids could sell any surplus power back to the regional power grids, they would reduce the need to put in additional large-scale power plants. Small gas turbines can even be used in place of a gasoline or diesel engine in gas-electric hybrid cars — yet another bridge technology that would allow us to reduce auto emissions and improve fuel economy dramatically while we wait for the hydrogen economy to arrive.

As important as all this versatility, however, is the way gas works as a bridge fuel to a more climate-friendly energy economy. Because methane contains less carbon than either coal or oil, it produces less carbon dioxide — about 50 percent less than coal and 33 percent less than oil — for the same energy production. This is crucial, because CO_2 is the most pervasive of the greenhouse gases that are raising global temperatures. Granted, gas is not the ideal climate-friendly fuel. Methane itself is a greenhouse gas, with a climate-changing impact roughly twenty times that of mere carbon dioxide.[16] Further, even if we managed to replace *all* current coal-fired power with gas-fired power, we would cut carbon emissions by only 30 percent: in other words, moving to a gas-fired economy will not solve our climate problems; but it would buy us some time — perhaps another five or ten years in the race to figure out some energy system not dependent on hydro-carbons.

In the meantime, gas is already eating into markets long dominated by coal and even oil. Cars can easily (if at some expense) be converted to run on natural gas, which is already the fuel of choice for many bus and taxi fleets around the world. Gas can also be converted into various liquid fuels, such as diesel and even synthetic gasoline, a capacity that could potentially end oil's stranglehold on transportation, although this gas-to-liquids technology remains primitive and uncompetitive. In fact, according to estimates (admittedly on the optimistic side) by the gas industry, in the United States alone, gas could replace as much as a third of the oil currently consumed, thereby dramatically reducing the nation's dependence on im-ported oil — and considerably improving energy security. As one somber gas industry magazine ad points out, "no wars will be fought to acquire it. No lives lost to protect its supply."

In May 2003, a select group of gas analysts and industry executives received an unusual summons from the U.S. Department of Energy. Even as indus-try experts were heralding the emergence of a gas economy, Spencer Abra-ham, the U.S. energy secretary, was calling an emergency "summit" to ad-dress what the department saw as a possible flaw in that grand and gassy vision: a shortage in domestic gas supplies. With the colder-than-expected winter and production problems in North American gas fields, U.S. gas re-serves were at their lowest level since 1976. Now, on the eve of the sum-mer cooling season, when a hundred million air-conditioning units would pump up demand for electricity, industry observers were forecasting a staggering shortage in gas supplies, and price spikes even higher than dur-ing the California crisis of 2000. Already, gas prices were hovering at six dollars per million Btu's — nearly three times the historic average — and some analysts were calling for twelve-dollar gas by summer's end.

Happily, twelve-dollar gas never materialized. The higher prices, by eventually encouraging more production, temporarily loosened the gas markets and eased fears of a disastrous winter shortage. Yet the speed with which the shortfall had appeared — only a few years before, the U.S. En-ergy Information Administration had been predicting virtually unlimited domestic supplies — led some astute market observers to wonder whether the much-discussed gas economy could actually develop, or whether gas, like oil, was a fatally flawed fuel.

As we have seen, worldwide gas reserves are extensive. Yet after decades of increasingly heavy use, much of the accessible gas has been depleted, and what remains is not always close to the markets that need it most. Whereas the cities and factories of Europe have ready access to gas from the Netherlands, North Africa, the Caspian, and especially Russia, other industrialized regions are not so lucky. The United Kingdom, long self-sufficient in gas because of its huge North Sea fields, will soon need to import gas from Europe, as these fields continue their decline. Similarly, the industrialized economies of Japan and South Korea, not to mention the no-longer slumbering Chinese economy, are increasingly reliant on gas imports.

As becomes clearer with every news cycle, an even more noticeable gap is opening up in North America, which burns through nearly one-third of the world's natural gas production, in large part owing to the aforementioned surge in gas-fired power plants.[17] In the United States alone, since 1999, more than 220,000 megawatts of gas-fired power capacity (roughly 30 percent of the nation's total electric supply) has been built, at a cost of some $143 billion, as investors and utilities have scrambled to exploit rising power prices.[18] What is more, over the next fifteen years, U.S. gas demand is expected to grow by another 50 percent.

Sadly, all the increase in demand has not been followed by a similar surge in supply. Canada, Mexico, and the United States together possess less than 2 percent of the world's natural gas deposits. The United States in particular, once the largest gas producer in the world, is now a "mature" province. Though the country remains the number-two gas producer, behind the former Soviet Union, American production can no longer meet American demand — and is in fact falling, even as demand skyrockets. As was true of oil three decades ago, the biggest gas fields in the United States have long since been tapped, and the fields now being discovered are smaller than before. In 1980, gas drillers working offshore in the Gulf of Mexico were routinely finding gas reservoirs of a hundred billion cubic feet or more.[19] By 2002, average discoveries had fallen to around a twentieth of that size.[20]

The result is a steadily eroding production base. In 1996, gas companies could manage twenty-five million cubic feet of new gas production a day. In 2001, that number had dropped to just fifteen million.[21] "We are drilling more and more and getting less and less, to the point where the new wells can't keep up with the fall-off from existing fields," says Merwin Brown, a gas analyst who consults for the U.S. government. "We are running up against a peak."[22]

The United States still has a great deal of gas left, but either it is in smaller, less economical fields that energy companies are loath to tackle, or it is located beneath national parks, offshore areas, or other protected government land. Lawmakers are pushing to open up off-limits lands to drilling. But given the likely environmental battles and the time it takes such projects to get under way, it could be at least a decade before substantial new sources become available. Nearby, in Alaska and on Canada's Mackenzie Delta, vast gas fields exist, but gaining access to them will require a twenty-billion-dollar, heavily subsidized pipeline to the United States — a project that has been delayed for more than a decade on account of the incessant political battles waged by Alaska, the Canadian provinces, and a host of aboriginal tribes whose lands the pipeline would traverse.

As a result, analysts say, the United States is facing a tighter supply "balance" than observers can recall until now. We have seen how tighter supplies increased the volatility in the oil markets, and U.S gas markets have similarly lost much of the excess capacity that once protected them from violent price swings. For example, because gas was historically burned as a heating fuel, demand was highest in winter. This situation allowed gas producers to run a surplus for the rest of the year and pump the extra gas back into the ground, where it was stored until needed the next winter. This extra supply insulated the market against sudden disruptions. If a winter was colder than expected, companies could tap into the stored gas and meet the demand before prices rose too high.[23]

Now, because gas is used increasingly to generate electricity, demand is also high in the summer, when air conditioner use places an extra burden on power supplies. That leaves even less time to build up surplus stocks, and since overall production is declining, it's even harder to keep inventories high. In April 2003, despite a warmer-than-normal winter, the volume of gas in storage was 40 percent lower than the historic average. "The U.S. market has never previously experienced a continuing decline in storage, relative to historical norms, of this magnitude," says Andrew Weissman, chair of Energy Ventures Group.[24]

What this means is that today there is no slack in the system, no extra gas to meet any unanticipated demand — and the market knows it. The slightest blip in demand, from a brief cold snap to a heat wave, or even a spike in oil prices, which encourages big consumers to switch to gas — anything that might conceivably increase need for natural gas or gas-generated electricity — sends prices skyward. They then plummet just as quickly. Such volatility is quite attractive to energy speculators who are willing to gamble large sums buying gas in the hope that prices will continue to rise. (In fact, it is widely speculated that the price hikes in the summer of 2003 stemmed in part from gas sellers' withholding supply to "squeeze" prices up at the margins.) Yet it is also true that such manipulations cannot occur in a loose market; they can take place only when supplies are tight, as appears to be the trend.

In such an environment, volatility is inevitable — and devastating, especially in an economy increasingly powered by gas-fired electricity. As gas prices rise, utilities are raising power rates. Industries like plastics makers that depend on natural gas for a "feedstock" are shutting down plants and moving overseas, where gas supplies are closer to hand. When U.S. gas prices spiked to ten dollars per one million Btu's in 2001, entire U.S. factories were shuttered, and at least two hundred thousand U.S. manufacturing jobs disappeared. This so-called demand destruction is one reason Greenspan warned in June 2003 that a gas shortage could effectively wipe out the struggling economic recovery. "Energy is rapidly becoming a major limiting factor on economic growth," Jeffrey Currie, senior energy economist for Goldman Sachs, told a congressional hearing last year. "If the core energy infrastructure in the U.S. does not improve, energy crises are likely to become progressively more frequent, more severe and more disruptive of economic activity."[25]

Without substantial new gas production, argues Weissman, "there is no readily apparent means to meet the incremental electricity needs of the U.S. economy over the next five to seven years — raising serious question as to how the growth of the U.S. economy will be sustained during the remainder of this decade, while new, longer-term sources of natural gas supply are being developed."[26]

Gas optimists — a breed of analyst related to oil optimists — say such dire predictions are overblown and are intended mainly to scare lawmakers into easing regulations on the gas industry and opening additional off-limits land to drilling. And, indeed, by the fall of 2003, the high prices of the spring and summer had finally begun to encourage fresh production, thus easing tight supplies in time for the 2003–2004 winter heating season. The relief is probably short-lived, however. Because existing U.S. reserves are still mature, and because overall production rates are still falling, experts say that tight markets will simply reemerge each spring, each perhaps worse than before. The current gas surplus "is temporary," says analyst Joseph. "It's taking more wells to produce what gas we're getting, while the decline rate of existing fields is accelerating."[27] Notes another analyst: "The days of two-dollar gas are gone. What we can look forward to instead is a price range of four to five dollars, with regional shortages and occasional spikes to six to ten dollars."[28]

In short, just as the United States, the largest, most sophisticated, and most politically influential country, is making a monumental shift toward a "natural gas economy," the country is waking up to the fact that it may need to procure that gas from someone else.

That brings us back to LNG and Costa Azul. If America is running short of gas, the rest of the world is not. Neighbors like Trinidad and Venezuela have massive reserves, which can be liquefied into LNG and tankered to the States, and the same is true for the huge gas fields in Siberia, Australia, and Southeast Asia. With gas prices expected to stay above $3.50 for the next five years and possibly beyond, energy companies will have every incentive to bring on additional LNG importing capacity, filling in supply at the margins and, with luck, adding enough to total supplies to keep the American economy from running out of gas.

Yet this hopeful scenario highlights what may ultimately be the biggest obstacle to a truly global gas economy: security risks. In many respects, gas is actually a less secure form of energy than is oil. Gas pipelines are as susceptible to terrorist attack or natural disasters as oil pipelines, and if anything, LNG terminals and ships are even more vulnerable than those used for oil. More to the point, since the map of gas consumers and gas producers is more or less identical to that for oil — most of the world's gas reserves are in the Middle East and the former Soviet Union, while the biggest users are the United States, Europe, Japan, and China — it is all but inevitable that gas will become as geopolitically volatile as oil is today, and just as central to regional and global conflict. In this sense, shifting to a gas economy simply means replacing the current social and political problems surrounding oil with a different set of problems that may be even more complicated.

Over the long term, switching to a gas economy could create a geopolitical dynamic similar to that of oil. More than half of the total known gas reserves are in just two countries — Iran and Russia — while most of the rest is in Qatar, Nigeria, Algeria, Norway, and Venezuela. And although Russia and the rest of the former Soviet Union are becoming more stable and more favorably disposed toward Western customers and ideologies, the reliability of other gas producers is simply unknown. Nigeria, Algeria, and Venezuela are all prone to civil unrest and anti-Western sentiments. Iran presents an even greater uncertainty. The country is so rife with political unrest and fierce anti-Western sentiment and so economically desperate that the prospect of price manipulation is not inconceivable — especially since America still treats Iran as an enemy and forbids trade. In fact, many Western analysts believe that U.S. animosity toward Iran is already backfiring. As Florence Fee, a former top executive with Chevron and Mobil, has noted, Tehran is carefully building an alliance with Moscow — an alliance that could control more than half the world's gas supplies — just as "U.S. dependence on imported gas supplies is growing markedly."[29]

In the meantime, global gas markets are witnessing the geopolitical equivalent of a horse race. Gas-rich Caspian countries are currently fighting over the possibility of building a gas pipeline across central Asia to the immense markets of India. An even fiercer competition is arising in the LNG markets, as the big energy companies team up with big gas-producing countries to compete for secure long-term delivery deals with the big consuming nations. Indonesia, Malaysia, Brunei, Australia, and tiny Qatar are all scrambling to sell their gas to China, Japan, South Korea, India, and, ultimately, the United States. Because the stakes are now so high — many LNG agreements will be worth tens of billions of dollars and lock countries and companies in for decades — the race for gas has touched off rivalries not only between companies, but between governments as well.

In 2001, for example, China announced that it would accept bids to supply LNG to Hong Kong and surrounding Guangdong Province for three decades. The prospect of so mammoth a deal touched off an intensive bidding war between Australia and Indonesia, and drew in politicians from

Britain and even the United States. As an Australian consortium, ALNG, appeared close to winning the thirteen-billion-dollar deal, other suitors brought in the big political allies. No less a spokesman than U.S. vice president Dick Cheney had lobbied Beijing on behalf of ExxonMobil, which hoped to sell gas from Qatar to the Chinese. British prime minister Tony Blair urged Chinese premier Zhu Rongji to choose BP and its gas from Indonesia. (Indeed, in May 2002, as the deal seemed to balance on a razor's edge, Australian prime minister John Howard, having flown to China to press the Australian case, checked in to his Beijing hotel to find a stern note from the deputy British prime minister demanding that the Aussies withdraw their bid.)

The competition was so fierce and prices were driven down so low, that many analysts began to question whether the deal would ever actually make any money. Beijing had negotiated brilliantly, playing the Australian oil companies against their industry rivals. They had also exploited Australia's desire to be a trading partner with China — to the point that the Australian government itself was pressuring ALNG to lower its bid.[30] In the end, ALNG priced its bid so cheap that some observers wonder whether the consortium will ever actually turn a profit over the twenty-five-year duration of the contract — or whether the deal might turn out to be simply the latest case of the West's blind obsession with the "China market." As one industry insider complained: "The Australian Government negotiated against the Australian consortium. . . . John Howard can crow all he likes, [but] he helped the Chinese get a good price."

Such trials and tribulations raise a critical question for those contemplating the development of a new energy economy. Because our energy circumstances have become so urgent, and because so many of the players have been driven to move so quickly, it is tempting to imagine that the emergence of a gas economy will occur fairly quickly and seamlessly. But that is hardly a safe bet. As we have seen, the oil infrastructure that many of us want to replace required nearly a century to build — and much of that time was spent in resolving the kinds of logistical nightmares that currently plague our gas warriors — problems ranging from tanker design to the scheduling of refinery operations to potential geopolitical rivalries. An entire system of financing had to be invented, based on the unique traits of oil as well as on the scale of the projects, and generations of engineers and operators had be trained and brought up through the ranks. Although to a certain degree the gas economy will simply build on the oil economy, much will have to be invented from scratch — at great expense, and probably after some delay.

This is not to argue that we should abandon the move to a gas economy. Quite the contrary: moving to a gas economy is probably the only feasible way for the world to delay the effects of a changing climate while we figure out how to completely revamp our energy economy. Nor is it to argue that the shift to gas won't happen — eventually. It is, however, to suggest that converting our energy economy to this "bridge" fuel will be slow, painful, and quite costly — and that moving beyond gas will be even more difficult. Whatever our "new energy" system ultimately turns out to be — hydrogen or some other fuel or, more likely, some collection of fuels — that future will require a reinvention of the energy business. This will add expense and delays to an enterprise that is, in many respects, already behind schedule and will almost guarantee a bumpy ride.

That, at least, is the case on the Baja Peninsula, where the LNG revolution is progressing by fits and starts. Last spring, in spite of numerous complaints by environmentalists and locals, Sempra received its environmental permit from the Mexican government, one of three permits required before construction can begin. Other energy companies are making progress on their LNG projects as well, and laying out a blueprint for a vast gas power system that will link Baja with the hungry markets of the north. All has not gone smoothly. Environmental groups insist that these plants will destroy local ecosystems. Residents insist that LNG tankers offer fat targets for terrorists. Activists, meanwhile, continue to complain that the United States is simply shipping its energy problems south of the border.

Predictably, those pioneers of the gas-powered economy are fighting back, with syrupy promises and relentless public-relations campaigns. To allay local concerns about its proposed LNG terminal and regas plant in Playas de Tijuana, fifteen miles from the U.S. border, Marathon Oil has offered to build a wastewater treatment plant and a desalinization plant. At nearby Rosarita Beach, Phillips Petroleum hopes to ease local concerns over a proposed regas facility by painting the storage tanks to look like works of art. As a hopeful company executive explained, "we think the storage tanks will be part of the attraction."

8

AND NOW FOR SOMETHING COMPLETELY DIFFERENT

ANASTASIOS MELIS is not someone typically described as excitable, nor is he given to wild statements or hyperbole. The tall, placid-looking Greek molecular biologist has made his name over the last thirty years by laboriously documenting the various ways green plants convert sunshine into chemical energy. When he offers his professional opinion, he usually does so quietly and cautiously, via a careful lecture or the final draft of a research paper. But when Melis walked into his laboratory at the University of California, Berkeley, in November 1999 to check on an experiment, he suffered a temporary lapse of cool. Some twenty hours before, Melis had filled a small flask with a colony of *Chlamydomonas reinhardtii*, a green alga with a peculiar talent for survival. Under normal conditions, *C. reinhardtii*, otherwise known as pond scum, behaves like other green plants, turning sunlight into sugar and oxygen via photosynthesis. When it finds itself in a dark, oxygen-deprived environment, such as the bottom of a pond, however, *C. reinhardtii* activates an emergency mechanism — an enzyme that generates a small ration of energy and, in the process, releases trace amounts of hydrogen.

Scientists and energy companies have known about *C. reinhardtii* since the 1940s and, given hydrogen's value as a potential fuel, have spent decades trying to induce the tiny plant to increase production — to no avail. The hydrogen mechanism, it turns out, is only temporary: as soon as the enzyme generates any energy, the alga releases oxygen, which automatically shuts off the enzyme. For sixty years, biochemists have sought to harness *C. reinhardtii*'s hydrogen-making powers with the same fervor with which alchemists once tried to transmute lead into gold — and with the same frustrating results. Melis, however, had an idea. In 1996, he began asking whether the hydrogen enzyme, known as hydrogenase, could be made to *ignore* oxygen. Specifically, Melis wondered whether by depriving hydrogenase of sulfur, a key nutrient for oxygen reactions in plants, he could effectively block the "off" switch and allow *C. reinhardtii* to produce hydrogen continually.

Melis began experiments in 1997 — and was promptly stunned by the results. Arriving in the lab on that November morning, he found his flasks so full of gas that he was sure it couldn't be hydrogen. "I thought it might be nitrogen or carbon dioxide or even oxygen, but not hydrogen," says Melis. "There was too much." Only after colleagues duplicated his results did Melis acknowledge the implications: "It's the equivalent of striking oil." A pondful of *C. reinhardtii*, Melis told a packed press conference in Washington in February 2000, could make enough hydrogen to power ten fuel cell vehicles. Looking forward, Melis envisioned vast networks of "bioreactors" — sealed plastic tubes filled with the algae — generating enough hydrogen to run cars and power plants, and someday supplying the world. He cofounded a company, Melis Energy, and set out to get *C. reinhardtii* on the market by 2005.

Three years later, Melis's enthusiasm has been tempered somewhat. Although he and colleagues have more that doubled *C. reinhardtii*'s hydrogen output since 2000, Melis estimates that photosynthetic hydrogen farms won't be commercially viable until he can increase the output by a factor of twenty. Researchers have identified three R & D pathways that promise substantial improvements, yet Melis doesn't anticipate attaining viability for at least a decade. Worse, while the technology has attracted interest and modest funding from the U.S. Department of Energy and even DaimlerChrysler, the big investors needed to take *C. reinhardtii* to the next level have been reluctant to jump aboard. Despite initial excitement, Melis's company, Melis Energy, has pulled in less than a million dollars in financing — only a fraction of the twenty million dollars that CEO Stephen Kurtzer believes would be necessary to get the technology rolling. Part of that is bad timing: Melis Energy went looking for money in September 2001; but Kurtzer also blames the low priority that clean energy gets in the United States. "We pay lip service," he says, "but the only real plan in this country is, 'Let's drill more oil wells.'"

~◦§§◦~

In many ways, the story of *C. reinhardtii* — a promising new energy technology that has been hobbled by market skepticism and doubt — drives

home one of the more central, and discouraging, realities of the next energy economy. For although we can be certain that our current, hydrocar-
bon-based energy system is slowly breaking down — and that problems
such as oil depletion and climate change will only worsen with time — we
have no such certainty about what fuels or technologies will come next.
Not only do we face grave problems with long-term supplies of oil and gas,
but even if reserves were infinite, the way we use hydrocarbons is destroying our climate. Gas holds out the hope of a transitional fuel, a bridge, between the current system and whatever is coming next. Yet the very term
"bridge" indicates the temporary nature of gas: sooner or later, humans will
have to take the energy revolution one step further and figure out how to
produce energy in new ways — energy that entails fewer problems of supply and political stability, that produces no carbon, and that can be put in
place cheaply enough and quickly enough to offer some hope of staving off
calamities like climate change.

How quickly? According to the U.N.'s Intergovernmental Panel on Climate Change, to have any hope of keeping our atmospheric CO_2 concentrations below the 550ppm red line, fully one-seventh of all our energy
must be coming from some kind of new, carbon-free technologies by no
later than 2030. By 2050, that share must be nearly one-third, and by 2075
more than half.

Thus far, our progress toward this goal could hardly be described as
encouraging. Despite an explosion of new energy technologies — everything from fuel cells and hydrogen-belching algae to technologies that
"scrub" the carbon from coal — and despite impressive growth rates in the
burgeoning wind and solar power industries, it is increasingly clear that
this "something new," this "alternative" carbon-free energy economy, won't
materialize overnight.

Hydrogen fuel cells and a ready supply of hydrogen to fuel them are
still decades away from mass deployment. Nuclear energy has so many
technical, economic, and political problems that its future is in doubt,
while *fusion* energy — the so-called good nuclear power — is by most accounts probably a century away from being feasible on a large scale. This
leaves the "renewables" — hydropower, solar and wind power, biomass,
and geothermal, tidal, and dozens of other intriguing technologies — and
the current picture here is even less encouraging. Today, renewables provide just over 8 percent of the total world energy supply. Of that, most

comes from hydropower (7 percent), and most of the rest from plants, crop waste, and other biomass, which are either refined into fuels like ethanol or burned directly in steam-power, or "cogeneration," plants. By contrast, the two technologies we most commonly associate with the alternative energy label — solar energy and wind power — together provide *less than half of 1 percent* of the world total. Indeed, if you add up all the solar photovoltaic cells now running worldwide, the combined output — around 2,000 megawatts — barely rivals the output of two coal-fired power plants.

Why, after three decades of effort, do alternatives claim such a tiny fraction of the energy market? One obvious reason, say many advocates of alternative energies, is that alternative energy must compete against an entrenched energy establishment. The industries that profit from hydrocarbons (and the politicians who profit from those industries) have zero interest in seeing the emergence of competing technologies or the new, more decentralized energy system these new technologies may make possible.

For decades, advocates say, the energy establishment not only has used its political leverage to exclude alternatives from the marketplace (with huge government subsidies and tax breaks that keep hydrocarbon fuels artificially cheap) but has used its great rhetorical authority to downplay any expectations for a renewable-energy economy. "Years down the road, alternative fuels may become a great deal more plentiful," Dick Cheney, the former oilman, conceded during the rollout of the White House energy plan in 2000, "but we are not yet in any position to stake our economy and our own way of life on that possibility. For now, we must take the facts as they are." And as far as Cheney and the rest of the energy establishment are concerned, "the facts as they are" means oil, gas, and coal.

But there are other reasons for the slow rise of alternative energy — reasons that go beyond the greed and duplicity of individuals or an entrenched system. For all their huge potential, most alternative technologies really aren't ready for prime time. Despite decades of research and development — and despite recent growth rates that rival that of computers and cell phones — nearly every major alternative technology still suffers from serious engineering or economic drawbacks. Automotive fuel cells are still many times more expensive than even a vintage gasoline engine, and they may require decades of work to be competitive. Solar power, even after nearly thirty years and many billions of dollars in R & D, still costs five times as much as coal-fired power. Beyond questions of cost, these technol-

ogies may still face inherent limits in the quality of the energy they produce, and where and when they can be used, that could keep them from assuming a dominant share of the future energy mix.

Eventually, these limits, too, may be surmountable. But as we explore
the ever-shifting landscape of new energy, it becomes clear that the much-
hyped, and much-needed, revolution in alternative energy remains as uncertain and risky as anything else in our energy future.

In the south of Germany, a few miles from the French border and smack in
the middle of the German wine country, the more technically oriented
tourist can hire a guide and spend the day touring Freiburg, the world's
first "Solar City." From the comfort of a chartered minivan, you can take in
the city's solar-powered train station, energy-efficient row houses, innu-
merable rooftop photovoltaic systems, and, high on a hill overlooking the
vineyards, the world-famous Heliotrop, a high-tech cylindrical house that
rotates to follow the sun — all while your guide (in my case, a tall, lanky
fellow named Jurgen), recounts how Freiburg gave birth to Germany's environmental movement nearly thirty years ago.

On the outskirts of town, on a gorgeous, tree-lined campus, Jurgen
brings the van to a halt in front of a gracefully curving structure of glass
and photovoltaic panels. This is the Fraunhofer Institute for Solar Energy
Systems, which as one of the top solar R & D facilities in Europe represents
the vanguard of the solar revolution. On any given day, several hundred of
the world's best scientists, engineers, and technicians can be found laboring
toward the breakthrough that will finally allow solar energy to live up to
its potential. One current project: a third-generation photovoltaic, or PV,
cell that will produce twice as much energy as any currently available PV
cells. "Of all the renewables, solar has easily the biggest potential," explains
Joachim Luther, the institute's ebullient, silver-haired director. "And the
reason is simple: sunlight is everywhere."

For advocates like Luther, solar energy is *the* pathway to the next energy economy. Solar energy is abundant. It emits no carbon dioxide or anything else. It also produces the most commonly used form of energy —
electricity — which makes it the ideal "clean" technology for the power
sector, where CO_2 and other emissions are the worst. Because solar works
on any scale, from rooftop units in Freiburg to mile-wide arrays of PV cells

in the Mojave Desert, it lays the foundation for a truly decentralized energy system, in which power production is handled locally, and even individually, instead of through centralized corporate or state-owned utilities.

Perhaps most important, solar energy is the one existing nonhydrocarbon technology that has any hope of filling the projected need for huge volumes of new carbon-free electricity — as much as twenty-eight thousand megawatts — or twenty-eight terawatts — by 2050. Hydropower, for example, has only limited potential for growth: most of the best hydro sites in the industrialized world have been exploited, and developing countries can rarely afford exorbitant construction costs for dams. Geothermal energy, which uses underground steam to generate electricity, is promising in certain places, like Iceland, but worldwide will probably account for less than 2 percent of global electricity by 2020. The quantity of energy that can be generated from biomass is limited by the amount of land needed to grow fuel crops. Solar energy, by contrast, is theoretically limited only by the amount of sunshine and land area: in theory, the sunlight falling on just 1 percent of the earth's surface would be sufficient to power most of the industrialized world. "Solar is the only technology that could provide twenty-eight terawatts of clean power by 2050 without breaking a sweat," insists John Turner, an alternative energy expert at the National Renewable Energy Laboratory in Golden, Colorado. "In fact, there is nothing stopping us from doing it right now — except that it would be extremely expensive."

Solar faces two related obstacles — prohibitive costs and a reputation so marred by a bad '70s experience that despite recent advances, the technology is a nonstarter in the eyes of most players in the energy business. Thirty years ago, at the height of the energy crisis, Western governments gambled heavily that solar power could liberate their industrial economies from dependence on Arab oil. Governments subsidized solar research. Citizens and companies were encouraged through tax breaks to buy solar equipment; homeowners whose new PV systems generated too much power could sell the surplus to utilities, which were required by law to buy it. With such assurances of a solar market, business, too, jumped in. Most major oil companies, including Exxon, Arco, and Mobil, invested heavily in solar energy, effectively ensuring that, if solar did succeed, Big Oil would own that new market as well.

By the early 1990s, however, the solar boom had gone bust. Despite investments of more than three billion dollars, researchers never got the improvements they needed from the photovoltaic cell, on which solar energy depends. The best PV cells on the market had efficiencies of barely 10 percent — meaning that only one-tenth of the solar energy falling on the cell was being converted into usable electricity. They had other weaknesses, too. The costs to manufacture the silicon-based PV cells remained incredibly high. More significantly, solar power is intermittent: it doesn't work at night or on cloudy days, and it does poorly at higher latitudes. Even in sunny regions, the average PV cell delivers its maximum capacity only 22 percent of the time, or about two thousand hours a year. By contrast, a coal-fired power plant can crank out its maximum wattage around 90 percent of the time.

All told, PV was nowhere close to being able to compete in the energy marketplace — despite the fact that solar "fuel" is free. By 1994, PV's "installed capital costs" — that is, the cost of buying and financing a PV system — was still around eight dollars for every watt of power that the new system could generate. (For example, the installed capital costs of a PV system capable of generating fifty kilowatts — enough to supply ten American homes — would have been four hundred thousand dollars.) To pay off that investment and the associated financing, it would be necessary to charge around forty cents a kilowatt-hour for solar electricity[1] — about fourteen times the cost of power from a coal-fired power plant (whose capital costs are only a dollar and a half per installed watt). As far as the power markets were concerned, although solar might be suitable for tiny niche markets where conventional power was unavailable, it would never compete with coal, gas, or nuclear power. By the mid-1990s, most of the energy industry had written solar energy off as a dead end.

Solar was about to get a huge boost, though. In 1995, Japan, home to some of the highest electricity rates in the industrial world, announced an ambitious program to subsidize and install millions of rooftop PV systems for homeowners. Two years later, the newly elected German Green party pushed through a similar law in Germany. Primed by these moves, the global solar market began to grow: Japan installed twenty-five thousand rooftop systems in 2002 alone; between 1995 and 2002, the number of PV systems installed each year jumped from eighty megawatts of total power to five hundred megawatts. Today, solar is growing at 30 percent a year — as fast as cell phones during their breakthrough period — and the energy industry has taken note. BP and Shell have made large new investments in solar, and Japanese electronics firms like Sharp, Kyocera, and Sanyo are vying for industry leadership.

Inevitably, as volume has increased, solar has slowly become more affordable. Every doubling in sales causes the costs to drop by around 10 percent, while breakthroughs in materials and design are yielding dramatically higher efficiencies. Companies are creating a new thin photovoltaic "film" that could be applied to windows and building sides — turning entire skyscrapers and stadiums into solar generators. The Fraunhofer Institute is working on the next generation, a "multilayer" PV cell whose efficiency is theoretically 40 percent — about twice that of the cells currently on the market.

Coupled with falling manufacturing costs, the improved efficiency is steadily improving the competitiveness of solar energy. Solar energy technology "is still too expensive by a factor of four to compete with nuclear and by a factor of three to compete with natural gas," says Paul Maycock, a former head of the Energy Department's solar division and now editor of the industry magazine, *PV Energy.* But, Maycock insists, the generation of PV technology that will provide "that 'factor of three' is in the lab right now and is virtually assured by 2010." According to a study by European energy company RWE, given the growing economies of scale, solar is fast approaching a capital cost of one dollar per watt of installed power, which in countries with sunny climates and low interest rates works out to about eight cents per kilowatt-hour. "That's very close to being competitive with gas," says Maycock, who believes that the remaining cost gap can be closed simply by volume sales. "I'm not talking about a technical breakthrough — just economies of scale — and it gets us PV that is very close to the economics of natural gas."

Fraunhofer's Luther agrees. If current trends continue, solar energy can be cost-effective — without government subsidies — in sunny regions, such as the Mediterranean, the Middle East, and the U.S. Southwest, by 2008. Glenn Hamer, executive director of the U.S.-based Solar Energy Industries Association, is even more optimistic. As the U.S. natural gas shortage began to be felt in 2003, Hamer did an analysis showing that by 2005, solar power "could mitigate nearly a third of the natural gas shortfall with clean, renewable power from the sun."

⁕

Forty miles west of the city of Walla Walla, amid the dusky sagebrush rangeland near the Washington-Oregon border, a left turn off the state highway and a mile's drive up a dirt track bring the lost or inquisitive motorist to the edge of one of the largest wind farms in the world. Begun in 2001, the Stateline Wind Farm today sprawls over seventy square miles and boasts 454 towers. Each is 160 feet tall and topped by a sleek, boxy Vesta V47 turbine and a gigantic three-blade rotor. When the wind off the prairie hits seven miles per hour, sophisticated sensing devices turn the turbines into the breeze and the 77-foot fiberglass rotor blades begin to rotate. At thirteen miles per hour, the transmission engages and the generator begins cranking out power. By thirty miles per hour, each turbine is putting out the maximum capacity of 660 kilowatts — enough juice to run 150 American homes, or 300 homes in Europe.

Technically, Stateline can generate a total of three hundred megawatts, a volume of production that has allowed its owners, Florida Power & Light, to make electricity for a wholesale rate of around three cents a kilowatt-hour — substantially more than coal, but a third less than gas-fired power. "Wind power is moving from the niche market to the mainstream," says Jan Johnson, spokeswoman for the Oregon-based PPM Energy, which buys power from Stateline and other wind farms and sells it to other utilities. "These are not tiny plants anymore — a three-hundred-megawatt wind farm is starting to look a lot like a [coal] power plant."

If solar energy is the energy alternative-in-waiting, wind is the alternative technology that is already making a difference. On hills in Spain, in the icy waters off the Dutch coast, on the Great Plains and in the mountains of California, and even in China, wind towers are new features on the landscape — and, increasingly, a force in the energy market. In 2002 alone, investors spent seven billion dollars to install equipment with seven thousand megawatts of wind power capacity — enough to supply 3.9 million homes in Europe. Although wind accounts for just .4 percent of the world electrical supply today, the wind market is doubling in size every two and a half years. By 2020, wind could be supplying 12 percent of our global power needs — all while generating an estimated seventy-two billion dollars in revenues for the wind power industry. "The 'alternative' tag for wind power should be ditched — it is old-fashioned and out of date," insists Corin Millais, the confident chief executive of the European Wind Energy Association. Wind power, Millais argues, is a "mainstream energy business."

The reasons for the success of wind power aren't hard to find. Whereas solar energy depends on finicky photovoltaics, the core technology for wind power is among the simplest in the world. In the standard, tower-mounted wind turbine, two to three large blades spinning on an axis drive an internal generator and produce electrical current — much like a hydroelectric dam. The harder the wind blows, the more electricity is generated.[2] No insurmountable technological barriers remain to be crossed. For the most part, the improvements made over the past decade have been mechanical: lighter blades, computers that turn the blades into the wind, taller towers to catch the stronger breezes, and ever-larger turbines. Whereas the average turbine from the 1980s put out 100kW, today's turbines average 1.2 megawatts — enough to supply 620 homes — and 4-megawatt units for offshore use are now under development.

In other words, whereas solar power must wait for both market growth and technological breakthroughs to bring its costs down, the costs for wind power depend almost entirely on scale — that is, how many units are being manufactured and sold; as the use of wind spreads, these costs will do nothing but drop. Today, for example, a watt of installed wind power costs twice as much as a watt of installed coal power, and more than four times as much as a watt of installed gas power. Yet this cost difference leaves out a critical part of the story. Coal- and gas-fired plants may be cheaper to build, but they cost a great deal to fuel: over the lifetime of the power plant, the fuel costs are usually about half again the construction costs, or more. Wind, by contrast, has no fuel costs: its up-front costs — manufacturing, installation, real estate, and financing — are its main costs.

What this means is that while gas- and coal-fired power can become more expensive over time (if fuel costs rise), wind power can only become cheaper and more cost-competitive, as greater manufacturing volumes continue to push the up-front costs down. Today, wind-generated electricity can be produced for around 4.8 cents a kilowatt-hour — around 2 cents more than the wholesale cost of electricity from coal, gas, nuclear, or hydro. But most experts say that this small difference will disappear as wind turbines move further into mass production. Wind turbines are already 20 percent cheaper than they were in 1998. By 2010, analysts say, manufacturing costs will have dropped enough to bring wind-generated electricity down to around 3 cents — at which point, wind can compete in nearly any market.

In the meantime, Germany, the United States, and other governments are keeping the momentum up with various subsidies and tax credits for utilities that sell wind power. In the United States, utilities and marketers that buy wind power get a tax credit of around 1.8 cents per kilowatt-hour — enough to make wind power competitive with many conventional power sources, especially given today's high prices for natural gas (which, in some regions, have already made wind power a bargain). Not surprisingly, with such assurance of profits, many utilities and energy investors have been turning to the wind sector, and wind farms have been popping up around the world like strange, gigantic forests.

Wind's attractiveness to industry goes beyond government subsidies. Because wind power has no fuel costs, wind farmers face little of the "price risk" that stalks conventional power vendors. Wind farm owners can thus offer their power to utilities and other customers in long-term contracts — ten, twenty, even thirty years — with a huge degree of confidence that they will earn a steady rate of return and won't be ambushed by price spikes in their "fuel." By contrast, during the thirty-year lifespan of a gas-fired plant, for example, its operating costs — the cost of gas — will fluctuate widely, as the gas market rises and falls and rises again — so much so that many gas-fired plants turn out to be far less profitable or cost-effective than their investors had hoped.[3]

Wind technology also offers far more flexibility than conventional power sources. Wind is the essence of modular energy production: a wind turbine will function just as well by itself as in a cluster or farm, and that factor gives utilities an amazing degree of flexibility. Whereas a gas-fired power plant must have a generating capacity of at least a hundred megawatts to be economical — and coal-fired plants a thousand megawatts — wind farms can be built on nearly any scale — from a single-turbine wind farm in Kiel, Germany, to the world's largest wind farm — the huge Stateline project. Again, this modular capacity is ideal for a decentralized energy economy: one can easily picture wind towers in a backyard, on the rooftop of a skyscraper; some people have even suggested building them on old oil rigs, to exploit forceful offshore winds.

Similarly, where a coal- or gas-fired or a nuclear power plant requires a massive up-front commitment of anywhere from four hundred million to two billion dollars (and, in the case of coal and nuclear, can take seven to ten years to license and build), wind power can be brought on quickly and incrementally, as market conditions warrant. "You can get turbines delivered in less than a year," says Chris Flavin, an expert on alternative energy sources at the World Watch Institute, a Washington-based environmental think tank. "It's more like ordering a refrigerator than a power plant." A wind farm planned initially for 300 megawatts could be scaled back to 150 or scaled up to 450 megawatts, depending on regional demand, and with fewer financial penalties for the utility.

And because wind turbines are emission-free, they are exceedingly attractive to utilities hoping to cut sulfur emissions or get credit for reducing emissions of carbon dioxide. In fact, says Flavin, in the current power markets — where demand for electricity is growing, but where high gas prices, environmental concerns, and fuel price volatility are making new gas- or coal-fired power plants unattractive, wind begins to look downright sensible. "If you're a utility needing to expand your supply and can't build coal anywhere, and nuclear isn't an option and gas carries a price risk, you don't have many options," says Flavin.

Given the advantages of wind power — and the continued existence of a hefty government subsidy — it's hardly surprising that new wind farms are being added more quickly than new gas-fired power plants — or that companies like General Electric and Vesta are ramping up production and coming out with new designs. Thanks to skyrocketing production numbers, unit costs are dropping so quickly that wind-generated electricity is expected to be cost-competitive with nearly any other power source except hydropower, without government subsidies, by 2008.

In some cases, wind power is already competitive without subsidies. During the U.S. power crisis of 1999 and 2000, when wholesale electricity rates climbed to twenty-five cents per kilowatt-hour in the western United States, many regional utilities launched ambitious plans for wind farms. As electrical prices fell, some of these projects were scaled back or put on hold. But today, with the prospect of high gas prices — and sharply higher power prices — for at least the next four years, wind is again looking exceedingly attractive. In 2003, the U.S. wind market grew by 25 percent, even as the rest of the power market remained flat. And if gas prices remain at four dollars per million Btu's, says Randall Swisher, executive director of the lobbying group American Wind Energy Association, wind energy could economically replace some hundred thousand megawatts of gas-fired power — enough juice for twenty-five million homes — by 2013. In the longer term, says Swisher, if gas prices remain at four dollars, the "total economically competitive U.S. wind resource is on the order of six hundred thousand megawatts" — enough for 150 million homes.

As wind takes a greater share of the global power market, an interesting dynamic kicks in. Growing numbers drive down costs, and growth itself becomes a target: a power company's profits come to depend more and more on how fast it can expand its wind portfolio by adding new machines and new farms. In this way, wind becomes yet another battleground where ruthless energy giants vie for market share, cut costs, and push wind power into new markets — just the impetus that this new power source needs. Already, big companies like General Electric are battling with niche players like Vesta to become the Boeing of the wind turbine industry, while power companies ranging from Pacific Gas & Electric to German-based RWE are fighting to become the next big wind power provider.

And as wind becomes more and more profitable, the wind industry becomes a potent political force. Wind lobbyists and trade groups are more able to compete with lobbyists from fossil fuel and nuclear industries for a greater share of political support and protection. They are also better able to win favorable legislation from lawmakers increasingly concerned about green energy — not to mention their own environmental images — which in turn encourages more growth. As one German energy expert told me, "today, you can't debate any energy policy or law in any German government without having the wind lobby show up and try to run the show." What results "is this powerful economic engine and a reinforcing political dynamic," says Flavin, who predicts that "we will soon be at a point where the huge players will have a huge interest in keeping that growth rate going."

With prospects like these, it's easy to see why advocates for both wind and solar energy have such high hopes for a "renewable" energy economy — and why they tend to discount the gloomier forecasts offered by conventional industry types. Groups like the European Wind Energy Association believe that if costs continue to fall, wind power will fulfill 12 percent of global energy needs by 2020. And some advocates of renewables say the number will eventually be much higher. Given current trends in cost and efficiency — and assuming a continuation of political pressure to replace carbon-intense energy systems with emission-free power — the potential of renewables is only now being realized. The wind market has been growing by a third every year — so fast that the amount of new wind capacity, in megawatts, being installed each year now exceeds the capacity of new gas-fired power plants. Solar energy, though well behind wind, has apparently hit its stride as well, posting gains of some 30 percent a year. And although these rates are not sustainable, even a more typical growth rate of 10 per-cent will still leave wind and solar vying for serious market share by 2020.

From that point on, says Jim MacKenzie, a renewables expert at the World Resources Institute, solar and wind not only will be adding to the world energy supply "at the margins" but should actually begin competing directly with conventionally produced energy, on a one-to-one basis, and especially energy that emits lots of CO_2, like coal. By around 2030, some ad-vocates believe, solar and wind together could be meeting one-fifth of power demand in the industrialized world — and could even be making inroads in the developing world, where renewables offer a way to get elec-tricity to remote areas. MacKenzie, a former official with the White House Council on Environmental Quality, speculates that by 2100, renewables could displace *all* conventional fuels in the United States — and help en-sure a peak in CO_2 emission before the middle of the century.

If renewable energy really can replace fossil fuels with clean, decentralized power, why is anyone still worried about energy security or climate? The reason, as even the most zealous renewables advocates will tell you, is that solar and wind are not without substantial limitations. Both solar cells and wind farms require space and resources. Wind farms provoke political op-position. And while some of the best solar and wind conditions are found in sparsely populated regions — the upper Midwest is known as the Saudi Arabia of wind power — the remoteness of those locations can actually be a liability, in that any generated power must be transmitted to markets over long distances. In addition, though renewable electricity may be clean, of-tentimes the equipment to produce it brings environmental costs. Photo-voltaic cells are essentially semiconductors, the manufacture of which can release cadmium and other toxic pollutants.

Other problems become more apparent when we look more closely at cost. Although wind and solar are getting cheaper, proponents often over-look the fact that their competitors are also getting cheaper and will con-tinue to do so. Just as fuel cell cars must compete with a constantly improving internal-combustion engine, wind and solar will have to battle with gas- and coal-fired technologies that will grow more efficient and less expensive and less polluting by the year. Renewables are also extremely vulnerable to energy price swings: if gas prices were to come down, for example, wind and solar power would lose much of their cost advantage. Renewables are politically vulnerable, as well: if wind or solar were to lose their government subsidies, the current boom in new installations would come to a screeching halt: the mere threat of such a loss has many potential investors looking elsewhere.

And these, it turns out, are the easy challenges. One of the main reasons that utilities prefer coal, gas, nuclear power, or hydropower is that these power sources are *dependable*. A coal- or gas-fired power plant designed to deliver 1,200 megawatts will, over the course of a year, deliver an average of 90 percent of its listed capacity.[4] A nuclear power plant delivers 80 percent. Such year-round, day-and-night dependability is why utilities tend to rely on gas, coal, nuclear energy, and hydropower for their constant, or "base load," requirements — the steady demand for power that exists twenty-four hours a day, 365 days a year.

By contrast, both solar and wind suffer from *intermittency:* they are not available twenty-four hours a day, nor do they always deliver their maximum power. A 1-megawatt wind turbine, for example, actually delivers 1 megawatt only during high winds; its average production will be considerably lower, because average wind speeds are lower. Factoring in this variability, a wind farm's average production, or "capacity," may be just 45 percent in high-wind regions like Spain or in Wyoming, but generally closer to 33 percent — or about a third the capacity of a gas-fired power plant. Thus, if a utility wants to add 100 megawatts of wind capacity to its portfolio, it actually needs to install closer to 250 megawatts in new turbines: a huge additional expense. Solar power has an even lower capacity — around 20 percent — meaning that to produce a steady 100 megawatts of solar power actually requires the installation of 500 megawatts of PV cells. This extra capacity is called overbuild, and it poses a huge problem for energy advocates — especially in an era of deregulated power sectors, where utilities are no longer required to carry so much surplus generating capacity.

Here we begin to see the first cracks in the rosy renewables scenario. Even if utilities were willing to overbuild, to cure renewables' many weaknesses would take more than building more wind turbines or PV arrays to compensate for this lower capacity. Both solar and wind power also lack a quality known as dispatchability: unlike a coal-fired plant, which can be called upon for power day or night, regardless of weather, neither wind nor solar is so reliable. Solar is simply unavailable at night or on cloudy days. Wind is even less dependable. Although meteorologists are getting better at forecasting the average amount of wind on a given day in a given region, "we still can't guarantee that it will blow at 10:00 A.M. tomorrow," says Tom Osborn, a renewables expert with the Bonneville Power Administration (BPA), a federal power supplier in the Pacific Northwest.

The BPA, which already owns or buys 200 megawatts of wind power and wants considerably more, has found ways to cope with some of wind's unpredictability. Computerized scheduling, for example, lets utilities delay a power delivery from a particular wind farm until thirty minutes before the scheduled delivery time. If the necessary wind speed is there, the delivery goes ahead; if not, the utility takes that power from some standby source instead — like a gas- or coal-fired plant or, in the Pacific Northwest, a hydroelectric dam. As a result, wind's unpredictability hasn't been as costly as many skeptics feared: Osborn says that BPA wind power sales miss their scheduled deliveries only 10 percent of the time.

Yet if you're a paying customer, 10 percent is too much. The BPA and other utilities still must maintain some kind of backup as insurance — typically in the form of base-load power plants that are paid for yet kept idle until needed. How much backup is required depends on the quality of the solar or wind resource and the overlap between the two. (For example, in some places, like northern Germany, the prevalence of nighttime winds can compensate for solar's after-hours weakness and smooth out the intermittency in power.) On average, analysts say, wind and solar renewables can provide a maximum of 20 percent of a region's power. Past that point, either the intermittency factor causes too many power disruptions, or the cost of maintaining so much backup base load becomes too high — a nonstarter for utilities trying to avoid blackouts, price increases, or anything else that might attract regulatory attention in the post-Enron era.

Some energy analysts, like Gerry Stokes, director of the U.S. Joint Global Change Research Institute in Maryland, worry that these kinds of limitations create a kind of natural barrier to the expansion of solar and wind. "We see wind and solar saturating the energy market at around 2030, constrained by their deployment and intermittency restrictions," he says.[5]

In this context, says Stokes, the main question about alternative energy is not which renewables technology to focus on or how quickly it can grow, but what to do about the 80 percent of the market that renewables cannot, on their own, supply.

This, of course, is where hydrogen comes back into the energy picture. In Chapter 3, we saw how hydrogen may ultimately replace oil in the transportation market. In that arena, hydrogen's value lies in its amazing ability to carry, or store, energy, and then deliver it in the form of electricity, via a fuel cell. This storage capacity, energy advocates say, provides the missing link between renewable-energy technologies and a renewable-energy economy.

Specifically, by using electricity from solar arrays and wind farms to make hydrogen, we could effectively neutralize intermittency. With hydrogen storage, utilities would essentially "overbuild" their wind and solar arrays to allow them to generate vast quantities of energy during periods of high winds or peak sunshine, and then, using industrial-size electrolyzers, convert this surplus of electricity into hydrogen for storage. The hydrogen could be run through banks of stationary fuel cells to produce electricity as needed for homes, offices, and factories — all far more cleanly and quietly than with coal, natural gas, or oil. The cost of overbuilding production would be offset by the sales of the "stored" electricity, the hydrogen itself.

Since the 1980s, this image of a hydrogen economy has tantalized energy advocates, who picture a global system of solar panels, wind farms, and other renewable energy sources, all feeding into a network of electrolyzers. In this scenario, the energy's inexpensiveness (at least, once the electrolyzers, fuel cells, and other infrastructure have been paid for) and abundance would simultaneously address the problems of pollution and dependence on foreign oil. MacKenzie, for example, has calculated that the development of hydrogen storage would allow solar and wind to grow fast enough that CO_2 emissions would peak by 2040.

Hydrogen storage would also be the first step toward a truly decentralized power system. Homeowners or companies with rooftop solar cells, for example, could make electrolytic hydrogen when the sun was shining and store it in underground tanks. The hydrogen could then either be used to power the home when the sun wasn't shining or be pumped into the fuel cell car. Such a setup could function as a stand-alone system or could be plugged into the local power grid — thereby effectively creating what Turner, at the National Renewable Energy Laboratory, calls a regional energy exchange.

In this model, my individual home becomes a kind of mini–power plant, with ultraefficient PV cells on the roof, an electrolyzer in the basement for making hydrogen, and a stationary fuel cell to turn hydrogen into electricity. During the day, when my household power demands are low, my home-based, Web-connected "energy system manager" takes any excess electricity from the rooftop PV unit and shunts it out of the house, through the wires, and onto the regional power grid, where it is purchased by the local utility to resell to industrial customers. During the night, when regional power demand is low, my energy system manager signs on to the Internet, checks the local power rates, which vary according to demand, and, if rates are low enough, automatically buys the utility's excess power, then converts it, via my small electrolyzer, into hydrogen, which is stored in underground tanks. I can then use the hydrogen to fuel my fuel cell car or send it to my stationary fuel cell to produce power for the house at night or on cloudy days, when the solar cells aren't working. And once my home hydrogen tanks get full, my energy system manager automatically starts converting the fuel back to electricity and selling it back to the utility during periods of high demand, when rates are high.

The impact of these mini–power plants would be huge. Even if only one in ten or even one in twenty homes participated, Turner says, in a large city, it would still create a tremendous power base. "Let's say you have fifteen hours of hydrogen stored, with a 3-kW fuel cell, multiplied by forty thousand homes — that's an enormous amount of available energy," says Turner.

Predictably, not everyone is so confident in that vision of a renewably powered hydrogen economy. Many veteran alternatives experts, and even some solar advocates, have serious concerns that PV technology cannot improve as quickly as it would need to, to compete effectively in the energy marketplace. They also express considerable anxiety over the sheer scale of a wind- or solar-based energy economy. Although the total wind potential of the planet easily exceeds the world's projected electrical demand in 2020 — around twenty-six million megawatt-hours — to produce even 12 percent of that amount will require the construction of more than a million one-megawatt turbines at a cost of some three-quarters of a trillion dollars.

Even if money were no object, the sheer physical scale of the enterprise would be daunting. While a six-hundred-megawatt coal-fired power plant requires a few dozen acres, a three-hundred-megawatt wind farm like Stateline may cover as much as seventy square miles. This is because wind technology has a far lower "power density" — that is, power produced per square foot of facility — than coal technology. Power density is, ultimately, one of the greatest weaknesses of renewables like solar and wind. With a power-dense fuel — coal, for example — you can generate an enormous amount of power quickly, in relatively small, centralized power plants, and then distribute it to urban consumers. By contrast, trying to power a large city with renewables would require huge tracts of land: a moderate-size city of a million homes would need as much as a thousand square miles of wind farms. And, as Vaclav Smil, an expert in energy economics at the University of Manitoba, points out, most of the world's population will live in urban areas of ten million or more, most often in high rises and densely packed housing. "Supplying those buildings from locally generated renewable energies is either impractical or impossible," Smil writes. The "power density mismatch is simply too large."[6]

Solar is even worse. MacKenzie, with the World Resources Institute, has calculated that on the basis of current PV technology, a solar-powered hydrogen economy in the United States alone would require the construction of tens of thousands of square *miles* of PV panels — at an astonomical cost — while the new electrolyzers would increase water demand nationally by 10 percent. "We could do it," MacKenzie told me, "but it would be expensive."

The weakest link in this vision of a hydrogen economy, say skeptics, is the cost of hydrogen storage. Even if installation costs for solar fall sufficiently to make PV competitive with, say, natural gas or coal, adding hydrogen storage — electrolyzers, pipelines, and special storage tanks — drives those costs back up again. "How many thousands of dollars per megawatt is it going to cost you to store that energy?" asks Stokes, who worries that the expense of hydrogen storage will torpedo renewable energy's chances in the marketplace — or its prospects for displacing hydrocarbons as the energy source of choice. Smil is much more definite. Given the already high costs of solar, the vast uncertainties of hydrogen, and the size of the existing fossil fuel infrastructure, "there is no alternative technique of nonfossil energy conversion that could take over a large share of the supply we now derive from coal and from hydrocarbons in just a few decades."

⁕

This is why a good many energy experts believe that our best bet isn't displacing hydrocarbons, but figuring out how to use hydrocarbons more cleanly — and, specifically, how to use them without releasing their carbon. For decades, energy companies and chemical makers have been "decarbonizing" natural gas — using superhot steam to split the methane molecule into hydrogen and carbon. The hydrogen is used in industrial processes (and a few fuel cell demonstrations), while the carbon is simply vented into the atmosphere as carbon dioxide. More recently, energy researchers have been experimenting with ways to capture the carbon and sequester it in safe places, such as abandoned mines and oil fields, or the depths of the sea, where it can't reach the atmosphere and contribute to climatic problems. Even if natural gas may be a fairly limited resource — especially in the U.S. market — carbon capture also works, at least in theory, with a fossil fuel that is extraordinarily abundant: coal.

Decarbonizing coal is a complex process, but it actually makes use of an old idea: turning coal into gas. Instead of being burned, as it is in most power plants, the coal can be refined first into a synthetic gas, much like the "town gas" that lit lamps a century ago. This synthetic gas, or "syngas," is a strange brew, composed of hydrogen, carbon dioxide, carbon monoxide, and steam, with trace amounts of methane, sulfur, and other pollutants. To complete the process, the syngas is scrubbed of sulfur and other contaminants and then subjected to intense heat and pressure. This procedure splits off the hydrogen molecules, which are stored for later use, and creates a separate stream of nearly liquid carbon dioxide, which is captured and pumped into storage.

Much of the technology for decarbonization is already available. Coal gasification, for example, has long been practiced, and power companies have already built a number of plants that can combine gasification and power generation. The process, known as Integrated Gasification Combined Cycle, or IGCC, begins by refining the coal into syngas and then uses this fuel for a standard gas-fired turbine. The turbine generates electricity, while the exhaust heat is used to make steam, which helps power the refining process. The exhaust itself, meanwhile, is put under high pressure, a process that causes it to separate into hydrogen and carbon dioxide, which in theory can then be sequestered.

Carbon capture technology — which will also work for other hydrocarbons, including oil, heavy oil, and tar sands — is in the early stages of development and has yet to be tried out on a mass scale. The process is energy-intensive — a decarbonzing IGCC plant, for example, must burn 20 percent more coal simply to generate the energy needed to run the carbon capture equipment, which itself is expected to be quite expensive. Moreover, no one is sure how to handle the staggering volumes of captured carbon dioxide. Because CO_2 contains both carbon and oxygen, it is actually three times as heavy and bulky as the original coal. In other words, for every freight car of coal delivered to an IGCC plant, three cars of captured CO_2 would need to be removed and somehow transported to a safe repository and placed underground — a task that, on a global scale, would involve handling a volume of waste material larger than the combined tonnage of the steel and iron industries. All told, decarbonization is expected to add perhaps 30 to 50 percent to the cost of electricity.

Given the novelty of capture technology, the few IGCC plants running today simply vent the CO_2 into the air. Once the technology is up and running — and once climate regulations make it necessary to capture carbon — IGCC plants are designed so that the capture equipment could be added on fairly easily. As demand for hydrogen rises, a growing share of the syngas will be split directly to produce hydrogen, instead of being burned for power generation. In the meantime, IGCC plants are anywhere from 20 to 40 percent more energy-efficient than existing coal plants (that is, they produce fewer CO_2 emissions for the same energy output), and they also supply hydrogen. A $1.2 billion IGCC plant in Italy, for example, turns sixteen million tons of heavy oil into 550 megawatts of electricity and several tons of hydrogen, which could be used to run fuel cell cars.

In spite of the huge uncertainties about carbon capture, support for the idea is growing — among energy companies, who see it as a way to preserve the value of their hydrocarbon assets in a "new" energy economy, and among governments, who regard the technique as a solution to a host of energy problems. To begin with, coal is astonishingly abundant: world reserves are estimated at nearly a trillion tons — enough to power the entire planet for more than 150 years. Heavy oils and tar sands, which can also be decarbonized, are similarly abundant: Alberta now claims to have tar sand deposits equivalent to more than a trillion barrels of oil. Reserves of coal and heavy oil also happen to be conveniently distributed within or near the world's industrial centers: the world's largest coal reserves are in the United States, followed by the Russian Federation, China, and Europe. Shifting to a decarbonized coal economy would thus dramatically improve the energy security of big energy consumers, like the United States and Europe, and would completely remake the geopolitics of global energy.

Just as important, advocates say, is that this so-called clean-coal technology provides another route to an alternative energy economy — one that can complement renewable-energy technologies like solar and wind, by providing a source of clean base-load power. Stokes refers to carbon capture and sequestration as a backstop technology — that is, a technology that can serve not only as the base load for the 80 percent the energy renewables cannot supply, but as a fallback if solar or wind, or any of the myriad other energy technologies now on the drawing board, for whatever reasons, does not advance quickly enough or become economical enough to meet demand and climate requirements. In Stokes's mind, the next energy economy will probably be a blend of technologies, tailored for different regions and sufficiently flexible to adapt to changing conditions and new technologies — but primarily built around two approaches: "I really think renewables and fossil fuels with carbon capture and sequestration are the two big dogs in the hunt."

Stokes isn't alone. According to the experts at the U.N.'s Intergovernmental Panel on Climate Change, whereas solar, wind, and other renewables, including hydropower, will account for less than 12 percent of the total energy mix by the end of the century, "clean" coal's share could be as high as 50 percent.

Not surprisingly, this concept of a hybrid energy economy, still heavily reliant on fossil fuels, is not universally praised. Renewables advocates are deeply suspicious of carbon capture. They are disturbed by the great technical uncertainties it presents, especially the challenge of transporting and permanently storing so much carbon dioxide, which, if it leaked, could pose tremendous health problems, not to mention add to global warming. Critics also fear that the costs of the technology are simply too high. As Turner, at the National Renewable Energy Laboratory, grouses, "a six-hundred-megawatt coal-fired plant will use 20 percent of its energy just to capture and sequester the carbon, which basically means I have to go out and build another power plant to make up for that loss." Studies by the NREL suggest that even with the greater efficiencies of the IGCC process, moving to an energy economy based on clean coal would boost overall energy demand by 17 percent. "We end up burning our fossil fuels at a higher rate in order to protect ourselves from the carbon they produce," says Turner, who derides carbon sequestration as a temporary fix, "duct tape on a very serious problem."

Turner is the first to acknowledge that the challenges attendant on a renewable-hydrogen economy are daunting, and that in order to meet emission goals and world energy demand, advances in solar, wind, and hydrogen technologies must continue to improve, at a rapid pace, for the next three decades. If anything, the growing focus on carbon sequestration makes that task even harder. "Every dollar we're talking about spending on sequestration," Turner says, "should be spent on renewables."

In a real sense, the debate over renewables is a series of arguments about the future of energy. If the consensus is that hydrocarbons can no longer be used as they have been for centuries and that replacement technologies must be found, we remain divided about what kind of technologies we should be pursuing. On one side of the debate are the proponents of a hydrogen economy powered largely by renewable energies. On the other side are advocates of a hybrid approach in which renewables are supplemented with a new hydrocarbon technology that, in theory, would do away with the problems of the old one. Clearly, whether we embrace one model or the other — or one that has yet to be invented — will have enormous ramifications for our energy future.

Yet in another sense, this debate masks an even more important question: whether we can produce enough energy by any means to provide a decent standard of living for the entire planet and at the same time satisfy our emerging criteria for climate and energy security. As we have seen, in all but the most optimistic scenarios, renewable-energy technologies are still seen as providing only a fraction of our clean-energy needs. Even the climate experts at the IPCC foresee solar, wind, and other renewables, including hydropower, as making up less than an eighth of the total energy mix by the end of the century.

Many alternative advocates say such pessimistic forecasts are a reflection less of true barriers than of political and cultural biases. Flavin, for example, argues that the main reason that prospects for alternative energy seem so bleak is that most of the forecasts come from a complacent political culture so accustomed to hydrocarbons that it is unable to believe that alternatives can exist anywhere but in the margins. "When you talk about how fast wind is growing, energy industry people will say, 'Who cares — you're starting with such a small base,'" Flavin says. But by that logic, he points out, IBM would still be the dominant force in computers. Just as the IBMs of the world had no way to conceptualize a threat like the personal computer, the energy establishment has no clue where the energy market is going, where the competition is coming from, or how we may be powering ourselves in thirty years, or even twenty. "If you had asked the computer industry in the 1970s where it was headed, you would have been told, 'Mainframes forever,'" Flavin says. "You wouldn't have heard about Bill Gates. Changes in the basic nature of the technology had already set up a new way to look at computing, but none of the big boys had figured it out. So while it's important to analyze what the big energy companies are thinking, ultimately, you may be talking to dinosaurs — creatures that are going extinct or that will barely survive, but only by completely changing their business model."[7]

Flavin's critique has merit. Proponents of coal were just as smugly convinced that an upstart fuel called "rock oil" would never unseat their own industrial model. Today, the political, economic, and even cultural inertia of the energy order remains firmly behind hydrocarbons: insofar as the energy order can even contemplate "something new," it is likely to be a derivative of what exists today — such as decarbonized coal or heavy oils. In reality, over the next few decades, we are very likely to see all kinds of technological advances that have nothing to do with hydrocarbons, or solar or wind, for that matter — advances that most of us, brought up in the age of oil, probably can't even imagine, breakthroughs, like the one with *C. reinhardtii*, that come from entirely unexpected quarters. To cite but one example, work by Craig Venter, mapper of the human genome, to design a microbe that can eat carbon and turn it into hydrogen, has raised great hopes that the future of clean energy may come from biotechnology — from living forms, not silicon panels or fiberglass wind rotors.

If the recent past is any indication, however, as we gain new technological powers, we will also gain a greater sense of the boundaries to those powers — and an understanding that humankind may be coming up against a fundamental limit in its quest to find new energy sources. For centuries, humans have advanced with the certainty that progress was inevitable. In the sphere of energy, especially, innovations in the fuels we used and the ways we used them have always ensured a steadily expanding energy supply. We might have had to use different fuels, or consume them in different ways, as we did when we shifted from coal-fired steam power to oil- and gasoline-powered internal combustion. Still, in the end we always had as much energy as we needed or wanted. This was the implicit assurance of the modern energy economy: we could always count on having some new technology or fuel arrive in time to maintain the energy status quo and let us go on living and working and consuming as we always had.

Since the 1970s, however, this assurance has become more explicit — yet at the same time less credible. Despite numerous promises that a replacement for gasoline (or oil or coal) would soon be available, we remain in the initial phase of the alternatives revolution. Renewables proponents like Flavin contend that the real problems are political and even cultural, and that that kind of institutionalized pessimism conveniently supports the hydrocarbon industry. But we may also have to confront the possibility that our innate energy optimism is itself obsolete, and that, in a future energy economy bounded by risks to supply, crushing energy privation, and a carbon ceiling, it will simply not be possible to continue producing energy in ever-increasing volumes. If that is the case, if there truly are limits to the size of our energy economy, then we will have to radically rethink not only the way we produce energy, but the way we use it.

9

LESS IS MORE

IN APRIL 2001, months before anyone outside Wall Street knew or cared about a company called Enron or the business of energy trading, Dick Cheney stood before a crowd of reporters and executives in Toronto and declared the United States to be in the throes of an energy crisis. Gasoline prices had reached record highs, the ex-oilman explained. The nation's reliance on foreign petroleum was nearly twice what it had been during the Arab oil embargo. Still more dramatically, California, the wealthiest, most populated state in the Union and the capital of its high-tech revolution, was besieged by rolling blackouts — a catastrophic power shortage that, in Cheney's somber view, was bound to spread to the rest of the country. "Without a clear, coherent energy strategy for the nation," Cheney warned, "all Americans could one day go through what Californians are experiencing now, or worse."

The problem, Cheney went on, was obvious. While America's energy needs had soared over the last decade, energy production had not kept up — not least because shortsighted politicians had failed to encourage more energy capacity. Under the Bush administration's much-anticipated new national energy policy, though, America would rediscover its supply-side roots. Oil companies would be encouraged to tap new domestic reserves — including those in the Arctic National Wildlife Refuge. The nation's energy infrastructure would be upgraded, for example, through construction of thirty-eight thousand miles of new gas and oil pipelines, dozens of new oil refineries, and as many as fifteen hundred new coal, gas, and nuclear power plants. "America's reliance on energy, and fossil fuels in particular, has lately taken on an urgency not felt since the late 1970s," Cheney said.

Few in the audience were surprised by the White House's heavy emphasis on fossil fuels, given the administration's well-known connections with the energy industry. Less expected, however, was Cheney's dismissal of nontraditional energy sources. After writing off alternative fuels, Cheney got downright nasty on the topic of energy conservation. Under the Bush plan, Cheney promised, Americans would not be exhorted to cut their energy consumption, to "do more with less," as they had in times past. "We all remember the energy crisis of the 1970s, when people in positions of responsibility complained that Americans just used too much energy," Cheney said. Even now, he warned, environmentalists were demanding that government "step in and force Americans to consume less energy, as if we could simply conserve or ration our way out of the situation we're in." Conservation, Cheney conceded, might indeed "be a sign of personal virtue, but it is not a sufficient basis for a sound, comprehensive energy policy." The Bush energy plan, Cheney said, "will recognize that the present crisis does not represent a failing of the American people."

Among energy experts, the "Toronto speech" is now regarded as one of the more revealing moments in the evolution of modern energy policy. Not only had the administration rather shamelessly exploited the California energy "crisis" — a crisis, it would later turn out, that had been largely manufactured by power companies and energy traders, many of them Bush's political allies — but the White House had apparently failed to understand how that crisis had actually been overcome: through conservation. In fact, even as federal and state politicians were scrambling to open new power plants in California that year, it was California's consumers and businesses that, by dramatically cutting power usage, dragged the state out of harm's way. When the crisis was finally declared to be over in late 2001, "it wasn't because the new power plants had come on line," argues Dan Kammen, director of the Renewable and Appropriate Energy Laboratory at the University of California, Berkeley, and an expert on the state's power problems. "It was that consumers immediately cut their power usage by 10 percent as soon as a crisis was declared." Yet the contribution of conservation was only belatedly credited. State officials and the Bush administration, Kammen says, "got a lot more political mileage cutting ribbons in front of new power plants."

Cheney's willful misapprehension of the California energy crisis reflects perfectly the confused disdain most modern consumers feel for the idea of conservation. To many of us, the word still evokes only the grim energy austerity of 1970s, when Europeans saw fuel prices quadruple and U.S. president Jimmy Carter, wrapped in an energy-saving cardigan, glumly asked his fellow Americans to drive less and turn down their thermostats.

In truth, conservation has always carried a broader meaning — one with great relevance to a new energy economy. When we talk about energy conservation, we mean not just using *less* energy, but using energy more efficiently — that is, squeezing more work, more goods and services, more wealth from each kilowatt-hour we consume. In this sense, conservation is less a question of morals or ethics than of sound business practices: maximizing the profit we can make for each dollar we spend on energy. Given that the main challenge in the next energy economy will be to find ways of creating more wealth without additional expenditure of energy, conservation — or, if that term is too embarrassing, the continual pursuit of better energy efficiency — would seem an obvious first step.

We often forget just how effective a tool efficiency has been. Today's cars travel twice as far on the same gallon of gas as they did in 1970. Today's appliances generate more comfort, entertainment, and other services than they did in 1970, for about half of the energy costs. Between 1975 and 2000, even as the American economy grew by nearly 50 percent, our "energy intensity" — the amount of energy needed to produce a dollar of GDP — fell by 40 percent, largely through improved technology, policies, and marketing methods.

And these gains are only the palest shadow of what could be achieved. Around the world, at every level of society, we squander an embarrassing volume of energy every day. Less than a quarter of the energy used in the standard stove reaches the food. Power plants in the United States discard more energy in "waste" heat than is needed to run the entire Japanese economy — and half the electricity generated in the United States isn't needed to begin with. Barely 15 percent of the energy in a gallon of gasoline ever reaches the wheels of a car — a missed opportunity that, if exploited, would completely rewrite the geopolitics of oil. As Amory Lovins, one of the world's most outspoken efficiency advocates, likes to point out, "just a 2.7 miles-per-gallon gain in the fuel economy of this country's light-vehicle fleet could displace Persian Gulf imports entirely."[1]

In fact, according to efficiency optimists like Lovins, the amount of oil, electricity, and other energy that could be saved through better efficiency in the United States alone — the so-called "efficiency resource" —

is actually larger than our physical reserves of oil and gas. In other words, it is now possible to *save* more oil than we could possibly find in the ground, and to do so at a per-barrel cost well below the average market price for oil.

In this context, aggressively improving energy efficiency would certainly seem as important as, say, researching hydrogen fuel cells, or building LNG liquefaction trains — and perhaps even more so. Because while we are accustomed to thinking of energy efficiency as optional — something we can *choose*, on the basis of the cost of fuel or our personal politics — it will soon become an absolute necessity.

Our rapidly growing population and economies will soon exceed our ability to supply that population with low- and no-carbon energy. This means we can expect a gap between the energy we need and what we can safely generate without permanently damaging our climate (or sowing more geopolitical discord or economy-wrenching price volatility). Optimistic forecasts show much of this gap being filled by new energy technologies — biofuels, solar power, clean coal, or hydrogen. On closer inspection, however, it becomes clear that most forecasters are counting on a huge contribution from conservation — both lower energy use and more efficient energy use. The reason: not only are the new energy technologies emerging more slowly than optimists had hoped, but many of the new fuels and technologies lack high power density and simply will not be able to deliver the same energy punch as the hydrocarbons they replace. To put it another way, within the next two decades, extensive and sustained improvements in energy efficiency will be not simply a sign of moral virtue, but an absolutely essential component of the future energy economy.

Yet whether efficiency is allowed to make such a contribution is increasingly in doubt. In many industrialized nations, and the United States in particular, any mention of efficiency is generally absent from energy debates. In spite of high energy prices and rising concerns about energy security, consumers and policymakers alike have all but stopped talking about the ways we use energy, how much we waste, and what might be changed. "We have an entire generation of policymakers, journalists, and consumers who either didn't live through the energy shortages of the 1970s or have forgotten everything we learned back then," Lovins told me. "We really need to go back and restart a thirty-year-old discussion."

Efficiency's lowly status is a recent development. Throughout history, humans have been obsessed with saving energy — not because it was morally correct, but because energy, whether extracted from coal, oil, or ox dung, was so costly that users had every incentive to use less. This impulse has been a great impetus for innovation, especially during the industrial revolutions. Today, a barrel of oil or a ton of coal produces five times as much in energy "services" — that is, goods, services, comfort, convenience, and other forms of material wealth — as it did a century and a half ago. This explains why industrialized economies are roughly three times as energy-efficient as industrializing societies: as economies advance and become more competitive, they are forced to use energy more efficiently.

By the mid-twentieth century, however, as the global oil economy matured, our obsession with efficiency faded. In the great postwar economic boom, as energy became more plentiful and less costly, emphasis naturally shifted from using less energy to acquiring more of it. This change was especially obvious in the oil-rich United States. Whereas Europe and Japan, anxious to reduce oil imports, continued to encourage conservation through high energy taxes, the United States saw no reason to conserve. As far as Americans were concerned, using less energy would actually hurt the economy — a fear that, apparently, still haunts policymakers in Washington.[2] When the oil shocks of the 1970s struck, driving oil prices skyward and forcing consumers to cut energy use, many economists believed that the world's industrial economies would be destroyed.

Instead, the world rediscovered efficiency. Energy researchers like Arthur Rosenfeld, a physicist turned efficiency expert from California, began to study just how much energy we had been wasting, and how much of the waste could be "harvested" through simple conservation measures. Within a short period, says Rosenfeld, who helped launch the efficiency movement and remains California's efficiency czar, "we realized we were discovering, or had blundered into, a huge oil and gas field buried in our cities, factories, and roads, which could be 'extracted' at pennies per gallon of gasoline equivalent."[3] By making basic improvements to cars and buildings, America could save the energy equivalent of twelve million barrels of oil a day — well over half the nation's total demand — thereby obviating the need for oil imports.

Even as many in Washington were talking openly about seizing Middle Eastern oil fields, Rosenfeld was arguing that "it would be far more

profitable to attack our own wasteful energy use than to attack OPEC."[4]
More important, despite fears that conservation meant giving up material
wealth — President Carter had called conservation "the moral equivalent
of war" and pleaded with Americans to make lifestyle sacrifices — Rosenfeld and other efficiency experts insisted that this new kind of conservation
could be largely "transparent" to the consumer. With the right technologies, regulations, and financial incentives, energy waste could be cut unob-
trusively, without affecting how people worked or lived. "The best form of
conservation is the stuff that makes so little difference to the quality of energy services delivered that you don't even notice it," says Rosenfeld. "You
get either the same energy services for less energy, or even a greater level of
energy services." Or as Rosenfeld is fond of saying, conservation "doesn't
mean putting on a sweater."[5]

If Dick Cheney is an icon for conservation skeptics, Rosenfeld would
be his ideological opposite. Soft-spoken and grandfatherly, with a shock of
white hair and a mastery of disciplines as diverse as building design, the
economics of power plants, and automotive technology, Rosenfeld is acknowledged as the "grand old man" of conservation. In the 1970s and '80s,
he and his colleagues helped turn California into a model of efficiency and,
ultimately, helped the United States and other big consumers among nations recover from the first oil shocks. "The irony of it all," says Rosenfeld,
"was that the Arabs, who didn't give a damn about energy efficiency, were
the ones who taught us we could get by with considerably less energy."

Between 1974 and 1986, Western economies, and especially the U.S.
economy, made enormous strides in conservation, often, as Rosenfeld had
predicted, in the form of efficiency improvements that were largely invisible to consumers. By government mandate, air conditioners were reen-
gineered to use less power, yet they suffered no loss of cooling capacity.
New building codes required double-paned windows, better insulation,
and more efficient heating systems. New refrigerators used only one-quarter of the power that a pre-1970s model had — a savings that, when multiplied by the number of U.S. households, helped avoid the construction of
forty new power plants. Most dramatic, however, was the improvement in
cars. Between 1977 and 1985, despite a booming U.S. economy that grew 27
percent, oil demand fell by more than one-sixth. As a result, the oil markets were glutted, and OPEC learned that the United States had a powerful
oil "weapon" of its own: conservation. All told, American energy intensity

— again, the amount of energy required per dollar of economic productivity — was falling by more than 3.5 percent every year.

Paradoxically, conservation's great success was also its downfall. As oil prices fell to ten dollars a barrel, few Western consumers saw any reason to continue conserving. In Europe and Japan, where energy security remained a critical issue, governments kept fuel taxes high, to discourage oil imports — a policy that has, by and large, worked. But in the United States, where raising taxes is anathema (and where the politically connected domestic oil industry was desperate to see demand increase), political leaders declared the energy crisis over. Conservative politicians like President Ronald Reagan regarded conservation not only as a governmental intrusion into the marketplace but as a surrender to the Arab oil embargo — an admission of U.S. geopolitical decline. Observes Denis Hayes, a top official in the Carter administration energy department, "in Reagan's view, America did not *conserve* its way to greatness. America was a modern industrial state, not a hunting-and-gathering society. It needed more and more energy every year, and the mission of government was to provide that energy."[6]

Arriving in office in 1980, Reagan tried unsuccessfully to kill many conservation regulations, while encouraging a massive buildup of power plants, new coal-mining operations, and new domestic oil production — essentially trying to reclaim our heritage as an energy giant. Reagan's efforts initially met with resistance, but as energy prices fell, political and popular support for aggressive conservation campaigns evaporated. In 1986, the Reagan administration froze the CAFE fuel standards that had been so effective, and within several years, American automakers were producing — and American consumers were happily buying — a progression of full-sized sedans, light trucks, and SUVs that grew larger and less fuel-efficient with each model year.

By the onset of the first Iraq war in 1990, American energy policy had come full circle. Instead of responding to this new threat to energy security through efficiency — Carter's "moral equivalent of war" — the United States, with the enthusiastic support of Europe and other importing nations, declared actual war. Even if no one officially referred to the "liberation of Kuwait" as a war for oil, the allied victory seemed to signal the end of the 1970s energy shortage, and the end of any need to save energy. Oil prices were down. The Middle East was stable. Arabs everywhere liked

Westerners. Why worry about oil or energy security or above all conservation?

For efficiency advocates, the most telling images of the first Gulf War, and the most irrefutable sign that conservation was passé, were the postwar TV shots of the victorious George Bush senior, roaring through the waves in a king-sized speedboat. If conservation's icon had been a glum President Carter, asking Americans to do more with less, to sacrifice, and to embrace austerity, the symbol of the movement's end was a relaxed and ebullient Bush, smiling in the sun, throttles wide open, burning through fuel as if there were no tomorrow. The message, says efficiency advocate David Nemtzow, was brilliant, powerful, and persuasive: "It was, 'You don't need to conserve. We'll go and *get* the oil for you.'"[7] By 1994, for the first time in its history, the United States was importing more oil than it could produce at home. Conservation was over, dismissed as a relic from the 1970s. Indeed, many efficiency advocates today are so fearful of scaring the public with glum '70s references to using less that they refuse even to use the word "conservation."

One explanation for the brief success and ignominious defeat of conservation is that the crisis truly is over, and with it any moral imperative to buckle down and save energy. Another is that energy efficiency no longer pays. Because energy prices have fallen steadily over time, while our economy has grown more robust, energy costs have become an ever-smaller share of a business's or household's expenses and have thus provided a declining incentive to use energy more efficiently. According to this viewpoint, conservation generally, and energy efficiency specifically, have ceased to be a viable economic proposition.

In fact, however, energy efficiency has hardly ceased to make economic sense, in that plenty of potential remains for energy savings. In the U.S. power sector alone, we could reduce our electricity rates by 40 percent and cut CO_2 emissions in half by upgrading power plants and transmission systems.[8] Replacing inefficient household furnaces with high-performance models would, within fifteen years, reduce gas demand in North America by nearly 25 percent. And, as we have seen, automotive fuel efficiency could be doubled through technologies that are already in use, thereby saving vast quantities of oil and, in theory, sparing us endless foreign entanglements.

What has changed, however, is consumers' capacity to recognize these economic benefits — largely because, in much of the industrial world, even educated consumers haven't a clue about what energy is or what part it plays in their lives and the larger economy. Beyond having an awareness of the cost of heating oil or gasoline (a cost so extensively covered by the media that it has taken on an almost religious significance), most consumers understand very little about the energy they use. Few can say how much they consume in the course of a day or a year, or where it comes from. (The classic illustration of this is the finding that a majority of U.S. consumers believe that most of their electricity comes from hydroelectric dams, when in truth most is produced from coal-fired and nuclear power plants.) A similar ignorance surrounds virtually every element in the energy economy: ours is a culture of energy illiterates.

This is not surprising. Whereas residents of poor nations are acutely aware of every aspect of their energy use, every stick of wood, every gallon of cooking fuel, in modern, wealthy societies, where energy costs are a small fraction of overall expenses, energy is not a hot topic of conversation. We may complain about the high cost of gasoline or castigate our leaders for making war for oil. Yet the nuts and bolts of energy — what energy is, where it comes from, how much we use, and how we might use less — are scarcely discussed, covered in the news, or taught in schools. In more affluent cultures, energy has become an invisible commodity, something we vaguely understand to be important on a national and international level, yet no longer fully recognize in our daily lives.

On those rare occasions when consumers in modern energy economies bother to look for energy information, it is extremely difficult to find. Our gas and electric bills don't show us how much energy we used for heating our homes, as opposed to cooking, or heating water, or which uses are most energy-intensive — that is, where a reduction in usage might produce the greatest energy savings. Instead, individual energy costs are folded into total energy costs, which for most of us essentially disappear into overall household or business expenses.

Even when energy information is broken out, consumers rarely make use of it. Many appliances today come with tags detailing their energy efficiency and giving the energy cost per year, ostensibly to allow customers to compare the yearly energy costs, or the costs over the lifetime of the appliance. Because ultraefficient appliances are often marginally more expensive to purchase, however, most consumers will buy the cheaper model —

even if they know that the cheaper model will cost more to run over its life-time than the more efficient model, owing to higher energy use. With energy as with most other factors affecting buying decisions, the modern consumer simply cannot embrace the notion of a long-term payoff.

We see this most clearly in car purchases. Western car buyers generally, and Americans in particular, focus almost entirely on the purchase price of the car, not the yearly operating costs — for insurance, repairs, and especially fuel. Such irrational decision making explains at least part of the trend toward ever more elephantine, less fuel-efficient cars and trucks over the past few years, in spite of higher gasoline prices — as well as the delayed "buyer's remorse" many SUV buyers have after a year of trips to the gas station.

If companies made purchasing decisions like that — looking only at up-front costs and ignoring costs over the life cycle — they would go bankrupt, or at least they would fire the purchasing manager. This observation brings us to a major fallacy about energy efficiency and about energy decisions generally — namely, that consumers can be counted on to behave rationally when it comes to buying energy or energy-consuming products. They can't. In theory, consumers behave much like little businesses, carefully gathering product information, comparing competing offers, and making decisions based on some sense of the total costs and total benefits of each option. In reality, consumers buy energy and energy-consuming products, such as cars or houses, with the same mix of rational and emotional criteria that they apply to any purchase: they are primarily trying to avoid up-front costs, even if the costs over the long term are greater. Given such tendencies, it is hardly surprising that we use energy so wastefully.

This energy obliviousness helps explain why we have so often misspent our "efficiency dividend"; for example, although today's lighting systems are dramatically more efficient than those of the pre–energy crisis era, any potential energy savings is offset by the trend toward homes with more lights: where a single light used to shine, many new or remodeled homes now have dozens of recessed or track lights. Televisions, too, have become more energy-efficient, but they have also grown larger — the big-screen home-entertainment center is increasingly the standard, especially as incomes rise — and far more numerous: most American homes have at least two or even three televisions. Our refrigerators may be four times as energy-efficient as a 1975 model, but many of us now have two refrigerators

— the shiny new efficient model in the kitchen and an older, less efficient model out in the garage for beer. All told, these trends help make clear why, despite great improvements in energy efficiency, demand for electricity jumped 15 percent in the United States and 17 percent in Europe in the last decade and is expected to jump by more than 50 percent by 2020.

More broadly, this trend helps us see why energy experts get so anxious when they begin calculating how the world is going to power itself over the next century. By most estimates, assuming that projections for future energy demand and population growth hold true — and that we maintain our current disdain for energy efficiency — by the year 2100, the world's ten billion people will need something on the order of fifty terawatts of electricity, or around four times what we produce today. That is a staggering amount of power. Generating it would require an energy infrastructure far larger and costlier than any that exists today, and it raises questions about not only the adequacy of our energy supplies, but the quality of that energy. By some estimates, given the slow success and low power densities of nonhydrocarbon energy technologies, we would not be able to meet all this new demand without using a lot of fossil fuels, which we've no way to ensure that we can burn cleanly. In other words, our unwillingness to take energy efficiency seriously enough to reduce demand may make it flat-out impossible to stay within any sort of reasonable carbon budget.

⊙∰∼

To a traditional economist, the bleakness of our record on efficiency is nothing to be ashamed of, but is instead natural and unavoidable, because efficiency is regarded largely as a one-shot deal. According to this view (which still holds sway with many policymakers), most of the possible gains from, say, building more efficient automobile engines or air conditioners or homes were already realized in the 1970s and 1980s, when energy prices were high. The big, easy improvements are gone, and the remaining savings will be marginal, especially now that lower energy prices have removed the big incentives.

Efficiency improvements can still occur, of course; but these will be the serendipitous results of other improvements in technology or business practices. Overall, the economy will experience a "spontaneous" improvement in energy efficiency of around 1.5 percent a year — basically, the historic rate at which economies have become more energy-efficient. Any efforts to boost efficiency beyond this spontaneous rate will be driven not by self-interested consumers, but by governments — an intrusive, unnatural, antimarket dynamic that all but guarantees failure. According to this traditionalist view, it would be far better to spend government money looking for new energy sources — that is, to increase the *supply* of energy, rather than waste public money trying to reduce energy *demand*.

Not surprisingly, such conventional wisdom is anathema to many of today's efficiency enthusiasts, especially Lovins, a fifty-six-year-old experimental physicist who argues that mainstream economics is blind to the true costs of energy and is thus unable to register the true benefits of energy efficiency. An intense, driven man, with a dark, thick mustache and a mile-a-minute speaking style, Lovins has spent much of the past three decades trying to correct this flaw of energy economics by reinventing the way we do our energy accounting. In lectures, interviews, guest editorials, and an endless stream of research papers from his Snowmass, Colorado–based Rocky Mountain Institute, Lovins attacks the pillars of traditional energy economics, among them, the sacrosanct belief that people and businesses save energy only when the cost of energy is high.

The truth, Lovins insists, is far more complex, but also more encouraging. Although price did to a large extent drive the conservation measures during the first era of conservation, such has not been the case recently. Between 1996 and 1999, industrialized countries made improvements in energy efficiency that were almost as dramatic as those made during the energy revolution of 1979–1985 — even though recent energy prices have been much lower. "Something else was getting our attention," says Lovins. That "something," Lovins says, was money — not the price of energy, but the money businesses were saving by adopting energy-efficient technologies and practices. The costs of improving energy efficiency, when it is undertaken correctly and systematically, are always *less* than the cost of the energy saved.

Lovins' favorite example of this occurs in the automotive sector. By conservative estimates, doubling the average fuel economy of American cars added around three hundred dollars to the cost of each car. But it also saved drivers anywhere from three hundred to five hundred gallons of gasoline a year. Assuming that the life of a car is ten years, customers ended up spending around a penny or less to save each gallon. In other words, the cost of efficiency (three hundred dollars) was dramatically less than if drivers had said no to fuel efficiency and had simply bought more gas instead.

And, says Lovins, you find this same kind of payoff potential in nearly every area of the energy economy. A more energy-efficient heating system installed in an office building, for example, usually more than pays for itself in reduced energy costs and thus creates a net gain in revenues for the company. The cost of installing more energy-efficient motors on an assembly line is almost always less than the cost of the extra energy used to power the older, inefficient motors. As David Goldstein, an efficiency expert at the environmental group Union of Concerned Scientists, told me, "anywhere companies have pursued energy efficiency, they have ended up making money, even if making money wasn't their initial goal."

The catch, of course, is that to achieve this kind of benefit, conservation has to be properly carried out, and this, too, requires a fairly dramatic shift in our assumptions about energy economics — in this case, how we use energy. Historically, mainstream thinking about energy has been oriented toward a supply-side model — that is, focused on producing primary energy, like coal or oil or natural gas or electricity — and getting it to customers. How customers used the energy — driving their cars, cooling their homes, running their factories: the so-called end use — was regarded as largely fixed. Engineers might be able to make an air conditioner run a bit more efficiently, but the essence of the end use — running an air conditioner — was not going to change. All you could really alter was the way you produced the power — that is, whether you used gas, nuclear, coal-fired, or solar energy: in short, the main variable was *supply*.

But the supply-side view, says Lovins, has it exactly backward. What we really need to be asking is, What kinds of things do we want to do with the energy we get from our raw materials, and how much energy do those things really take? "Why do people want energy in the first place?" Lovins asks in one of his trademark editorials. "Customers don't want lumps of coal, raw kilowatt-hours, or barrels of sticky black goo. Rather, they want the services that energy provides: hot showers and cold beer, mobility and comfort, spinning shafts and energized microchips, baked bread and smelted aluminum." The point, Lovins says, is to look at the desired end use and then determine how to achieve it as efficiently as possible. If the desired end use is a well-lit reading space, then the quantity and quality of light should be the criteria, not whether a certain number of light bulbs are installed or whether the power plant is burning coal or gas. Likewise, if you want a cool office space in the summer, then air temperature should be the criterion, not the size of the air-conditioning unit.

The point, says Lovins, is to start with the end use, define the energy services required, and then look for the most efficient means to provide them. Thus, cooling an office space becomes more than simply installing an air conditioner and ducts; it calls for designing an office with special heat-proof windows, effective ventilation, passive channels that draw up cool air from underground, and other low- or no-energy approaches to cooling. And here's the bonus: when correctly undertaken, these whole-system approaches to efficiency cost no more, and often even less, than the systems they replace.

What quickly becomes clear, says Lovins, is that energy efficiency is inseparable from other kinds of efficiency. For example, new machinery, systems, or processes that happen to be more energy-efficient are often also more efficient in other ways. They usually cost less to operate, and they also boost productivity. In fact, spending money on energy efficiency can be a lucrative investment. Every dollar spent retrofitting an old office building with more energy-efficient lights, heating and cooling systems, and windows typically nets the owner savings of $1.20 or more — a 20 percent return on investment that easily beats the Wall Street average, but with far less risk. Energy experts like Lovins say this side benefit is a big part of the reason that the global economic growth was so robust during the first conservation era, when energy intensity fell sharply, and it is also why the economy continues to hum. Individuals, companies, and entire nations are getting more productivity from even less energy and investing the savings elsewhere.

This is why efficiency advocates have long argued that saving energy at the end use is always cheaper than adding more supply. Today, generating a kilowatt of electricity at a power plant costs, on average, just under three cents. By contrast, by the time the electricity has reached the home or business, its cost has climbed to 8.5 cents a kilowatt, owing to operational expenses and waste; up to half of all electricity is lost from transmission lines in the form of heat. So every kilowatt of power spared at home saves eight cents, whereas generating another kilowatt costs three cents — money that businesses or homeowners can spend elsewhere, boosting economic growth while lessening demand for new power.[9]

Saving energy is also faster than producing energy. In most cases, energy can be "produced" more quickly through improved efficiency than through building more power plants or drilling more wells. Changing to energy-efficient light bulbs, for example, yields an instant reduction in power demand — essentially, other users have access to electricity without the power stations' having to add a single kilowatt. This was a painful, if rarely discussed, lesson of the 1980s. Convinced that the energy shortages were entirely an issue of supply, U.S. policymakers dismissed conservation and chose instead to encourage increased production — especially more power plants and oil wells. Yet by the time many of the supplemental barrels and kilowatts became available, conservation had already "produced" the needed energy. That much of the new oil was now unnecessary contributed to a market glut and the great price collapse of 1986. Similarly, some industrial countries found themselves with so many surplus power plants (many of them overpriced nuclear plants) that many were mothballed — left idle while utilities tried to get taxpayers to bail them out. "Efficiency had actually captured the market that the suppliers thought was theirs," quips Lovins, clearly relishing the irony. "Efficiency got there first."

The best example of efficiency's fleetness is California. Before the 1973 oil embargo, per capita energy use in California was growing at 4.5 percent a year. But starting in 1977, the state embarked on a comprehensive and ambitious campaign to reduce energy consumption. Efficiency standards were adopted for appliances and buildings. Utilities, through a novel incentive program, were actually paid by the state for every kilowatt their customers conserved — in effect a reversal of the traditional incentive structure that had encouraged greater demand.[10] The effects were dramatic. By the mid-1990s, California's per capita energy growth was flat. As a result, the state avoided building dozens of power plants (and coping with the emissions) that would have been inevitable had pre-1977 demand continued. As one California efficiency advocate boasted, "if the rest of the United States got even half as aggressive as California did, we'd basically solve most of our energy problems."[11]

Given such successes, it becomes clear why efficiency experts believe that industrialized societies have realized only a small fraction of the total energy savings that would be possible if efficiency were approached not simply as an afterthought but as a core element in industrial design. For example, reengineering the entire car concept around fuel efficiency — that is, focusing not simply on building better engines, but also on making lighter, more aerodynamic bodies — could yield gasoline-powered cars that get not just forty miles per gallon but sixty miles per gallon or even eighty miles per gallon and, as a result, could dramatically reduce CO_2 emissions and cut oil demand. Introducing vehicles like this on a global scale would save as much oil as is produced by all the members of OPEC combined — and effectively "conquer" the Gulf-dominated oil order without firing a single shot.

This is not just a question of liberating the West from the thrall of the oil order. Although most discussions about efficiency focus on industrialized nations and their oil addiction, efficiency's biggest payoff may come in the developing world, which is now desperately trying to figure out how to attain anything approaching a modern energy economy without having to spend billions on a new energy infrastructure. Efficiency may be part of the answer. Programs to distribute or subsidize ultraefficient compact fluorescent light bulbs, for example, would give third-world consumers the light they need, without having to add as many expensive and polluting power plants — thus freeing up money for schools, health care, water systems, and other services that are fundamental to bringing a society into the modern era. Expand that idea from light bulbs to cooking stoves, heating systems, communications, and transportation, and suddenly it begins to seem possible to provide a decent standard of living to the billions of people who today lack access to even the most basic energy services.

If conservation is so beneficial, why hasn't it become more of a commodity? A classically trained economist will tell you that if big opportunities remained where efficiency could be exploited cost-effectively, the omnipresent market would have noticed, and investors and businesses (which, unlike individual consumers, do make rational buying decisions) would already be on to them. The argument, in Lovins' summation, "is that if there was any more efficiency worth buying, it would already have been bought, because we all live in a perfect market."

In reality, when it comes to energy efficiency, the market is far from perfect. Information about energy efficiency is neither widespread nor clear, so most businesses simply are not aware of the potential cost savings. As a result, those which do embark on efficiency retrofits often do so incompletely, say, by replacing a single component — a furnace, for example — but leaving in place leaky ducts and old, single-pane windows that waste half the heat the new furnace produces.

In many cases, businesses are encouraged *not* to invest in energy efficiency. Many buildings and much of our transportation infrastructure were designed with an eye to minimizing construction costs, not energy use. Electrical systems are a prime example. Contractors know that thicker-gauge copper wire conducts electricity more efficiently, with less energy lost through waste heat, than does thin wire. The difference is big enough that using a thicker wire, though more expensive to install than a thinner wire, will pay for itself through lower energy bills in less than five months. Nevertheless, contractors rarely use thicker wires, because electrical work is usually done by the low bidder, whose goal, not surprisingly, is to minimize up-front materials costs, and who doesn't care about the "life cycle," or the operating costs of the building. The thinnest wire allowed by law is invariably installed, and the house or business essentially throws away much of the electricity before it reaches a light bulb or appliance.

Or consider the color of your roof. Because dark colors absorb more heat than light colors, people in hotter climes used to paint or tint their houses and especially their roofs white. In the West, however, architects shy away from white, in large part because it shows dirt. "So they choose an earth tone," says Rosenfeld. "They think they're being 'environmental,' but really what they are doing is creating a roof that gets ninety degrees hotter than the surrounding air, instead of the fifteen degrees of a white roof." This extra heat soaks into the house, forcing the air conditioner to work 20 percent longer and use a fifth more power. Worse, when you have a large number of dark-roofed houses and buildings in close proximity — as in a city, for example — you create what Rosenfeld calls a heat island, raising the outside air temperature by several degrees and forcing cooling systems in buildings and in cars to work even harder. In Los Angeles, the combined effect of so many million dark roofs, as well as dark asphalt roads, forces the city to use up an extra 1,500 megawatts of power cooling itself — the equivalent of one-and-a-half power plants — or about 3 percent of California's total summertime power load.

Negative incentives like these work against efficiency at all levels of business and industry. Because landlords buy appliances, for example, but tenants pay for the electricity they consume, landlords buy the cheapest models available, which usually are the least energy-efficient. In short, businesses and consumers often have incentives *not* to pursue conservation. Multiply these disincentives by all the apartment buildings, office towers, and factories around the world, and you waste a staggering amount of power.

Reversing such disincentives won't be easy. Governments would need to rewrite everything from building codes to tax laws, in order to encourage investment in efficiency upgrades. Industries would need to rethink the way they do their energy accounting and, in particular, incorporate life-cycle energy costs into the bidding process for capital projects. But the pay-off would be enormous. If national governments resumed the aggressive approach toward energy efficiency that was so successful in the 1980s and began reducing energy intensity by 2 percent a year (which is actually less than the United States has been achieving without really trying, over the last decade), world power needs in 2100 would be cut to around *half* of current demand. If we reduced energy intensity by 3 percent a year, we could meet world demand in 2100 with around a *quarter* of the energy we use to-day. In other words, improving efficiency only slightly faster than is already happening "spontaneously" in the United States would mean that within a century ten billion people could be enjoying a modern level of energy services for less than a fourth of the energy used today.

The implications are obvious. Given the volume of energy services we will soon have to provide, given the pressing need to reduce CO_2 emissions, and given the slow pace at which carbon-free energy sources are likely to become available to displace carbon fuels, the only plausible solution is to reduce substantially the rate at which our overall energy demand is growing. And since no one envisions reducing energy demand by cutting economic growth (at least, not yet), the only way to maintain our standard of living without using more energy is to become more energy-efficient.

How much more efficient? According to the U.N. Intergovernmental Panel on Climate Change, by the end of the century, more than 30 percent of our energy demand will be met not by new clean technologies, but as a consequence of energy conservation. And many energy advocates believe that this share must ultimately be higher — again, because the new low-carbon fuels and energy technologies will lack the power density of the hydrocarbons they will replace. "Of the energy we use today, two-thirds and maybe even three-fourths will be replaced by energy efficiency, and only one-third to one-fourth will be replaced by new supplies and technologies," predicts Lee Lynd, the Dartmouth College biofuels expert.[12]

To be sure, alternative fuels, like hydrogen or biofuels or solar or wind, will be critical to the future energy mix, but their high profile has tended to obscure the role that efficiency must play. "You can't simply rely on the hope of new supply — you have to have efficiency," says Lynd. "But if you are willing to consider the idea of new supply in conjunction with efficiency, then suddenly your supply options open up substantially." Fuels and technologies that couldn't power today's cars and houses are suddenly able to do the job in a future where cars and houses require only half or a quarter of the energy. "If we are willing to see these two ideas together," says Lynd, "we can come up with some extremely attractive scenarios that just might get us out of the box we've gotten ourselves into with oil."

As a possible catalyst for the next energy economy, conservation finds itself in an awkward position, caught between its great potential for saving energy and the equally great obstacles to achieving that potential. It's abundantly clear that dramatically better energy efficiency not only is possible but will be essential to any long-term efforts to keep the world supplied with clean energy. Yet much like the alternative energy industry, the efficiency sector faces a daunting array of obstacles, ranging from consumer ignorance and prejudices to a market and political system that still assign greater value to producing energy than to saving it.

Many of the modest successes that governments have had in promoting energy efficiency have been offset or wiped out entirely by larger political and economic trends. Low energy prices, for example, destroy the incentive for efficiency: if oil or gas prices were to fall for several years, any campaign to improve the efficiency of cars or home heating systems could easily be scuttled.

Efficiency can also fall victim to entirely unrelated trends, such as the wave of deregulation that has swept state and national power systems over the past decade. Proponents of deregulation argue, quite reasonably, that traditional power sectors (as state-protected monopolies) are bloated and inefficient, resulting in high power rates that hurt the economy and waste energy to boot. By opening the power sector up to competition, proponents argue, utilities would have to streamline operations and become more efficient, thereby bringing power prices down. Whether that will eventually happen remains to be seen. In the meantime, however, one unintended consequence of streamlining has been the abandonment of many state conservation campaigns, most notably, programs like California's, which paid utilities to encourage consumers to save power. Had that single program remained in place, says Rosenfeld, California would very likely have reached the year 2000 with a power surplus large enough to avoid the blackouts entirely.[13]

California is not alone: of the ten states that rewarded utilities for discouraging consumption in the early 1990s, only two — Oregon and California (which only recently restarted its program) — do today. The rest, says Lovins, operate like regular businesses — they are rewarded for selling as much power as possible, penalized for any action that cuts customers' bills, and thus "unenthusiastic about energy efficiency that would hurt their shareholders."

Of course, Lovins' complaint highlights one of the main flaws of the efficiency program: that most energy vendors, like any business, don't want to sell less of their product. While greater energy consumption imposes extra costs on society, such as supply risk and climate, it means nothing but profits for energy producers — even more so when supplies run short and prices spike. Reversing this powerful incentive, as we'll see in later chapters, will require innovative policies and some tough political choices.

The larger problem is that efficiency, by itself, is only half a solution. No matter how efficient we become, if we want to reduce CO_2 emissions and other negative effects of energy use, we must somehow alter the historic trend whereby any gains made through energy efficiency are more than wiped out by a corresponding jump in overall energy consumption. And to date, that trend shows few signs of changing. Except in rare instances, every significant improvement in efficiency has eventually preceded — and perhaps produced — an offsetting increase in overall consumption. Our car engines became more efficient, so we made them larger and more powerful, or we drove more miles or made more trips. We learned to build houses that used less energy per square foot, then built bigger houses and filled them with more gadgets. As energy historian Vaclav Smil points out, "whatever the future gains may be, the historical evidence is clear: higher efficiency of energy conversions leads eventually to higher, rather than lower, energy use."[14]

For now, this trend may be tolerable. With energy prices low, and with no other real restrictions on energy use, we face no immediate penalty for misspending our efficiency dividend: it is still possible simply to use more energy and get away with it. But from everything we have seen so far, such forgiving circumstances may be short-lived. Ultimately, we may find ourselves living in a world where simple "transparent," painless efficiency no longer suffices. Thirty years from now, for example, all our cars may get twice as many miles to the gallon as they do today. But if there are four times as many cars by then, we'll still be using twice as much fuel, and putting twice the load on an energy economy and a natural environment that are already under stress today. As Andrew Rudin, a California-based energy consultant, puts it, ultimately, "our environment does not respond to miles per gallon: it responds to gallons."[15]

PART III

INTO
THE BLUE

10

ENERGY
SECURITY

IT'S NOON on a broiling hot August day, and I'm cruising the streets of Wenatchee, Washington, listening to Golden Oldies and engaging in that peculiarly American pastime, the gasoline bargain hunt. Though we're still a week away from Labor Day, the traditional peak of the U.S. driving season, a series of shocks to the global oil system have already boosted prices. Here in the States, refinery outages, a ruptured pipeline, and low oil storage levels have pushed prices past two dollars a gallon, provoking the usual claims of collusion and price gouging from a nation of permanently indignant motorists. Yet out in the markets, the oil traders seem far more anxious about events beyond U.S. borders — and beyond American control. Market analysts are nervously watching Nigeria, America's fifth-largest supplier, where violent ethnic unrest and offshore oil piracy have cut exports by as much as three hundred thousand barrels a day. In Venezuela, the number-four U.S. supplier, oil production has yet to recover from a national strike in 2002 and opposition groups are gearing up for another showdown with President Hugo Chavez.

The worst news, as always these days, comes from Iraq, where saboteurs have just blown another hole in the six-hundred-mile-long pipeline between the oil fields in Kirkuk and export facilities on Turkey's Mediterranean coast. The pipeline is one of Iraq's main oil export routes, and a centerpiece in U.S. efforts to revive the Iraqi oil industry and fund postwar reconstruction. Each day the pipeline stays empty, the Iraqis lose another $6.25 million in vital oil revenues.[1] Meanwhile, world oil markets lose confidence that Iraqi oil will be a major force in the energy economy — or an effective weapon in Washington's campaign to undercut OPEC and stabilize global oil markets. Oil prices that were falling for the first time since

Saddam fled in April are rising again past thirty dollars a barrel — and reviving concerns about the global economic recovery.

Press accounts that morning have Thamer al-Ghadaban, Iraq's beleaguered acting oil minister, vowing to restore pipeline operations quickly, but the markets are not soothed. Security analysts believe sabotage will hamper the Iraqi oil industry for the foreseeable future, thereby making it impossible to know when the country can resume prewar exports of 2.5 million barrels a day — oil that energy forecasters have been counting on to meet unexpectedly high demand from the United States and China. Even al-Ghadaban himself seems doubtful that much can be done to protect the pipeline in the postwar chaos. "In the past regime, we had the oil police, the army, and the cooperation of the tribes, as well as what we call internal security," al-Ghadaban tells reporters.[2] "Now all this has disappeared. There is a void in security."

Iraq is simply the latest reminder, if one were needed, that in a global economy dependent largely on a single fuel, "energy security" is a thin fiction. Since September 11, we've all become much more aware of the vulnerability of the sprawling energy infrastructure that moves oil, gas, and power around the planet. On TV talk shows and in magazines and newspapers, a parade of experts on terrorism and sabotage have described in gory detail the devastation that would be caused by a well-planned al-Qaeda–style attack on any of a dozen "choke points" in the energy order — the trans-Alaskan pipeline, for example, or the huge oil ports in Rotterdam, or worse, Ras Tanura, the massive Saudi export facility that handles six million barrels a day.

Yet if Iraq's oil problems show us anything, it is that the real threats to energy security go well beyond sabotage and dirty bombs. At its most basic level, "energy security" is our ability to meet immediate energy demand — that is, to produce adequate volumes of fuel and electricity at affordable prices and to move that energy to the countries that need it, when they need it, to keep their economies running and their people fed and their national borders defended. A failure of energy security means that the momentum of industrialization and modernity grinds to a halt, and survival itself becomes far less certain.

Energy security is where the rubber hits the road in global energy policy, the harsh reminder that crises like air pollution and energy colonialism and even climate change are by no means the most serious ones facing the energy economy. For if we can't meet basic energy demand — demand that, under current trends, is growing so fast that it will double by 2035 — none of these other things will matter. And energy security goes well beyond mere questions of supply. No matter how much oil or gas we can find, this supply is worthless unless we have in place the physical infrastructure, the political stability, and the financial and technological resources to get it to those who need it — criteria that are growing more and more difficult to meet.

Today's global energy system, the massive network of production and delivery, is meeting the needs of the industrial world — but barely. Many non-OPEC oil fields are in decline, and even if OPEC has vast reserves, the cartel itself may lack the political stability and financial ability to exploit that crude as quickly as skyrocketing world demand will shortly require. The main alternative to oil — a gas economy — faces similar political and financial challenges, as do, by extension, the world markets for electricity.

And this is the situation in the *advanced* energy economies of the industrialized world, where technology, politics, and market forces are said to operate with a high degree of sophistication, efficacy, and harmony. In rapidly industrializing, or "transition," economies, like India, China, South Korea, Brazil, and Malaysia, demand for oil, gas, and electricity over the next several decades will be almost unimaginably high — yet no one seems to have a clear idea how that energy is to be delivered. Even worse are the energy problems of the developing world. More than 1.5 billion people — a fourth of the world population — not only lack access to the most basic of energy services but lack any realistic hope of getting those services, short of some massive bailout by the industrial world. Such energy poverty raises the specter of yet another global divide, between wealthy and poor nations, and sets the stage for a new kind of conflict: the energy war.

The widening gap between our demand for energy and our ability to meet it is already emerging as a powerful force in the shaping of the next energy economy, a force that could easily override other priorities and undermine prospects for a cleaner, more sustainable energy economy. It may be beyond dispute that protecting the climate is a long-term imperative. It may be widely understood that an urgent need exists for crash programs to develop alternative fuels and technologies and promote energy efficiency.

Yet in light of the grim and immediate realities of energy security, these goals begin to seem like luxuries — as if we had a *choice* about the kinds of energy we produce and consume, or the way we use it. Given the enormous task of simply getting energy to those who need it today, the most pressing question to be asked about any future energy system may no longer be whether we can produce the *right* kind of energy, but simply whether we can produce *enough* energy.

Just outside Dhabol, India, residents of this impoverished region can peer through a padlocked chain-link fence and see one of the biggest examples of the modern energy gap. In the early 1990s, the Enron Corporation and several partners ponied up more than two billion dollars to build a mammoth gas-fired power plant in the coastal Indian city. The flagship project in India's campaign to privatize its ailing power sector, Dhabol was a test case for a brave new, market-oriented approach to bringing energy to countries that simply could not afford it themselves — a "win-win deal," to use the saccharine term of art. Under the agreement, India would get a huge volume of dependable, high-quality energy for its booming industry and exploding population, without having to beg for financing from the World Bank. Enron, for its troubles, would get fabulously wealthy from long-term contracts that guaranteed unusually high power prices for a developing economy. Better still, because Enron was a gas trader, it could buy fuel for the power plant very cheaply — an advantage that allowed the company to cash in bigtime on the "spark spread."

That, at least, was the theory. In practice, Enron found itself trying to sell overpriced electricity to a third-world energy economy with little taste for capitalism. Like many developing nations, India's power sector has no market-oriented tradition. Tens of millions of Indians have no access to electricity at all, and most who do are accustomed to electricity rates kept artificially low by government subsidies. Many consumers do not pay their bills at all, and the utilities rarely try to collect. When utilities could not honor their long-term contracts for Enron's overpriced power, Enron took them to court. As that strategy faltered, Enron executives begged their friends in the Bush White House to lean on the Indian government, to no avail.[3] The Dhabol deal bogged down in the courts, the plant was padlocked, and Enron moved one step closer to bankruptcy.

To many U.S. observers, Dhabol seemed yet another example of Enron's greed and hubris. But to the rest of the world, and particularly to India, where the collapse of the deal has further delayed the slow rise from poverty, Dhabol is much more a lesson about the sheer challenge of energy security in the developing world. In India, rural China, and Bangladesh, in large parts of Southeast Asia, Latin America, and the Caribbean, and in most of Africa, 1.5 billion people still rely on wood, dried animal manure, or other so-called biomass for nearly every calorie of energy used for cooking, heating, or lighting. Another 500 million people burn coal — not in furnaces, but in cooking fires and braziers — producing poor-quality heat and constant clouds of asphyxiating soot. In all, some 2 billion people — almost a third of the world's population — rely on energy systems that fail to meet even the most basic human needs. As developing nations have the fastest population growth, energy poverty — the slow-motion failure of energy security — is sure to be one of the most serious problems of the next several decades.

On the surface, energy poverty is simply another measure of the generally poor economic conditions the developing world faces. Like water or food, energy is a resource that is in chronic short supply. Yet because energy is so interconnected with all other aspects of life, energy poverty tends to play a more significant and central role, one that creates a ripple effect through a developing economy and has an inordinate impact on living standards, which can all but destroy a population's move toward modernity.

Families that heat and cook with wood or dung spend hours each day hunting for fuel; in many communities, villagers will travel for days to find enough fuel for the week, only to repeat the process the following week. Wood-based energy economies also precipitate rapid deforestation of entire regions, causing erosion, landslides, and other environmental problems. Yet these effects pale by comparison with the direct human costs. Wood fires produce clouds of toxic smoke, the leading cause of respiratory illness in the rural third world, and especially among women and young girls, who do most of the cooking, and infants, who tend to be with their mothers.[4]

Even when time might permit women and children to read or write, the light cast by a wood fire is so weak — typically less than that from a flashlight — as to be useless for the purpose. "There are hundreds of millions of children who are trying to learn and study under the dim light of a kerosene lamp," said Kurt Hoffmann, director of Shell's corporate giving. More generally, according to one aid agency report, reliance on wood and other "traditional fuels . . . barely allows fulfillment of the basic human needs of nutrition, warmth and light, let alone the possibility of harnessing energy for productive uses which might begin to permit escape from the cycle of poverty."[5]

Conversely, even a tiny improvement in the level of energy services tends to raise living standards remarkably. Replacing wood-fired cooking stoves with kerosene stoves means that families that once spent hours or days gathering wood can now devote that time to earning more money, producing more food — even getting an education. "When you can alleviate even some of the most basic energy needs, a lot of the usual stressors disappear," says John Steinbruner, director of the Center for International and Security Studies in Maryland. "Their world is fundamentally improved."[6]

Even more benefits flow when communities switch from biomass or liquid fuels to electricity. Electricity solves many indoor air problems and eliminates or minimizes the need to gather fuel. It provides adequate indoor lighting, "extending the day" for education and simple leisure. If electricity is available, consumers can install basic appliances — televisions, of course, but also refrigerators, which dramatically improve food preparation and safety, and water pumps, which make possible a supply of fresh drinking water — a huge benefit for communities now decimated by water-born illnesses. Electricity powers irrigation pumps that improve crop yields; it powers radios and telephones, lights for schools and hospitals — in short, it can raise living standards immensely.

With electricity, billions of people now mired in a preindustrial existence could find themselves living somewhere in the early twentieth century — perhaps not an ideal life, but certainly one far better than they have now. As Amy Jaffe, an energy consultant from Texas, told a conference on sustainable development, "if we can solve the energy problem, we can solve other problems, such as food supply, water, poverty, and also ease the pressure on greenhouse gases."[7]

Even more amazing is how little extra energy is required to produce this improvement in living standards. By one estimate, the amount of electricity needed to bring the entire developing world up to minimum energy standards would be around one thousand terawatt-hours — or roughly the amount of electricity used by the United States; and many energy experts believe the developing world can do significantly better than achieve "minimum energy standards." By adopting coherent energy policies that integrate cutting-edge energy technologies with aggressive conservation and efficiency programs, developing countries could essentially "leapfrog" directly into a twenty-first-century energy economy, one that provided a high level of energy services, while largely avoiding the messy "smokestack" phase that the industrial world had to pass through.

Unfortunately, given current trends in the energy economy, such a hopeful scenario seems only remotely possible. Developing nations lack not just modern fuels or electricity, but the capital and know-how to build and maintain a modern energy infrastructure — the pipelines, refineries, power plants, and transmission lines that permit the move into modernity. More to the point, because energy systems are inextricably linked with the larger economy, attempts at building a modern energy economy must wait until a country has addressed such broad issues as economic reform and overpopulation. Without a strong economy, developing nations cannot afford even basic energy services, much less a more advanced "smart" energy economy. And until they have population growth under control, developing nations will of necessity be far more concerned with the *quantity* of the energy they produce than with its quality.

This situation is not likely to change soon. Historically, developing nations' energy budgets are almost entirely funded by outside players — development agencies like the World Bank, or big donor nations like the United States, France, or Saudi Arabia, or even big multinational energy companies that are looking to build markets.[8] Yet sadly, the energy systems that evolve in such a "donor" environment are often quite dysfunctional: poorly built power plants and dams, inefficient transmission technology, corrupt management, and heavily subsidized power and fuel rates for a population too poor to pay anything approaching a market rate.

Predictably, many outside contributors have lost interest in the third world. After disasters like Dhabol, few private energy companies are willing to risk billions of dollars building power plants or other energy infrastructure in countries whose utility customers can't pay. "Today, the only energy projects that anyone wants to finance are projects that get energy *out* of the developing world so it can be sold to the developed world," says Ira Joseph, a global gas analyst in New York. "There is no money in supplying energy *to* the developing world." Tellingly, the largest U.S. investment ever to be made in Africa is a $3.5 billion pipeline that ExxonMobil, Shell, and Elf-Aquitaine will use to pump oil from Chad out of Africa to world markets.[9]

Private energy companies are not the only ones rethinking their third-world investment strategies. Foreign aid from the United States, Europe, Asia, and the Middle East has long been key to building new energy systems in the developing world, but recently, U.S. and European funding strategies have shifted to reflect the new realities of energy security: as the big consuming nations seek new sources of oil and gas for their own energy economies, they are channeling aid dollars more strategically — to favor countries with energy reserves. "There has definitely been a shift of U.S. aid to regions like the Caspian, West Africa — places that are reliable energy suppliers or could be," says one U.S. aid consultant who specializes in energy issues in the developing world. "The United States still spends lots of money where we have no huge strategic energy interest, like South America, but less than in a region where we can build an energy relationship."

Even aid agencies are thinking twice. After years of criticism from environmental groups and human rights organizations, big donor organizations like the World Bank are reluctant to finance such large energy projects as hydroelectric dams, power plants, or pipelines, which not only tend to wreak environmental and social havoc but often turn into huge slush funds for corrupt regimes.

More generally, donors are simply less willing to throw good money after bad — that is, to fund a new energy project when the rest of a poor country's energy system and overall economy remain so inefficient and corrupt. The big push in the 1990s to introduce market reforms and capitalism to inefficient third-world energy sectors has been widely, if quietly, written off as a colossal bust. "If you've got a country where prices have been kept artificially low for decades, raising prices to market rates without first figuring out how to raise people's incomes is a recipe for disaster," says one U.S. economist who advises aid agencies on energy issues in the developing world. Big aid agencies "are really getting sick of constantly bailing these countries out, yet having so little to show for it. They are taking a big step back and reevaluating how they are spending money."

The result is a growing energy gap between the developed and developing worlds. As the energy economies in the industrialized West become larger and more efficient, those in places like Africa, South America, and rural Asia become poorer and less effective. In places where energy resources simply do not exist, the lack of energy continues to ripple through the economy, depressing living standards and exacerbating malnutrition and disease — effects that not only defeat the chances for a modern state, but increase the risk of conflict. Indeed, competition for energy in poorer countries may be the spark for a new category of war in the developing world: the energy war. To the extent that such conflicts threaten Western access to gas and oil, as may be the case in strife-ridden, oil-rich West Africa, they can only add to the industrial world's own concerns about energy security — and may in fact offer a foretaste of what the industrial world itself will face in the global competition for energy.

In a terrifying way, this is the good news. For if poor nations do find ways to obtain adequate energy — by which I mean simply raising their energy standards to today's global average, or the equivalent of France's in the 1960s — the results for the world would be staggering. Nearly two billion people in effect live outside the modern energy system: they subsist on wood and other biomass and have little influence on the global dynamics of supply and demand. To begin to bring even a fraction of these people up to modern energy standards — by providing them with coal-fired power plants, for example, or steady supplies of diesel or stove fuel — would add enormous stress to the global energy system. To bring all of them along would change the world in ways we have trouble imagining.

Not only is the population of the developing world growing rapidly, but people there have the furthest to go to reach even a modest living standard — a fact with two disheartening implications. First, even achieving an adequate energy standard for the existing population of the developing world would require more energy than our current system, or any easily constructed system, could produce. Second, even if population levels off or begins to decline, world energy demand — driven mainly by growth in the third world — will continue to climb inexorably, as all the poor continue to push for a twentieth-century energy existence. In a strange way, energy security is analogous to the climate problem: just as CO_2 concentrations in the atmosphere will continue to rise for decades, even after we stabilize emissions, demand for energy services will probably climb until the end of this century, regardless of what happens with population or energy technology, as the developing world "catches up" to the industrialized West.

What becomes clear is that the quest for energy security, because it ties in so directly with economic survival, supersedes most other concerns. Nations that lack adequate energy must find it, while those which do possess energy resources will be forced to exploit them — by whatever means necessary, and with little regard for the impact. The classic example is China's sprawling Three Gorges Dam, which will provide badly needed, pollution-free electricity, but whose construction has flooded millions of acres of farmland and displaced a million Chinese. Yet in comparison with the kind of energy development that is coming to the developing world, Three Gorges Dam will in hindsight seem almost benign.

In today's economy, clean, sustainable energy is a luxury reserved for the richest nations. In countries staggering under high population growth, the drive for energy security rarely means "leapfrogging" to a sophisticated, clean technology. Instead, these nations tend to take the easiest, fastest, and cheapest path possible — which usually means technologies that are obsolete, low-quality, and highly polluting. We have already seen how China's near-desperate push toward industrialization is producing inefficient, polluting cars and adding to an urban air-quality health disaster of epic proportions. The effects on China's power sector will be worse. As in the West, China's fastest-growing energy market is for electricity. In the first eight months of 2003, power consumption jumped 16 percent — about four times the amount Western analysts had predicted — and China is building new power plants at a staggering rate.[10] Yet whereas new power plants in the West are almost always fueled by gas, in China, a gas economy seems even less likely. Like the United States, China possesses only small gas reserves of its own — just 1 percent of the world's total proven reserves — and most of these are located in the nation's central and western regions, far from the big markets in the east.

As in the West, Chinese officials are working hard to build pipelines and LNG terminals in the densely populated coastal cities, but outside these pockets of affluence, gas is a long-term goal at best in the rest of China. The huge costs of such projects — coupled with investors' anxiety over just how much Chinese consumers can afford to pay for energy — leave the prospects for such critical infrastructure in doubt. Even if China had abundant gas, the country lacks the technology to use gas as fuel. China simply cannot manufacture or afford to import the small, highly efficient gas turbines that Western utilities now rely on for relatively clean power generation. As a result, gas, which currently supplies just 3 percent of

China's total energy, is expected to provide only 6 percent by 2010 and perhaps 12 percent by 2020 — compared with a 25 to 30 percent share in the rest of the industrial world.

Instead, China will solve its looming energy security problems in the worst way possible: through coal. China is, in fact, well on its way toward becoming the world's largest coal economy. According to one forecast, to meet its demand for electricity, China must build as many as sixty 400-megawatt electric power plants every year for the next decade, and most of them will burn coal. Despite an apparent decline in coal use during the 1990s (which Western analysts optimistically attributed to improved energy efficiency and a shift toward gas), Chinese coal consumption is again rising — by nearly 8 percent in 2002.[11] All told, the demand for coal in China, and in neighboring India, which is on a similar coal track, will account for more than two-thirds of the growth in world demand for coal. By 2050, more than a third of the energy consumed by China and its neighbors will come from coal. "The real question isn't whether China is going to use its coal," warns Reid Detchon, a former energy official in the first Bush administration, "but whether China will use its coal cleanly."

At this point, the answer seems to be no. China is so poor that it simply cannot afford the kind of cutting-edge IGCC technology needed for a "clean-coal" energy economy. Instead, Beijing is relying largely on the same obsolete coal-fired technology that plagues the West. Indeed, many of China's existing coal-fired power plants are so ancient they lack emissions-control technology and waste most of the energy they generate. The result is a power sector that is horribly polluting and so inefficient that, to meet the nation's rising energy demand, it has been forced to to build new plants faster than if it used a more efficient power technology, like gas — thus committing China to burn even more coal and emit even more pollutants.

The consequences aren't encouraging. China is already the leading emitter of sulfur dioxide, the component in coal smoke that causes acid rain, which is ravaging China's cities and nearly 40 percent of its forests and farmlands. Whereas many Western coal-fired power plants must install sulfur-"scrubbing" technology, most new coal-fired power plants in China do not — not because the Chinese like acid rain any more than Americans or Europeans do, but because scrubbers add 30 percent to the cost of a new power plant — the difference between building four new power plants and building only three. In electricity-starved China, where blackouts are still common in most cities, the choice isn't hard.[12] And if China can't afford sulfur scrubbers, it is almost impossible to imagine how, or why, Beijing would spend billions of dollars installing clean-coal technology.

The climatic consequences of China's coal-fired drive for energy security are staggering. Today, China is the second leading emitter of carbon dioxide, right behind the United States — despite the fact that China's per capita CO_2 emissions are just one-eighth those of the United States. Given China's current energy trends, it should occupy first place before the end of the decade. Between now and 2030, China's CO_2 emissions will increase as much as those of the entire rest of the industrialized world.[13] What is truly alarming here is that, despite all the new growth in power usage and in construction of power plants, China's per capita consumption of electricity is still less than a *tenth* of the average for industrialized countries.[14] What this suggests is not only that China still suffers from chronic energy poverty but that, once China starts to lift itself out of that poverty and approach a Western level of energy use, its energy needs will exceed the capacity of any global system that currently exists.

In countries like China and India, lack of energy security is clearly inseparable from the larger economic problems. Developing nations are simply too poor to attack energy security by any means other than brute force, even if doing so carries enormous social and environmental costs. Yet we should not imagine that lack of energy security is a problem of the poor alone: even a nation as spectacularly rich and clever as the United States is having serious trouble shifting its energy economy to gas. Worse, these difficulties are spreading rapidly throughout the industrial economies, disrupting energy markets and delaying the emergence of the gas economy that is supposed to serve as a bridge to the energy economy of the future.

Again, the problem isn't so much one of total global supply — by even conservative estimates, world gas reserves could last through 2050 — as one of access, of linkages. In the United States, for example, gas production shortfalls have prompted a rush to build a massive, multibillion-dollar gas import infrastructure, with a giant (but unbuilt) gas pipeline from Alaska, and intricate systems of terminals in places like Baja, to bring in tankerloads of liquefied natural gas. Elsewhere in the world, and especially in industrializing Asia, gas demand is encouraging a plethora of large-scale LNG projects and pipeline proposals, including for an outsize line from Siberia to Japan.

The actual logistics of bringing in so much gas are daunting, even in wealthy markets that are eager to pay for it. Because gas is so much harder to move than oil is, gas projects are far more capital-intensive — up to four billion dollars for a single LNG liquefaction "train," not including the fleet of special refrigerator tankers and the huge new regasification terminals in every receiving port. Over the next twenty years, according to the U.S. Energy Information Administration, investors will need to spend eighty billion dollars to build the gas and LNG infrastructure the United States will need by then. Considerably larger investments will be needed to expand the gas economy worldwide — and many energy companies are reluctant to put up that kind of cash.

Why? If future gas demand is expected to be so high, you would expect investors to be knocking one another out of the way in their rush to finance the world's gas infrastructure. In reality, the gas business is so volatile and uncertain that funding it poses huge risks for investors. In the American gas market, for example, many investors and companies actually fear that gas prices may be *too* high or, more precisely, too vulnerable to wide price swings. Speculators love volatility, but large energy companies and institutional investors do not, since it makes it hard to know what long-term gas prices will be, and thus whether they can invest safely.

This is one reason the big energy companies are reluctant to build a gas pipeline from Alaska to the lower forty-eight states: they won't commit to a twenty-billion-dollar investment without a guarantee that, if the market becomes glutted and prices fall below what the companies need to make a profit, the U.S. government will step in to pay the difference. Many energy analysts believe that, as the scale of gas and other energy projects rises, such government guarantees will become a standard feature of the energy business — and a critical component of energy security.

Globally, the economics of LNG are even less certain. The key to a worldwide LNG economy is mobility: being able to move gas from, say, Indonesia, to markets in Japan or Europe, or America. But LNG is costly to move: the special gas tankers are more expensive than oil tankers. Every extra mile LNG must be shipped adds considerably to the overall costs of an LNG project and thus eats into its long-term profitability. In this sense, gas is still a stranded asset, a factor that explains why energy investors have shown greater enthusiasm for shorter-distance LNG projects, such as selling Australian gas to China, but are dithering over longer-distance endeavors. Consider the much-hyped development of a Baja LNG energy hub.

Here, energy companies say they can tanker LNG from South America, Australia, and Indonesia and sell it to the fast-growing U.S. market. Yet while many analysts believe gas can be profitably shipped into the United States from South America, they are far less sanguine that companies can make money bringing gas tankers all the way from Australia or Indonesia — at least until North American gas prices rise substantially.

Of course, demand will eventually attract supply. With a market as large, as lucrative, and as apparently endless as North America's, investors will find a business model that makes long-term price risk tolerable. Yet it is not simply financial risk that is holding up a new gas infrastructure, but political risk as well. Pipelines are so enormous and involve so many political players that delays are inevitable. The Alaska gas pipeline has been held up repeatedly, as U.S. and Canadian politicians, not to mention a variety of native tribes whose lands the pipeline must cross, bicker over whether the pipeline should follow a longer, southern route that takes it through more U.S. territory or a more northerly route that is shorter, but crosses more Canadian land.[15] Similar fights are breaking out around the globe, as big gas producers like Russia or Turkmenistan try to build expensive gas pipelines to burgeoning markets in Asia and Europe.

Liquefied natural gas is even more susceptible to political risk. As much as industrialized economies need more gas, many individuals within those economies regard the big LNG ships and regasification terminals as safety risks, especially with the threat of terrorism. Even before the September 11 attacks, few American communities wanted LNG facilities nearby, out of fear that LNG is polluting and likely to explode — despite the fact that LNG, being frozen, is far less flammable than gasoline. This is one reason that, in the thirty years since LNG became technically feasible, the United States has built only four LNG receiving terminals — capable of handling less than 2 percent of what the country will soon need — and why energy companies have been so keen to travel to sunny Baja.

Again, demand invariably attracts supply, and most industry analysts I've spoken with see these infrastructural and political hurdles as temporary. The political urgency of the gas shortage, not to mention the sheer profit potential of the U.S. gas market, will drive politicians and energy companies to make gas infrastructure a top priority. Even industry optimists, however, admit that the rise of the U.S. gas economy will take place slowly — much more slowly than advocates of a transitional energy economy powered by gas had hoped for; in the meantime, we are left with an economy powered primarily by coal and oil.

Ultimately, the question of energy security in the modern world comes back to the place it started: oil. In time, a gas economy will probably emerge, followed by new kinds of fuels and energy technologies. In the here and now, though, oil remains the single most important fuel and in many ways the least secure. True, the infrastructure for oil is already in place, and its price (if we overlook the recent prewar spikes) is near the historic average — two factors that have helped maintain the fiction that oil is our safest form of energy. Nevertheless, this sense of security is only temporary, for oil's dependability is fading by the month.

Oil production in non-OPEC countries, for example, is falling with surprising speed. Between June of 2002 and June of 2003, according to the International Energy Agency, oil production in three of the four biggest non-OPEC states — the United States, Norway, and the United Kingdom — reported a net loss of output of nearly a million barrels a day, mainly through depletion of reserves. Much oil still resides outside the OPEC countries, but, except in Russia, most of this oil is harder to get at and thus more expensive to drill for and transport: indeed, as with gas, many projects to extract oil are so costly that they can be handled only by consortiums of oil companies. Even then, much of this difficult oil is simply not economical to produce, and it won't become economical unless oil prices rise significantly. Barring that outcome, analysts believe that non-OPEC oil production will fall each year by as much as a million barrels a day. With world oil demand expected to increase each year by nearly two million barrels a day, the global oil industry must somehow add another three million barrels of oil daily just to keep markets happy.

In theory, such production increases are possible. Russia, West Africa, and the Caspian have a lot of oil, at least in the near term, and the Middle East has more than eight hundred billion barrels — more than half the world's total. As we have noted, though, the main challenge to energy security is not simply procuring enough oil, but spiriting it out of the ground and into the tanks of those who need it. And here is where the trouble starts. Over the next three decades, according to the International Energy Agency, the oil industry will need to invest $1.7 trillion simply to maintain its current oil production levels — that is, to find new oil fields fast enough to replace those now in decline or soon likely to decline. On top of that, oil companies will need to spend an additional $600 billion to meet all the *new* demand, especially from booming Asia. Taken together, that means $2.2 trillion in oil investments — a pile of money, even for oil companies and petrostates — and it's not at all clear where it will come from.

Consider the situation in OPEC. By 2020, according to most forecasts, rising world demand, coupled with declining non-OPEC production, will make it necessary for OPEC to more than double daily production, from around 26 million barrels today to as much as 54 million barrels by 2020 — more than half of which will be shipped to the rapidly growing economies in Asia. Indeed, by as early as 2009, OPEC must be pumping an additional 5.1 million barrels a day to meet world demand.[16]

Finding five million new barrels a day may not seem tough, especially in today's healthy oil markets. Yet signs are growing that the days of loose oil markets are fading and that even mighty OPEC will struggle to increase its production. Venezuela and Nigeria, two of OPEC's biggest producers, remain highly unstable, and that instability means that the investment needed to expand production may stay away. Iraqi production is recovering far more slowly than optimistic U.S. officials had predicted — because of sabotage, but also because of the four hundred thousand barrels of crude a day that the U.S. occupation forces themselves consume.[17] Even OPEC producers that have not been affected by regime change or civil war are increasingly short of the funds necessary to expand their oil production capabilities. Despite decades of lavish oil export revenues — as much as three trillion dollars since the 1970s — many Middle Eastern countries have so mismanaged their economies that they can no longer afford to expand their production capacity, or even maintain what they currently have.

Saudi Arabia is a case in point. It is widely known that the Saudis are barely keeping pace with government expenses, despite the recent high oil prices, and now lack the cash flow to finance a major program to install wells and pipelines. To be sure, Riyadh and other Arab governments could easily attract the necessary investment: Western oil companies, kicked out in the 1970s, would be only too happy to bring their capital and expertise back to the world's largest and cheapest oil fields.[18] Still, Arab leaders, who also know that a renewed Western presence could anger many Muslims, are desperate to avoid any kind of civil unrest that might ultimately force them to reform their corrupt government, share political power, or flee for their royal lives.

Barring significantly higher oil prices, OPEC is likely to struggle to expand its production capacity sufficiently to meet long-term demand. Even in the short term, some analysts say, cartel members will be hard pressed to add even another 2.5 million barrels of production by 2009 — about half of what the global economy will need from OPEC by then.[19] This puts most of the production burden on the Caspian and Russia. Although these hot new oil regions will have less trouble attracting oil investment, they, too, will struggle to step up production fast enough — or for very long, given suspicions that Russian oil production could peak as early as 2015.

What all this means, some analysts believe, is that the global oil business is in the midst of a sea change. After twenty years of overabundant capacity and relatively low prices, the market is moving inexorably into an era of ever-tighter supply and significantly higher prices — risky business for a global economy built to run on cheap oil. How high prices will go is an open question; but some estimates suggest that a new price range of thirty-five to forty-eight dollars a barrel may be necessary, simply to finance all the additional oil production that our spiraling demand will require (an increase that could shave economic growth in consuming countries by a hefty 1.5 percent). How long prices would remain high is also uncertain. Typically, high prices depress demand and encourage new production, which raises supply and brings prices back down — a mechanism that once helped ensure long-term energy security. But given the possibility that large swathes of the world's oil production base are approaching production peaks and that the remaining oil will be increasingly expensive to produce — and in light of the voracious demand we expect when China and India emerge as true car cultures — a high-price "regime" may become a fact of economic life in the twenty-first century. As *Arab Oil and Gas* magazine has noted, "the five-year perspective for world demand, growth at current rates, and accelerated depletion in key oil provinces indicate a rapid and certain end to Cheap Oil."[20]

Worse, high prices will be accompanied by higher volatility. The supply buffer that major oil companies once held was abandoned in the cost-cutting campaigns of the 1980s and 1990s. Now the remaining buffer against volatility — OPEC's spare capacity — will also vanish. As OPEC struggles to meet demand, the necessity for countries like Saudi Arabia to deploy their excess production capacity will dramatically reduce the critical ability to compensate for supply disruptions and unexpected spikes in demand. Like the United States before it, Saudi Arabia will lose its mantle as swing producer and savior of world markets.

In the absence of such a savior, even "normal" fluctuations in supply or demand — for example, a cold snap in New England that unexpectedly drives up the demand for heating oil, or a hurricane that turns back oil tankers coming from Venezuela — could lead to dramatic spikes in price. And these are small disruptions. A large disruption, like a revolution in Venezuela or Nigeria, would mean a loss of three million barrels a day. In the kind of tight, volatility-prone oil markets that may emerge in the next five to ten years, "it is highly unlikely that substitute supplies could be made up," warns *Arab Oil and Gas* magazine, going on to predict that during such a disruption prices could easily be bid up past sixty dollars a barrel and kept there for months.[21]

Serious as the loss of Venezuela or Nigeria would be in this tight market, it would be peanuts by comparison with the loss of the big one: Saudi Arabia. Various grim analyses by intelligence experts portray the kingdom as teetering on the edge of a rebellion by fundamentalists who have little interest in continuing to serve obediently as oil pump to the West. Granted, even fundamentalists need revenues. Sooner or later, according to conventional wisdom, even a radical government in Riyadh would be forced to sell oil, and if the United States is the Great Satan, it is also the greatest, most lucrative market. (Iran, for example, would be delighted to sell oil to the Great Satan, if it weren't blocked from doing so by U.S. trade sanctions.) Some Western observers, however, argue that such confidence is misplaced: many Saudis are now said to be so disgusted with the corruption that oil has supported and so desperate to purify their countries of any Western influence — especially American — that such economic imperatives may no longer apply. As Robert Baer, a former Middle East expert with the U.S. Central Intelligence Agency, has written, "Saudi Arabia is more and more a breathtakingly irrational state. For a surprising number of Saudis, including some members of the royal family, taking the kingdom's oil off the world market — even for years, and at the risk of destroying their own economy — is an acceptable alternative to the status quo."[22]

Energy and security experts continue to debate the plausibility of such a nightmare scenario. In the meantime, the United States, Europe, and other big energy importers are now so concerned about large-scale disruptions that they have quietly stepped up efforts to fill up their strategic petroleum reserves. (Government oil purchases were so large in 2002 that they helped keep oil prices high after the Gulf War.) Yet even when these emergency reserves are topped off — at around 1.5 billion barrels — the world's backup oil supply is only a stopgap, which could cover the loss, say, of Venezuela or a "reconstructed" Iraq for less than fourteen months, or of Saudi Arabia for less than eight.

Given that exposure to disruptions is likely to continue, it is no surprise that the United States is now scrambling to develop a more "diverse portfolio" of oil suppliers — or that American foreign policy is as focused on oil as it has ever been. In 2002, for example, U.S. intelligence agencies provided support to Venezuelan military personnel who had briefly toppled President Hugo Chavez; Washington hoped that by replacing Chavez with someone less anti-American, the United States could better ensure that Venezuelan oil would keep flowing to American refineries.

The United States is also building up a military presence in and around oil-rich Africa. Pentagon officials are already planning to deploy small "rapid-reaction" teams to unstable areas and, according to one account, are exploring the possibility of stationing troops at camps and airstrips in Africa. "I think Africa is a continent that is going to be of very, very significant interest in the twenty-first century," General James Jones, head of the U.S. European Command, explained to a Senate panel in 2003. State Department officials insist that oil "is not the driving force," and that the primary aim is to fight terrorism in Africa.[23] Still, it is clear that the United States, which is working with American oil companies to accelerate development of West African oil fields, would be less concerned about terrorism in West Africa if the region had no oil: witness how reluctant the United States was to send troops to oil-less Liberia.

How far nations will go to protect their access to oil is impossible to predict, but energy security is likely to emerge as the newest pretext for geopolitical conflict. Both Gulf Wars were in part campaigns to defend the energy security for the entire industrial world, even if they were led by the United States and served U.S. interests. These days, oil security is taking on a more competitive aspect. The past five years have witnessed a race by

China and Japan and Europe to lock in a share of new production in West Africa, but also in Russia, South America, and the Caspian. Ostensibly these moves are meant to "diversify" sources of supply, much as the United States is doing, as insurance against a meltdown in the Middle East. In fact, Tokyo, Beijing, and other importing governments now worry about one another as much as they do about the Middle East. The campaign for oil diversity masks a budding competition among the world's big consumers for the last oil.

Thus far, the primary battleground in this competition has been Central and East Asia. For much of 2003, Japan and China were locked in a high-stakes bidding war for access to Russian oil. Japan, which depends entirely on imported oil, is pushing hard for a 2,300-mile pipeline from Siberia to coastal Japan, to carry a million barrels a day. But China, which sees Russian oil as critical to its own burgeoning economy, wants the Siberian crude to flow instead via a 1,400-mile, six-hundred-thousand-barrel-a-day pipeline south to the Chinese city of Daqing. The competition became so intense in summer 2003 that the Japanese offered to not only finance the five-billion-dollar pipeline, but to invest seven billion developing the Siberian oil industry, and another two billion in Russian "social projects"[24] — despite the certainty in Tokyo that if China loses out, Chinese-Japanese relations could plunge to a new low.

An even more intense oil battle is unfolding in the Caspian region. Since the late 1990s, China, America, Russia, and Iran have waged a diplomatic war to control the flow of oil out of Kazakhstan and Azerbaijan. Each country has not only proposed a different route for a Caspian pipeline but in many cases worked to undermine competing route proposals. The somewhat ironic result is that, so far, only a trickle of the Caspian's much touted "big oil" is getting out — much to amusement of the "unstable" Saudis. As one Saudi official smugly told me, "we've been hearing about 'Caspian oil' for ten years, and it's still not producing."

Predictably, this new oil competition is fostering all manner of geopolitical animosities and alliances. Beijing is firmly convinced that China is being intentionally shut out of world oil frontiers by a "conspiracy" of Western oil companies and Western governments fearful that China may become too important in oil geopolitics. As a result, China is pursuing new oil alliances in West Africa, South America, and, of course, the Middle East. In 2000, Chinese Prime Minister Jiang Zemin paid high-level visits to Iran and Libya — two countries the United States refuses to do business with.

China also has a major interest in Iraqi oil, which helps explain China's intense reluctance to support the United States–led war in 2002. "Basically we will choose those countries not enjoying a good relationship with the United States," says an official at a Chinese oil company with deals in Iran, "because there is almost no hope for Chinese companies to go into those pro-U.S. countries."[25]

China's biggest, and most controversial, oil alliance is with Saudi Arabia. In the classic win-win deal, the Chinese gain access to the world's largest oil reserves, while the House of Saud gets a foothold in what many regard as the largest potential oil market in the world. So intent is Saudi Arabia on securing a share of the Chinese market that Riyadh has been willing to offer Beijing special incentives, including below-market oil prices and special access to the kingdom's higher-quality, low-sulfur crude — even to the point of depriving existing customers in Europe and the United States. As one senior executive of Saudi Aramco explained, "we need the Chinese market, and we're going to get it just as we got Japan and the United States — through aggressive marketing subsidies."

In some ways, China is a more ideologically suitable partner for Saudi Arabia than the United States has been, particularly as anti-Saudi sentiments rise in the United States. With China, the Saudis get a giant new customer whose government, unlike Washington, won't chide Riyadh for its record on human rights or its links to religious extremists and terrorist groups. Best of all, for its oil, Riyadh gets access to sophisticated Chinese arms — including ballistic missiles and other hardware that even the Saudis' Western allies, such as the United States and Europe, will not sell to them. As one former high-level Saudi intelligence official put it, the Chinese are so anxious for Saudi oil that, "at the end of the day, we know that the Chinese would not have a problem selling us any kind of weaponry — as long as we can pay for it."[26]

What is so alarming about this more intense push for energy security is that it must ultimately fail. No matter how successful the United States is at building a military presence in West Africa, the fact remains that West Africa's known oil reserves of sixty-six billion barrels are around a tenth of those in the Arab Middle East — and can thus only temporarily delay the day when the United States and other big importers must return to the Middle East and all its instabilities. Similarly, no matter who wins the bidding war between China and Japan for Russian oil, the victor earns only a reprieve from energy insecurity. Fast-growing China, in particular, will eventually need more oil than even Russia can provide and will, sooner or later, have to look somewhere else for it — an eventuality that must give rise to other, perhaps less diplomatic, conflicts.

Of course, the United States, with its vast economic and military power, would be likely to win an overt battle for resources, including oil. But U.S. officials do worry that a growing rivalry among other big consumers will create conflicts that will both require U.S. intervention and destabilize the world economy upon which American power ultimately rests. As John Holdren, a Harvard energy economist and former energy adviser to the Clinton administration, told Congress recently, "a plausible argument can be made that the security of the United States is at least as likely to be imperiled in the first half of the next century by the consequences of inadequacies in the energy options available to the world as by inadequacies in the capabilities of U.S. weapons systems."

Clearly, traditional approaches to energy security are no longer viable. For more than a century, maintaining energy security has meant expanding the existing system incrementally — drilling more wells, building additional pipelines or power plants. From now on, however, adding supply will be less straightforward and the payback less assured. Gas projects are so gargantuan and expensive that they will require a novel form of government-corporate alliance, whose political and economic complexity may actually slow down deployment. Oil is even more problematic: in a world of declining production, unstable suppliers, and unprecedented demand, the struggle to maintain oil security will only become more challenging over time and absorb more resources and political attention.

Energy security, always a critical mission for any nation, will steadily acquire greater urgency and priority. As it does, international tensions and the risk of conflict will rise, and these growing threats will make it increasingly difficult for governments to focus on longer-term challenges, such as climate or alternative fuels — challenges that are in themselves critical to energy security, yet which, paradoxically, will be seen as distractions from the campaign to keep the energy flowing. This is the ultimate dilemma of energy security in the modern energy system. The more obvious it becomes that an oil-dominated energy economy is inherently insecure, the harder it becomes to move on to something else.

11

THE INVISIBLE HAND

ON THE BANKS of Clinch River, thirty miles east of Oakridge, Tennessee, the twin smokestacks of the Kingston steam plant rise a thousand feet over the Cumberland foothills, funneling coal smoke into the upper stratosphere and serving up a potent reminder of the weight of the energy order. In the 1990s, Kingston's owner, the Tennessee Valley Authority, realized it might have to replace the 1,700-megawatt plant. Built in 1955, Kingston was obsolete and extremely inefficient. Its cooling systems poured so much hot water into the river that nearby roads were often shrouded in a dangerous fog. More seriously, Kingston was one of a dozen TVA coal-fired plants whose high-sulfur emissions were destroying the region's forests with acid rain and choking the Eastern seaboard with smog. Summertime visibility in nearby Great Smoky Mountains National Park had declined by 80 percent since 1950. By building a new facility, Kingston would not only cut its sulfur emissions by 85 percent but, because greater efficiencies would produce the same power with less coal, cut carbon emissions as well.

There was just one catch — Kingston makes extraordinarily cheap electricity. Like many older U.S. coal-fired power plants, Kingston's initial construction costs have long since been paid off. Running the plant is mainly a matter of buying coal, which, in Tennessee coal country, is so inexpensive that Kingston can generate electricity for less than two cents a kilowatt-hour, well under the going market rate. That sweet deal would end, however, if TVA were forced to tear down and replace Kingston. Under the federal Clean Air Act, all new coal-fired power plants need expensive, state-of-the-art pollution scrubbers on their smokestacks. Because it was an older plant, Kingston was exempted from the scrubbing requirements; however, if TVA replaced Kingston with a new plant, the new emission control technology would run construction costs to $2.1 billion, effectively doubling the cost of Kingston's electricity. A new gas-fired power plant would cost only half as much — and would produce considerably fewer emissions — but the gas itself would be three times as expensive as TVA's cheap coal, so electricity costs would be just as uncompetitive.

Faced with two losing propositions, TVA did what most utilities were doing at the time. For around four hundred million dollars, TVA performed a kind of stealth-retrofit on the Kingston plant: tearing out and replacing the old steam boilers, burners, and pipes, and adding taller smokestacks, to better disperse the smoke. In effect, TVA rebuilt the plant from the inside out without technically replacing it — and, more to the point, without tripping the law's expensive scrubbing requirements. Today, the half-century-old Kingston is still generating 1,400 megawatts of cheap electricity — and emitting a hundred thousand tons of sulfur and nearly four million tons of carbon every year.

Kingston is a classic example of what might be called the Myth of the Perfect Gadget. Embraced by some environmentalists, energy executives, and policymakers, this fallacy holds that solving global energy problems is mainly a matter of waiting for the "right" energy technology — the right engine, the right fuel, the right scrubbing device. Once that miracle technology is available, the fantasy runs, success is assured — barring the stupidity of consumers or the greed of dark and powerful corporate forces.

Yet if we've learned anything from thirty years of global energy consciousness, it ought to be that the next energy economy will *not* be driven by technology alone. Although nearly every advance in our energy evolution has indeed centered on some new machine or process — the steam engine, the oil lamp, internal combustion — what ensured its success in the end was raw economics. No matter how clever a new technology might be, no matter how neatly it might fit into someone's vision of the ideal energy economy, if the gadget didn't pan out economically — if it couldn't do something better or faster or more efficiently or conveniently or cheaply than an existing technology — it didn't last. The future will be no different. Although factors like oil depletion or climate change or energy security may push us to champion new fuels or different emission policies, the fundamental question remains the same: Does the innovation help turn a profit?

This is by no means just free-market folderol. When we talk about building a new energy economy, consider the scale of our task: we need to take all our current energy assets — our coal-fired power plants, our oil pipelines and refineries, our tanker ships, our trains, and planes, and, of course, our automobiles — worth well over ten trillion dollars, and replace them all with an equally colossal and interwoven system of technologies, processes, and networks (many yet to be invented), which by 2050 must be efficiently producing enough energy for nine billion people, their companies, and their lifestyles, all while emitting half the carbon per capita than is currently the case.

We are talking, in short, about something so vast and complex and dynamic that it cannot be launched by a single technology but must be built, one transaction at a time, by the same relentless economic engine, the same competition between technologies and ideas, the same ruthless pursuit of profit, that built our old energy economy.

This is not to excuse the greed and shortsightedness of energy companies and their political allies, who often view a "new" energy economy as either a threat to their profits and power or an opportunity to sell old technology under a green label. It is, however, to recommend that we no longer be shocked, *shocked* at such self-interested behavior. The competition that is already shaping the next energy economy is occurring not only between rival technologies and ideas, but between the people, companies, and countries that have staked their existence on those innovations — and that will, quite reasonably, fight like hell to see their investments pay off. For at the root of every political conflict over energy, and every political debate over the best energy policy, is a conflict between economic propositions.

I am not advocating that we simply turn the task over to the market and cross our fingers. Our wonderfully efficient market has some astonishing blind spots and will require innovative political action to ensure that the energy economy we get is the one we truly want. I do, however, want to argue that until we gain a clearer understanding of the economic risks and rewards in the energy economy, we — and our policymakers — have very little hope of preventing the next energy economy from simply repeating the mistakes of the last one.

One of the clearest examples of this conflict between economic propositions occurs every year in Washington in the form of a noisy and somewhat disingenuous debate over automobile fuel-efficiency standards. Since 1987 — when Ronald Reagan froze the Corporate Average Fuel Efficiency (CAFE) standards (and terminated a decade of dramatic improvements in fuel efficiency), lawmakers have convened each spring to discuss whether those standards should be updated. And each time, after heavy lobbying by the auto industry, the answer has been a resounding No, with predictable consequences. In the absence of tougher standards, fuel efficiency in cars sold in this country has steadily worsened: today, the average new vehicle gets 20.8 miles per gallon — down 6 percent from the industry high of 22.1 miles per gallon in 1988.

Efficiency advocates complain that automakers, especially U.S. automakers, have the technology to make even their largest and heaviest SUVs far more fuel-efficient — and should have done so years ago.[1] That is undoubtedly true — but consider the economics of the proposition. First, for an automaker to retrofit, say, a fifth of its production lines to build highly fuel-efficient gas-electric hybrids would require the company to reengineer entire assembly lines, develop more efficient and cost-effective hybrid-engine systems, redesign body styles for aerodynamics and weight savings, then retrain thousands of workers and redevelop its network of parts suppliers. All this would could cost billions of dollars, at a time when Detroit is barely holding its own against Japanese rivals that *are* selling more fuel-efficient cars and thus would have an even *greater* advantage under tougher CAFE standards. Second, however bad they may be for the climate and energy security, large, inefficient cars have been awfully good for the auto industry. The profit margins on SUVs and trucks are roughly ten times as fat as those for smaller, more fuel-efficient sedans — and are widely credited with having almost singlehandedly kept American carmakers out of bankruptcy court.

Third, it is far from clear that American consumers even *want* fuel-efficient cars. For all the griping about conniving carmakers hypnotizing otherwise sensible motorists into buying four-mile-per-gallon Humvees, the *real* reason the auto industry is selling so many Ford Excursions and Dodge Rams is that consumers in the United States, and, increasingly, in other countries, *want* them. Despite higher gasoline prices — despite the many stories linking U.S. oil imports with worldwide terrorism — U.S. motorists have shown only marginal interest in getting more miles per gallon. Instead, big, gas-hungry trucks and SUVs grow more popular by the year and now account for more than half of all new cars sold in the United States, and even a growing share in European markets, like England (where higher fuel taxes were supposed to nip such absurd behavior in the bud). By contrast, although car manufacturers offer more than thirty car models with fuel economy of thirty miles per gallon or better, the ten most fuel-efficient models sold in the United States make up just *2 percent* of the sales. "In addition to addressing environmental concerns, we have to balance what customers want," explains Toyota vice president Donald Esmond, "and many of them want SUVs."[2]

Critics blame this trend toward lumbering inefficiency on the brilliance of automotive marketing and the abject stupidity of consumers, and these criticisms have merit. Yet it is also the case that, for many college-educated consumers, paying for fuel efficiency simply isn't an economically tempting proposition; for example, if I drive a midsized SUV with a fuel economy of twenty miles per gallon, and if gasoline averages $1.50 a gallon, I'll spend around $1,125.00 a year in fuel, or $93.75 a month.

If I trade in my SUV for a new Honda Insight, with its aerodynamic curves, ultraefficient gas-electric hybrid system, and fuel economy of fifty miles per gallon, I'll suddenly find my monthly gasoline bill falling by fifty-six dollars — which is, to use the technical term, chump change. If I'm the kind of guy who likes a larger, more powerful, more luxurious ride, fifty-six dollars a month isn't going to be enough to persuade me to switch to a smaller, more spartan chariot. As one auto industry analyst puts it, "when I pay twenty thousand or thirty thousand or forty thousand dollars for a vehicle, I'm not that worried about fifty cents or a dollar more for a gallon of gas."[3]

The point here is not that recalcitrant automakers should be forgiven (they shouldn't be), but that their behavior is entirely predictable. We may be appalled by Detroit's disdain for small, fuel-efficient cars — or the fact that automakers have spent hundreds of millions of dollars to defeat every attempt to raise CAFE standards — but we should not be surprised: in purely economic terms, fuel efficiency makes no sense for American automakers, or for many drivers. Like any self-interested party, automakers are squeezing as much as they can from their existing assets — their production lines, their technology, their current work force, and their current customers — and they will do so unless it becomes more profitable or less expensive to do something else.

This tendency might be thought of as asset inertia, and it is one of the most powerful forces guiding — or impeding — the evolution of the energy economy. The more a company has invested in a particular system, the more reluctant it will be to put those assets at risk — even if that means prolonging the existence of a system that is inefficient, noncompetitive, and, in the long run, likely to cost the company its market share.

In Detroit's case, for example, one could argue that it would be a better economic proposition over the long term to build a new kind of car that would dominate the type of market we know we will face: one with higher fuel prices and more environmental regulations. This is what Detroit should have done in the 1970s, when skyrocketing fuel prices suddenly opened the U.S. car market to smaller, more efficient Japanese and German cars. Instead of competing head-on, though, U.S. automakers, reluctant to give up their big-car production assets, waited years to produce fuel-efficient cars, thereby effectively offering the foreign carmakers a marketing beachhead they have never relinquished. In the last seven years alone, despite the popularity of the SUV and the pickup truck, American automakers have seen their control of the U.S. car market fall from 73 percent to 63 percent, and that share is still falling, because Japanese and European carmakers, having already won the battle for smart, fuel-efficient cars (Japanese gas-electric hybrids are nearly a decade ahead of American models), are now going after the lone category in which Detroit still dominates: pickup trucks and SUVs. Indeed, one the splashiest debuts at the 2003 New York International Auto Show was the enormous Nissan Pathfinder Armada. Nearly as massive as Ford's Excursion (currently the world's biggest SUV), the Armada offers seating for seven, all-wheel drive, the largest tires in its class — and enough horsepower to tow three and a half Honda Insight hybrids.

Because asset inertia operates most powerfully in those sectors with the greatest capital outlays, it is a hallmark of the energy business. Consider the worldwide "dash to gas." One of the main reasons energy companies have been so cautious about investing billions of dollars in new pipelines and LNG facilities is not that they doubt that gas has a future, but that so much of their money is tied up in other, more traditional energy assets, like oil or coal. "Every one of these oil companies wishes it could have a far greater position in gas," argues Fadel Gheit, senior energy analyst at Fahnestock in New York, "but the fact is that they have existing assets in oil and they just can't walk away from them. They would go out of business. It's almost like parents who have boys but wish they had girls."

Energy companies are, in other words, limited in how much they can spend building the new energy economy, in part by how much they have already invested in the old one. Whereas many companies believe a hydrogen economy will arrive eventually, they have no clear sense of how to prepare for a hydrogen future, in terms of their assets. How much should a company invest, and where and how soon? In theory, says asset expert Rick Gordon, a farsighted oil company could easily invest today in a mix of new technologies, assets, and expertise that would give it a clear advantage when the hydrogen economy finally emerges. In the meantime, however, the risks to that company would be horrendous, because in preparing to compete in a future hydrogen economy, it would be sacrificing some of its ability to compete in the oil economy today. In fact, such a right-thinking company would in all likelihood never live to see the hydrogen future, but would instead be driven into bankruptcy by its more conservative competitors, all of which were still positioned to succeed in the traditional, hydrocarbon energy economy of the present. "Companies are trying to keep their options open" for new energy sources, Gordon told me, "but without overcommitting to them."

Thus, while some companies — BP and Shell, for example — have been willing to make relatively small investments in solar power and hydrogen research, others, like ExxonMobil, believe that the smartest move is simply to wait to see how and when a new energy market develops — even if that means delaying the global transition to a cleaner energy system. "Our friends at ExxonMobil would rather keep their heads in the sand," a top executive at a European-based oil company with a big emphasis on renewable energy told me. "But it won't hurt them. ExxonMobil is the largest, most profitable company in the world, and if the 'new energy economy' comes about, they can simply buy their way into as much of the new market as they want to. But they are not going to create that position before they have to."

This asset-driven reluctance is most clearly evident in the power sector. In a perfect world, power generation is the first place to implement a new energy economy, because it's where the biggest impact on climate and energy supply can be realized. Not only is power generation appallingly inefficient, but it produces three-quarters of all CO_2 emissions. Yet the power sector also suffers from the most serious asset inertia. With few exceptions, the power industry has always been characterized by immense components — power plants, hydroelectric dams, and transmission lines — each costing hundreds of millions and even billions of dollars and each requiring twenty, thirty, or even forty years to pay off fully.[4]

For most of the twentieth century, this business model functioned adequately, especially for utilities with coal-fired power plants. Even though the plants themselves were polluting and terribly inefficient (more than 70 percent of the coal's energy goes right up the smokestack), coal-fired electricity was the cheapest power on the market. This was especially true in North America, which has more coal reserves than any other country, as well as in much of Europe, Russia, and Asia. (Even today, though most Americans believe they get their electricity mainly from hydroelectric power and natural gas, more than half comes from the nearly nine hundred U.S. coal-fired plants.)

By the 1970s, however, the asset inertia of the coal-fired power sector began to collide with another powerful force: environmental politics. After European governments forced utilities to scrub emissions from their coal-fired plants, many power companies simply abandoned coal and switched to gas, which was then plentiful, or nuclear power. In the United States, it was a different story. Although coal-related pollution was at least as severe as in Europe (the rainwater in the eastern United States, forty times more acidic than normal, was corroding buildings and killing entire forests), U.S. coal remained plentiful and dirt-cheap. Just as important, the large U.S. "fleet" of coal-fired plants was nowhere near the end of its useful life. By the time the Clean Air Act was being proposed, in 1977, many of the nation's coal plants still had years to go before they reached the end of their normal thirty-year payoff periods. Forcing the utilities to "retire" these assets early, before they had been paid off, and then replace the coal-fired plants with even more expensive scrubbing facilities, would have meant hundreds of billions of dollars in losses — which TVA and other utilities, not surprisingly, were not keen on absorbing.

Instead, the utilities asked for a stay of execution. Promising to retire these old coal plants as soon as they had been paid off, American utilities persuaded U.S. lawmakers to "grandfather" all coal-fired power plants built before 1985 — and to exempt them from the more stringent clean-air regulations. Most of these exempted plants would be required to take modest steps like building taller smokestacks to disperse the pollution better, or switching to a lower-sulfur coal. As long as the old plant technically remained intact, however, it escaped the tougher emission laws.

Environmentalists and energy efficiency analysts were furious. From the standpoint of efficiency and emissions, "those plants all should have been retired right then," says Bill Chandler, an energy economist with the Pacific Northwest National Laboratory. Yet from an economic standpoint, Chandler says, the utilities were being eminently reasonable. Investors, when they initially decided to build those plants, expected to have thirty years to recoup the investment. Had government forced early retirement of the plants before the payoff was complete, Chandler quips, "there would have been a revolution."

Still, as economically sensible as the exemption may have been at the time, it has not made sense in terms of energy efficiency or reduction in pollution. Despite the promise to retire their old coal plants after payoff, most utilities, by retrofitting from within, have managed to extend those plants' operational existence well beyond their normal thirty-year economic lifetime. (TVA's plants, for example, are an average of forty years old, and the utility has no plans to retire any of them.)

Moreover, most of these power plants have been substantially expanded. Just as TVA upgraded Kingston, other U.S. electric utilities have, by steadily improving their grandfathered fleet, almost doubled power output, and coal input, while avoiding the expense of installing scrubbing technology. In 1980, the U.S. coal fleet burned five hundred million tons of coal annually. By 2000, even though relatively few new coal plants had been built, the coal used in power generation nearly doubled, to nine hundred million tons annually. In effect, the same plants are now producing nearly twice the electricity — and, of course, twice the emissions. Today, U.S. coal-fired plants provide 320,000 megawatts of power — a little over half of total American electricity production — and of that, two-thirds is generated in unscrubbed plants.

Thus far, of course, we're talking only about "traditional" pollutants like sulfur. Scrubbers can't remove carbon dioxide; as a result, every year that a coal-fired plant has stayed in operation, every upgrade that has expanded its coal-burning capacity, has represented another increase in overall emissions. Today, coal-fired power generates more than one-half of all U.S. CO_2 emissions, and roughly one-eighth of the world's CO_2 emissions. To put it another way, had U.S. utilities been forced to replace those older power plants with, say, gas-fired generating stations, carbon emissions globally would be around 12 percent lower than they are today.

The situation is unlikely to improve. Absent any new regulation — a federal lawsuit over the retrofits has yet to be settled — utilities have absolutely no incentive to retire their plants. "These aging coal plants have a tremendous economic advantage," agrees Neville Holt, an analyst with the Electric Power Research Institute, a utility-funded research organization, told me. "To a large extent, their capital costs are paid off, so the cost of electricity is largely the cost of coal and the cost of maintenance." Because coal-fired power is so cheap — still around two cents a kilowatt-hour — utilities use their coal plants to provide base load and run them twenty-four hours a day, seven days a week. When more power is needed, a utility turns to other sources — often more expensive gas-fired electricity, which can cost four cents a kilowatt-hour. But every year these coal-fired plants can be kept running is another year of cheap, base-load power. "The economic imperative," says Holt, "would be to keep them running as long as possible." Coal is, in fact, so much cheaper than other fuels — especially cleaner-burning gas — that most forecasts have U.S. coal consumption actually rising by 25 percent between now and 2020, by which time the U.S. energy economy, ostensibly the most sophisticated in the world, will still be making 44 percent of its power from a coal-fired power sector whose core technology is more than a hundred years old.[5]

It is no surprise that climate policymakers focus so intently on American coal-fired power. Not only are these plants major CO_2 emitters, but upgrading or replacing them would yield huge gains in energy efficiency while creating new demand for alternative power sources, such as renewables or decentralized microgrids. Yet the economics make such action unappealing to investors. Building a new coal-fired "scrubbed" power plant, for example, would boost electricity costs to about 4 cents per kilowatt hour — killing coal's competitive edge, without fixing the CO_2 problem, since scrubbers remove only conventional pollutants — sulfur, nitrous oxide, and mercury. Removing CO_2 is an entirely different, and more expensive, matter. For every ton of coal burned, around three tons of CO_2 are produced — far too much to be scrubbed economically. In fact, according to studies by the Electric Power Research Institute (EPRI), if utilities were forced to capture carbon from existing coal-fired plants, technology costs would raise electrical rates to at least 7.5 cents a kilowatt-hour, at which point the older coal plants would simply no longer be worth running. "There is no economical way to control carbon emissions from existing coal plants," says the institute's Kurt Yeager.[6]

Instead, the only reasonable way to produce carbon-free power from coal is to fundamentally change the way we "burn" coal. As we saw in Chapter 8, using a so-called Integrated Gasification Combined Cycle, or IGCC, process, coal can first be refined into a synthetic gas. This syngas can be burned in a gas-fired power plant, much as natural gas is today, and the exhaust can be run through a secondary process that extracts carbon and pumps it away to be sequestered, presumably underground.

Yet the capture and sequestration process not only is unproven but is likely to be incredibly expensive. Research by EPRI and others suggests that the capital costs of replacing an existing coal-fired power plant with a new IGCC facility would raise the costs of electricity to 4.5 cents a kilowatt-hour. Bear in mind that this is simply the power-generating part of the equation. If we add the secondary process that actually captures the carbon and sequesters it (again, assuming that these technologies can be developed), we add another 2 cents per kilowatt-hour. All told, utilities would be spending about eighty to a hundred dollars to capture and sequester each ton of carbon. This would raise electricity costs to 6.5 cents, again, effectively pricing coal-fired power out of the marketplace.[7]

As a consequence, not only are most utilities profoundly disinclined to adopt any kind of cleaner, climate-friendly coal technology but, from a political standpoint, they have every incentive to fight climate policy.[8] Given that coal produces the lion's share of CO_2, and given that decarbonizing coal is not economical, any meaningful climate policy is, by definition, one that is anticoal. This is why, worldwide, the utility industry, the coal-mining industry, and essentially any industry or state with a stake in coal (China, India, and West Virginia, for instance) have been willing to spend hundreds of millions of dollars to defeat, delay, or weaken climate legislation, as well as to discredit climate science and persuade the public that climate change is mere conjecture. Every year that the coal lobby can stave off any kind of enforced CO_2 reduction is another year not simply of avoiding costs, but of avoiding costs so high they could put most of the coal-fired utility sector out of business.

These concerns are well known to politicians around the world. When German lawmakers tried enacting laws to reduce CO_2 emissions in the 1990s, they were nearly defeated by opposition from Germany's highly subsidized coal industry. In America, fear of alienating the "coal states" — Virginia, West Virginia, Kentucky, and Tennessee — helped persuade the Clinton administration to delay its support of Kyoto, and it is clearly a factor in the energy and climate policies of the Bush administration.

Climate advocates, environmentalists, and others with even a passing interest in prudent energy policy are understandably dismayed by the market-driven Realpolitik of coal-fired power, which they often point to as proof of the greed, shortsightedness, and generally wicked character of the energy companies. Yet apt as these descriptions may be, we would have been stunned if energy companies had instead embraced climate policy. In light of the overwhelming influence of asset inertia, moving to a new energy system will exact a considerable cost from some players in the existing energy economy.

To divest in any real way from fossil fuels, or at least to change how we use them, will entail enormous changes for companies that produce hydrocarbons, those which consume large quantities of hydrocarbons (such as utilities), and those whose products now burn hydrocarbons (such as automakers). True, some players will make the transition profitably — oil companies, for example, are already investing in gas and alternative-energy technologies — but many more will not. In some cases, the failure will stem from a simple lack of capacity to change: the factors that gave a company an advantage in the old energy economy — its technology, its business relationships, the expertise and experience of its work force — may simply not apply in the new energy economy. In many cases, however, companies contemplating a move to the new economy must cope not only with their own weaknesses but with the realization that the new energy economy may not be a very profitable place.

Indeed, a central obstacle to a new energy economy, over and above the fear and greed of the current stakeholders, is a profound uncertainty

about the entire economic picture. For years, alternative-energy systems —
everything from nuclear power and solar energy to hydrogen fuel cells and
clean coal, along with other, even more exotic technologies — have been
discussed as a kind of natural successor to fossil fuels, the last in a steady
progression of energy transformations. Just as wood was displaced by coal,
which was later displaced by oil and natural gas, now all fossil fuels would
give way to hydrogen, or solar, or something entirely different.

There are, however, significant differences between these earlier shifts
and what is happening today — differences that are critical to understanding why alternatives are struggling, why the existing energy companies are
so reluctant to participate, and why it will be so hard and take so many
years to move beyond oil and other hydrocarbons. In both the earlier shifts
— from wood to coal and coal to oil — the emergence of the alternative
energy was fundamentally an *economic* event. In the eighteenth century,
coal displaced wood as the dominant energy source after trees became
scarce in Europe; indeed, most Europeans resisted coal for years, because
of the smoke it produced. Two hundred years later, the dominance of
coal was broken by oil and then by gas, not because coal was scarce, but
because oil and gas had higher energy content and better handling charac-
teristics that more closely matched the needs of a maturing industrial
economy.

In other words, although each transition may have been sparked by an
accidental discovery or a single enterprising individual or a government
policy (for example, the English navy's strategic switch from coal to oil), ultimately, the shift from old fuel to new fuel was driven primarily by market
forces, and specifically by the competitive advantages of the new fuel over
those of the old. To have continued using wood or coal when your competitors were using oil or gas would have put you or your company or nation
at a serious competitive disadvantage.

This is not what is happening today. The hydrocarbon economy suffers from no direct competitive disadvantage that, say, wind turbines or solar arrays or fuel cells or some other noncarbon energy technology can easily exploit. Oil is not scarce. It will become more so in the near future, but
this risk is not yet apparent to the market. Instead, what the market "sees" is
that hydrocarbons are still vastly cheaper than alternative energy.

Hydrogen, for example, is routinely touted as the inevitable successor to gasoline because hydrogen contains more energy and burns cleanly.

But hydrogen also costs about three times as much as gasoline today, and it is not likely to become significantly cheaper anytime soon, even if a global hydrogen fueling system is deployed. According to Joan Ogden, a researcher at the University of California, Davis, and one of the leading analysts in alternative-energy economics, even when hydrogen is produced in the cheapest way — from natural gas — it still costs about $2.20 for the energy equivalent of a gallon of gas, compared with ninety-five cents for an actual gallon of gasoline. (And Ogden's studies, it should be noted, were conducted when natural gas prices were relatively low.) Making hydrogen from a renewable source, such as methanol or solar power, is even more expensive. With a cost advantage like that, gasoline has nothing to worry about: hydrogen is not an economic proposition.

Similarly, most of the machines that burn hydrocarbons, such as gasoline engines or gas-fired power generators, are much more competitive than their clean alternative challengers. Compare the gasoline-powered internal-combustion engine to the automotive fuel cell. Nearly a century of continual refinement of the ICE has created what may be the best-designed, best-engineered mechanical device in the history of the world — a machine that not only is vastly more powerful and efficient than it was even a decade ago, but each year becomes cheaper to manufacture. The fuel cell, by contrast, is still ten times as expensive as an equivalently powered ICE and is nowhere near as reliable.

Fuel cell advocates counter that these disadvantages will fade with time and research. They contend that the current cost disparity between a fuel cell and an ICE stems mainly from the low numbers of fuel cells being produced: once mass production takes over, costs will fall to the point where a fuel cell can go head to head with an ICE.

Still, researchers like Ogden, who carefully break down the costs of competing technologies, are not so certain of the fuel cell's eventual superiority. According to Ogden's research, even in the best-case scenario for the future — where fuel cell vehicles are being mass-produced at industry-standard volumes of three hundred thousand units a year in 2020 — a fuel cell car would still cost about 65 percent *more* than a car powered by a gasoline ICE — in part, Ogden reminds us, because the ICE itself will continue to improve. By 2020, she says, the "standard" internal-combustion engine will be significantly lighter, cleaner, and probably almost twice as fuel-efficient as today's version — and even more so if the auto industry moves to a gasoline-electric hybrid model. "Today's internal combustion engines average twenty-seven miles per gallon, but that's not what the fuel cell will be going up against" in 2020, says Ogden. "Fuel cells will be competing with advanced internal-combustion engines that gets forty-five miles per gallon and perhaps even more."

All told, if we compare the so-called life-cycle costs of the two vehicles — that is, if we look at the cost of manufacturing each car, plus the cost of fueling that car over its useful lifetime — the fuel cell car will be nearly 60 percent more expensive than an equivalently powered car with an internal-combustion engine — hardly an incentive. Indeed, in the context of this poor economic prognosis, it's unclear not only why any company would want to build such a car, but why any consumer would want to buy one.

In fact, the only real advantage that fuel cells and other alternative-energy technologies have over their hydrocarbon counterparts is that they emit less carbon or none at all — and as far as the market is concerned, that is no advantage. Not only are most new carbon-free energy systems vastly more expensive than the technologies they're supposed to replace, but it's hard to see how investing in them will be particularly profitable or advantageous — mainly because there is, at present, no economic *disadvantage* to emitting CO_2. Putting out a ton of carbon doesn't make you or your company less competitive or less profitable — whereas cutting CO_2 emissions almost always will, in terms of additional technology costs and lost productivity.[9] "Right now, carbon simply doesn't pose a business risk," says David Victor, the Stanford University expert on climate and energy economics. For most U.S. energy companies, Victor says, "carbon doesn't even appear on the radar screen."

In this context, carbon-free energy technology is truly a radical proposition, since it aims to attack hydrocarbon's dominant position at a time when, as far as the market is concerned, there is nothing wrong with the hydrocarbon economy.

⊘§§~

There is, of course, *plenty* wrong with the hydrocarbon economy. Among other things, burning hydrocarbons imposes terrible costs in the form of air and water pollution, smog-related illness, and above all, global damage due to climate change. In China, for example, rising urban air pollution from coal-fired power and the growing number of cars costs several billions of dollars a year in lost productivity and increased medical expenses. The projected effects of climate change — from rising sea levels to an increase in storms and droughts — are likely to be even higher. In short, every gallon of gasoline or ton of coal we burn imposes an economic cost — one that is not included in the price we pay for our energy, but which is real and, increasingly, can be measured.

Ogden, at the University of California, Davis, is one of the leaders in this emerging field of energy cost accounting. In the 1990s, curious to know whether hydrogen fuel cells could ever compete with the ICE, Ogden and her colleagues developed a clever method to calculate the hidden costs of driving a gasoline-powered vehicle — costs that, if unmasked, might seriously undermine the ICE's mantle of economic superiority and perhaps give alternatives like hydrogen a boost. Borrowing from existing research, Ogden was able to catalogue precisely how much pollution each gallon of gasoline is responsible for, from the time the oil is produced and refined to the moment it is burned in the engine — the so-called well-to-wheels analysis. Next, Ogden collected data on the various known health costs attributable to gasoline-related pollution, including medical costs, sick days, and premature death due to respiratory illness.

Combining these data sets, Ogden determined that even a super-advanced gasoline-burning car will, over its lifetime, cause an average of $1,162 in health-related damage associated with air pollution. Similar calculations for climate-related effects found that the same car will produce $846 in extra costs: everything from flood damage to crop losses due to drought. Ogden found that all told, even the best ICE car yields $2,006 in "external" costs — costs that aren't included in the retail price of gasoline or cars, but must be borne by society. By comparison, pollution and climate damage from a fuel cell car running on hydrogen made from natural gas come to just $736, or even less — $225 — if the carbon is sequestered from the natural gas. In this context, the problem is not that alternative fuels or technologies are too expensive, says Ogden, but that hydrocarbons are far, far too cheap.

The external costs of hydrocarbons aren't limited to health or climate. When we factor in the vast amounts of money that America, Europe, and other oil importers currently spend on energy security — mainly in the form of a military presence in the Middle East — the true costs of gasoline become even higher. Calculating on the basis of the amount of oil imported around the world, Ogden estimates that even an advanced car with an advanced internal-combustion engine incurs another $1,571 for the cost of keeping the U.S. Fifth Fleet in the Middle East, along with other military expenditures associated with protecting oil supplies, whereas a fuel cell car incurs no such expense.

By Ogden's calculations, if the price of gasoline truly reflected all these external costs, it would add at least a dollar to the pretax price of a gallon, bringing it to around $2.00, compared with hydrogen's cost of $2.21 to $2.46. Factor in the fuel cell's greater fuel efficiency, and suddenly, the fuel cell car is about 25 percent cheaper than today's gasoline-powered cars, and just slightly more expensive than the advanced ICE we might see in two decades. When the total costs of gasoline are considered, the fuel cell car is finally within striking distance of being competitive with an ICE car.

The issue of hidden costs is not unique to energy. For decades, the retail price of a pack of cigarettes never included the health costs that the average smoker incurred. Instead, the price of cigarettes reflected only traditional, or "internal," costs, like manufacturing, tobacco, and marketing. As the external costs from smoking, such as medical expenses, grew larger, governments sought to "internalize" those costs, mainly by taxing cigarettes heavily and using the proceeds in part to pay medical expenses.

This idea of internalizing hidden costs hasn't quite hit the energy business. The costs of pollution, climate damage, respiratory illnesses, and so on, are still not reflected in the price of a gallon of gasoline. Instead, these cost are paid by "society," in the form of higher medical insurance premiums or higher taxes for defense budgets. Because motorists don't pay these costs directly, at the gasoline pump, they have no incentive to make different energy decisions — to use less gasoline, for example, or switch to a cleaner alternative fuel. External energy costs, in other words, are costs that the normally very efficient market cannot see and thus cannot allocate correctly.

If, however, we could internalize these hidden energy costs for the ICE, much as we have done with cigarette taxes, things would change considerably. If carbon were to become an expense — a cost, like materials or labor, that a company, a country, or even an individual consumer sought to avoid — the energy economy would be transformed. Suddenly, every gallon of gasoline, every barrel of oil, every ton of coal would carry an economic penalty, with the result that we would either try to use less of them, through conservation or more efficient technologies, or find some other fuel or energy technology that produced less carbon. After nearly a century of running blind, the global energy industry would have some hint, some signal how it should treat carbon.

This is why, for decades now, energy experts across the political spectrum have been calling for some form of tax on each ton of carbon emitted — a penalty that reflects the damage that carbon does through global warming. In most such proposals, fuels would be taxed according to their carbon content, with coal being penalized the most and gas the least — an effective acknowledgment that the "cheap" cost of hydrocarbons has come about because the market was never forced to recognize the hidden economic impact of fossil fuels.

This idea is not new. In the United States, coal-fired power plants already pay a penalty for each ton of sulfur dioxide they emit — a requirement that has dramatically reduced sulfur emissions and the acid rain they cause. A similar system for carbon would be even more transformative. As carbon began to represent a cost to be avoided, consumers, companies, and entire industries would shift their business strategies, investment patterns, and technology programs to minimize carbon consumption and emissions. A carbon tax would rectify the myriad perverse incentives that today not only encourage wasteful building, driving, and other inefficiencies, but give hydrocarbons an advantage over other energy technologies, such as hydrogen or renewables. If, for example, it cost a hundred dollars for every ton of carbon emitted, utilities might find that their older coal-fired power plants were no longer such a bargain and that all at once a portfolio of renewables, gas, and IGCC carbon capture was looking cost-effective. Consumption patterns would shift dramatically: as the price of gasoline or coal-fired power rose to reflect carbon capture, consumers and businesses would move toward more efficient cars and appliances.

Stanford's Victor, for example, believes even a modest carbon tax — starting at ten dollars per ton, then growing each year predictably — would be sufficient to alert industry that a new variable was entering the market. The initial cost would be low enough to avoid causing emitters severe economic pain. Yet the "signal" to emitters would be firm enough to set market forces in motion, encouraging companies to rethink investments and schedules for retiring assets. Others have suggested starting the tax at only around ten dollars a ton and holding that level for fifteen or twenty years, before raising the final cost to around eighty to a hundred dollars per ton. "You'd keep the cap 'soft' for a long introductory period, then ramp up very dramatically at a certain date," explains Reid Detchon, a former Energy Department official who now crafts climate policy at the Energy Future Coalition. "This allows companies to adjust their capital stock replacement to prepare at minimal cost."[10]

At the same time, a tax would help create demand for carbon-free energy technologies — everything from solar and wind to processes for capturing and sequestering carbon. Observes Victor: "Even a small signal that we're going to have some limits on carbon in the future will, if it's credible, release a lot of innovation." Such a tax would not be difficult to implement: most countries already tax fuels. The cost to business and consumers could be offset by the reduction or elimination of some other tax. (U.S. analysts have suggested eliminating the capital gains tax, the idea being that we should be taxing the things we want less of, like carbon, not those we want more of, like capital investment.)

How high the tax should be allowed to rise — and proposals range from eighty to two hundred dollars per ton — is the subject of strenuous debate, since we don't yet know how much it will cost either to prevent climate change or to cope with the unavoidable damage. Many analysts believe that since we can't know the actual costs associated with climate change, we'll never be able to set an accurate carbon tax. "You are valuing the carbon tax based on an estimate, a best guess, of what you think emitting that carbon into the atmosphere will cost, in terms of climate impact," says Jim MacKenzie at the World Resources Institute. "If you make the value too high, you destroy the economy. If you make it too little, then you haven't made the penalty high enough to discourage polluters and curb emissions before they cause damage to the climate."

Many climate experts argue that taxing carbon may ultimately be too complex and expensive. An alternative approach, being implemented in the European Union under the Kyoto commitment, is to create a so-called cap-and-trade system. Under this system, governments set an overall emissions cap for a country or an industry or even an individual company. A given utility might be told it could emit a total of no more than 120 million tons of carbon between now and 2012. If, as 2012 approached, the utility realized it was going to exceed its carbon budget, it could buy carbon credits from other utilities or other energy companies or even other governments that had reduced their emissions more quickly and thus had surplus credits.

Such a cap-and-trade system is already being used in the United States to reduce sulfur emissions: companies can choose either to cut sulfur emissions themselves or to buy emissions credits from other companies.

Each credit might be good for one ton of carbon emissions, and the price would vary, depending on how many credits were available. Early in the program, when all utilities had many tons of carbon left in their carbon budgets, credits would be fairly cheap, since few utilities would need to buy them. As time went on, though, and as individual utilities began to use up their carbon budgets, demand for credits would rise, as would price. And here is where the market mechanism comes into play. As the price per credit rose, companies would face a crucial financial question: Should we purchase the credit and go ahead and emit a ton of carbon, or has the credit price gotten so high that we are better off simply paying for new technology that cuts our emissions? Should we install some clean technology now and hold some credits till later? Or should we use all our credits now, and then, at the last minute, upgrade to cleaner equipment? In one scenario, a utility might build an IGCC coal-gasification power plant but delay installing the more expensive equipment for carbon capture and sequestration until the cost of buying a carbon credit became more expensive than the cost of capturing and sequestering the carbon. In other words, companies could choose when to move into a cleaner energy regime; but the market would ensure that, eventually, they *would* move.

Carbon trading is in its infancy. In Europe, governments are launching a cap-and-trade system in 2005, starting with the heavily emitting corporations. A great many questions still need to be answered, about how the law will be enforced and how much each carbon credit should cost. Already, though, the European business community is responding pretty much as expected to the appearance of the new cost. A carbon emissions market has emerged, complete with emissions brokers and daily price updates. Bankers, investors, and market analysts have begun assessing utilities and other emitters on the basis of their carbon strategies, and the companies themselves have begun including the carbon costs in their long-term plans and are already making critical investments with an eye toward lowering carbon.

Ultimately, what a carbon-trading system reflects is the understanding that markets aren't perfect. Indeed, were markets allowed to dictate entirely the shape of the new energy economy, industrialized nations would use oil until it ran out or became too costly and then, to avoid replacing the sprawling hydrocarbon fueling infrastructure, would simply move to an economy based on synthetic fuels distilled from coal, tar sands, or other carbon-dense stock — regardless of the impact that might have on climate. In the developing world, meanwhile, coal, not gas or even oil, would become the fuel of choice and, worse, it would not be burned cleanly. As far as carbon-free renewables technologies were concerned, a market-driven energy economy would relegate any nontraditional energy technologies to the margins.

In other words, much as we will depend on market forces to build our next energy economy, the market's "invisible hand" will need a little guidance — mainly in the form of political action to assign a cost to carbon emissions. In Europe, governments have already taken this action, and the energy industry is already responding with tentative moves to reduce emissions. It remains to be seen, however, whether such an idea can cross the Atlantic and overcome the traditional American disdain for paying for pollution, something that has been free for centuries. In the mid-1990s, an attempt by the Clinton administration to enact a minuscule four-cent-a-gallon energy tax, in hopes of gradually phasing in a carbon-reduction regime, caused a revolt in Congress so severe that Clinton essentially shelved the idea, and the Bush administration seems even less inclined to grapple with it. True, during the months before the September 11 attacks, the White House is said to have been considering a carbon "registry," where utilities and other carbon-emitting companies would establish an emissions baseline, or a "share" of a carbon budget, from which a cap-and-trade system would be built. The White House abandoned that plan, however, in the interests of avoiding political conflicts with coal states, and it shows little signs of resurfacing.

Interestingly, in spite of political inaction, a shift toward this new low-carbon economy may be occurring anyway within the United States–based energy industry. Many multinational companies that must do business in Europe now face the prospect of having to cope with two fundamentally different markets — one in the European Union that penalizes carbon and one in the United States that doesn't — and may lobby Washington to reconsider its carbon policy. Oddly enough, some of the strongest pressure is likely to come from American power companies. Despite all the resistance, all power companies badly want to know whether their aging "fleet" of power plants, many of them coal-fired, can simply be updated or need to be replaced altogether with plants that emit no carbon. Such companies aren't necessary morally superior. They are, however, looking to position themselves, with regard to investment, acquisition, and technology, to survive the transition to the next energy economy — not only to minimize their costs, but perhaps even to profit under the new green regime. In other words, the attempt to deal constructively with climate could, in some small way, be evolving into an economic proposition.

"The utilities will never admit this in public," says one climate analyst who has worked closely with the power sector, "but if you talk one on one to senior guys from the power industry and you ask them whether they think that at some point in the next five to ten years there will be a significant limit on carbon, they will all say yes. They know this is coming, and they are investing in little clean technology things on the margins. But until they see what the limit will be, what the carbon market actually is, they can't move."

12

DIGGING IN
OUR HEELS

IN SEPTEMBER 2002, a small army of diplomats, business leaders, and advocates of various stripes converged on a conference center in Johannesburg, South Africa, for what some were billing as a first step toward a new energy economy. The occasion was the United Nations World Summit on Sustainable Development, a ten-day policy marathon on alleviating third-world poverty, and much of the agenda was to focus on the role that energy, and particularly "sustainable" energy, might play in moving developing countries into the modern age. Alternative-energy technologies had top billing. Between sessions on new energy systems, delegates could watch virtual demonstrations of renewable energy facilities and ogle a prototype of BMW's hydrogen-powered 735i. But the political highlight was to be a proposal by the United Kingdom, Germany, and other big European states calling for a commitment by all nations to boost renewable energy's share of the global market to 15 percent by 2010.

This was an incredibly ambitious idea. Given that nonhydro renewables now account for less than 2 percent of the world's energy, getting anywhere close to 15 percent by 2010 would require vast sums of money and technology — most of it from the wealthy developed nations — and an aggressive program for implementation. Yet as far as European delegates were concerned, such ambitions were warranted: trying to provide energy to the burgeoning developing world by using only fossil fuels would not only push the existing energy infrastructure to its functional limits but undermine any hope of reducing CO_2 emissions in time to avoid catastrophic global warming. "Solving the climate change challenge," U.N. officials had bluntly declared, "means reducing global dependence on fossil fuels."

Not everyone at Johannesburg was pleased by such talk. Whereas European states would gain from such a commitment to renewables (most wind-powered technology, for example, is manufactured in Europe), other countries saw the proposal as an economic disaster. Diplomats from OPEC oil states, for example, made it clear that they regarded any measure to boost renewable energy as a direct threat to their own market share. Delegates from developing nations, meanwhile, complained that such a commitment might reduce available supplies of cheap hydrocarbon energy, which many poorer nations now depend on.

Some of the most strenuous opposition, however, came from the United States. American negotiators were justifiably concerned about the aggressive deadline for a 15 percent renewable share: eight years was too short a time to expand renewable energy so broadly. But it was also plain that the Bush administration feared how a commitment to renewables would play politically with two key constituencies back home: political conservatives and the U.S energy industry.

American conservatives have never been keen on commitments to foreign aid, especially when aid is tied to some dubious left-wing notion of "green" energy. Energy companies, meanwhile, dislike being pushed to sell renewable energy, which tends to be less profitable for them than hydrocarbon energy. More to the point, many U.S. energy companies and utilities (some of which happened to be major contributors to President Bush's election campaign) see the third world as a potential market for traditional oil and gas projects and large-scale power plants and have no interest in seeing American foreign aid spent instead to build renewable "off-the-grid" energy systems. As one U.S. energy analyst put it, "this is not an administration that wants to spend foreign aid dollars for energy on anything other than oil and gas development."

As summit negotiations wore on, backers of the renewables provision became aware that Washington was lining up allies to quash the proposal. Among them, Canada, Japan, and, to no one's surprise, OPEC, which, despite generally tepid relations with the United States, had not been pleased by such phrases as "reducing global dependence on fossil fuels." As the days went by, delegates from Saudi Arabia and Venezuela could be seen slipping into meetings with U.S. officials and later, in what one critic called "the axis of oil," joined the Americans in a threat to scuttle the entire summit unless the renewables goals were dropped. Desperate to salvage some agreement on world poverty, the Europeans caved in. "We tried until the very end," Yvon Slingenberg, a senior negotiator for the European Union, wearily told the press. "At a certain stage there is a point of exhaustion."[1]

The exhausted Slingenberg could just as well have been describing any part of a long and rancorous political debate over the future of the energy economy. Energy is not the only critical challenge humans face. Yet because energy connects so centrally to everything else that is important to us, overhauling our energy systems is guaranteed to be among the most challenging political tasks of the twenty-first century. Even if every country and company agreed on what the new energy economy should look like, building it would be enormously daunting. As it is, such consensus is almost impossible to imagine. As events like the Johannesburg conference demonstrate, little agreement exists about kinds of energy technologies or policies we should pursue or how they should be paid for or how quickly they should be rolled out — or even, in some cases, whether there is a need to change anything all. The OPEC cartel is hardly the only energy player with a desire to freeze the world in a perpetual state of hydrocarbon dependence: nearly every participant in the modern energy economy, from individual consumers to multinational oil companies to superpowers, is so deeply vested in the status quo that any fundamental change poses enormous political and economic risks.

These risks, and the ways countries and companies try to manage them, will define the coming political battles over energy. Some players will seek to avoid risk altogether by obstructing or delaying changes to the energy system. Others will try to control the pace and direction of change — or exploit the desperation of other players — in such a way as to profit from the new energy order. Whatever the particular strategy, all participants can be counted on to influence the process every step of the way. Ultimately, it is this complex amalgam of competing self-interests that will rule the coming energy transformation and determine everything from the way we describe the problems with the current system to the kinds of solutions we pursue to the timing of the change itself. This may be the most important aspect of the coming energy revolution. Although the transformation of our energy system has been set in motion by such "hard" factors as declining oil production, energy security, climate change, and the availability of new energy technologies, where the transformation takes us will be largely a question of politics.

The world map of energy politics is dominated by five major players, each with its own agenda and, more to the point, its own role in the building of the next energy economy. The largest and most volatile player is the developing world — mainly the nations of Africa, Asia, and South America. It is a highly diverse group, ranging from severely impoverished nations like Nepal and Sudan, which will have little influence over the evolving energy system but will depend upon it entirely, to emerging giants like India and China, which will soon be the world's largest energy consumers and whose appetites are already warping the politics of supply and demand.

Because developing nations are poor, their governments make energy choices mainly on a "least-cost" basis, giving priority to such short-term objectives as energy security, while marginalizing such longer-term concerns as air quality, water pollution, or CO_2 emissions. As a Brazilian diplomat noted at a recent conference on poverty, "we are not as concerned about putting food on the table seventy-five years from now as we are about simply putting it on the table today." The developing world looks at energy in the same way.

Politically, the energy poverty of the developing world will influence the transformation of the energy economy in important ways. Because developing nations currently have little choice but to use the cheapest energy available (coal, in China and India), they regard policies to reduce carbon emissions as undercutting their own efforts to escape energy poverty and to modernize. By the same token, because developing countries rely on "dirty" energy, any success they have in achieving economic growth will doom global efforts to shift toward cleaner energy. This implicit threat gives developing nations a surprising measure of power over such global energy decisions as climate policy. Countries like China, India, and even Russia, with its obsolete and energy-intensive industrial base, will refuse to support global initiatives like CO_2 reduction unless wealthy developed nations promise financial and technological aid. But developing countries will also become political pawns as industrialized nations — mainly the

United States and Europe — maneuver for advantage on such issues as climate policy.

The next major player on the map of energy politics is Europe. The member states of the European Union have emerged as the collective leaders in the push for a new energy economy, particularly in the areas of climate policy and alternative energy. This is a sign of neither altruism nor moral superiority. In Europe (as well as in Japan), decades of dependence on imported energy have enforced a culture of conservation (including a historic tolerance for tiny, energy-efficient cars and high energy taxes) and helped support the shift toward renewable-energy sources. Countries like Germany, Denmark, and the United Kingdom, as well as Japan, are the fastest-growing markets for renewable energy and are establishing themselves as leaders in manufacturing clean-energy technology, especially wind power. Japan, meanwhile, is now the undisputed leader in solar technology. This is one reason many developing nations regard European proposals for "green" energy with considerable skepticism. As one European energy consultant told me, "developing countries are never sure whether a renewable policy is best for their environment, or simply best for the economy of some developed country."

On global energy policy, Europe finds itself playing the political broker between the developing world (including Russia), with its unrestrained financial needs and rising political acumen, and the United States, which sees energy policy, and especially climate policy, as an economic weapon that Europeans and developed nations are using against Americans. (This is less the case with Japan, which tends to follow Washington's lead on energy issues.) Europe's position as middleman has added a new twist to an already complex transatlantic energy relationship. The United States and Europe remain highly dependent on imported oil and still coordinate efforts to stabilize world prices; however, Europe has stopped waiting for the United States to join in on climate policy and is implementing programs to reduce emissions on its own, though it has little hope of bringing developing nations aboard without American assistance.

Third on the map of energy politics are the energy producers themselves. This group includes the big international energy companies: OPEC and other oil-producing states; the coal industry; the power sector; and the massive pyramid of financial, trading, and other ancillary services that support the energy industry. The producers are mainly invested in a hydrocarbon energy economy — oil, gas, and coal. They are thus heavily biased toward an energy economy in which hydrocarbons remain dominant, and they have the political influence to fight for that outcome.

This influence has several sources. World energy production, especially in oil and gas, is controlled by a relatively small number of oil states and companies. The top six oil producers in the world — Saudi Aramco, the National Iranian Oil Company, Mexico's PEMEX, Venezuela's PdVSA, ExxonMobil, and Shell — together control nearly one of every three barrels of oil consumed on the planet.[2] This concentration of wealth and power means that, when confronted by unfavorable laws, a competing energy technology, or even a threatening idea (climate change, for example), producers collectively bring to bear enormous financial, political, and even diplomatic resources to defend themselves. Such collectivism is most evident in the oil sector. Shell and ExxonMobil, for example, may compete brutally for market share, but they have in common concerns about climate policy or energy regulations that could affect oil consumption, and they have, in times past, pooled their considerable lobbying power to delay or defeat those policies. As one economist who advises the U.S. Energy Department put it, "a few CEOs represent most of the non-OPEC production, which means that oil can make its voice be heard in Washington or anywhere else much more easily than, say, conservation, or any of the new energy technologies, or even natural gas."

Because energy is critical to national power, producers have traditionally enjoyed close ties to national governments, and thus have been able to shape national energy policies. Whereas the United States might otherwise regard Saudi Arabia, with its anti-Western attitudes and its links to terrorist elements, as a legitimate political enemy, the kingdom's vast oil reserves and especially its enormous surplus capacity have for decades ensured that Washington would overlook such criminal behavior. As one political analyst told me, "the fact that a U.S. president can call up the Saudis and say, 'Something major is going to happen tomorrow and we desperately need you to pump more oil to reassure the market' has given the Saudis a level of access in Washington that is pretty much unparalleled."

Such influence isn't likely to diminish anytime soon. Because hydrocarbons will play a central role in any transitional energy economy, and because producers alone have the capital and resources to build the next energy infrastructure, they will also have considerable say in when and how quickly we move to a new energy economy.[3] In fact, many oil-producing states have worked assiduously to prevent any change, either by trying to keep prices low (not always successfully) or by attacking competing energy sources. The Saudis, for example, have gone so far as to file complaints with the World Trade Organization claiming that European programs to cut CO_2 emissions unfairly constrain the Saudi oil trade. "We are against any policy that unfairly discriminates against oil," one top Saudi oil official told me bluntly.[4] "We want to keep oil the fuel of choice."

After the producers — and often in direct opposition to them — come the advocates, the diverse community of activist groups, nongovernmental organizations (NGOs), and international agencies that are dedicated to changing some aspect of the existing energy order. Ranging from environmental groups like Greenpeace and Friends of the Earth to think tanks specializing in energy security and even oil depletion (the Association for the Study of Peak Oil & Gas, for example) to "official" organizations such as the U.N.'s Intergovernmental Panel on Climate Change, the advocates exert a moderate but steady force for change in the energy economy — and they are often the only constituency willing to take on the producers.

Lacking true economic power, advocates rely instead on persuasion and coercion and often browbeat other energy players into taking action. This has been critical in getting some producers, like BP, to change their policies and practices, but it has led to excesses. Too many advocacy groups, in choosing issues for their publicity potential, exaggerate various energy and environmental calamities and ignore the economic realities of the new energy economy. In the climate debate, for example, environmental groups are among the strongest advocates for making ultradeep cuts in CO_2 emissions quickly — even though this approach may be so costly that it ultimately defeats longer-term efforts to reduce emissions. Villains are also often an essential element in most advocates' rhetoric. Some U.S. and European advocates, for example, say the only reason we don't have a hydrogen economy now is that automakers and oil companies are wicked and greedy — but these claims conveniently omit any mention of the huge financial risks and engineering uncertainties inherent in shifting to a hydrogen economy. In the same way, some NGOs, in criticizing developing countries for using fossil fuels instead of renewables, ignore the expense of renewables and fail to acknowledge that often the quickest way to alleviate energy poverty is not with a solar panel but with a truckload of stove oil. Notes one expert on world poverty, "Most of the NGOs come from the north — Europe and the U.S. — where energy issues look a lot different, and a lot simpler."

Yet without advocates and the sense of alarm they bring to the debate, our energy economy would be in far worse shape, with far less hope for transformation. Not all advocates play fast and loose with facts. Many do their work the old-fashioned way, through carefully researched analysis of energy problems and sensible solutions, and have played a critical role in influencing government policymakers and even energy companies to change course. The United States–based Natural Resources Defense Council, often one of the shrillest voices in Green politics, has been one of the staunchest advocates for the eminently sensible cap-and-trade system for pollution trading.

Last and certainly not least on the map of energy politics is the United States. As we have seen, American prowess in both energy consumption and CO_2 emissions is second to none, and the U.S. role as self-appointed policeman of global energy markets is beyond dispute. What matters equally, however, is the enormous ability the United States has to influence change in the global energy system. The giant U.S. car market, for example, could be a catalyst for a cleaner auto industry. Likewise, even a small move by the United States toward improved energy efficiency in the American power sector could set off a revolution that would utterly remake global energy politics. Undoubtedly, with its unrivaled economic muscle and technological capabilities, the United States could anchor any international initiative to reduce CO_2 emissions, while using its vast political and diplomatic influence to help ensure that other nations stuck to their reduction goals.

Nonetheless, the sword cuts both ways. The United States is heavily invested in the hydrocarbon energy economy: it not only requires vast amounts of oil, gas, and coal for its own economy but derives much of its wealth from a global economy that is even more dependent on fossil fuels. Thus, just as any threat to global energy security is a threat to American economic and political power, any effort to move away from a hydrocarbon energy system — or worse, to use less energy — poses alarming economic and political risks to the United States — a reality that tends to reinforce the status quo. This consideration helps explain why the United States has doggedly remained an ally with oil-rich Saudi Arabia, despite mounting evidence that top Saudi officials have been involved in anti-U.S. actions.

Adding to the political inertia, American consumers and businesses, accustomed to decades of the cheapest energy in the world, have traditionally shown little support for the kinds of energy policies that are standard practice in Europe and Japan. As one European energy advocate puts it, "I think it's very difficult for most Americans to even imagine what a different kind of energy system might look like."

Just after dark on a chilly December evening, the wind picks up off the Baltic Sea and begins turning the blades of Brar Riewerts' two-hundred-kilowatt windmill. With each rotation, the wiry northern German farmer earns a few fractions of a cent from his local utility, while Germany moves a few electrons closer to the kind of clean-energy economy that advocates in Johannesburg believe could one day power the world.

In most energy markets, Riewerts' single wind tower couldn't compete with enormous coal-fired generators and the nuclear power plants, or even be allowed to feed into the grid. Since 1990, however, when German energy politics underwent a reformation, utilities here have been required to buy power from anyone who produces it, while the German government has paid wind farmers like Riewerts a hefty subsidy — around seven cents per kilowatt-hour — and has even financed construction of wind towers.

Such inducements have touched off a boom in the German wind market. Big energy companies and utilities are investing heavily in wind, and thousands of German farmers can now make a tidy profit on every kilowatt of electricity they sell to the grid. Riewerts, whose weather-beaten face betrays not so much the zeal of a clean-energy advocate as the anxieties of a struggling small farmer, regards the environmental contribution of his wind farm as a secondary benefit. "What does a farmer count these days?" he asks me, shouting over the steady *chuff-chuff* of the ice-covered rotors. "It used to be, 'one cow, two cows, three cows.' But since we got wind power, it's 'one euro, two euros, three euros.'"

If one were searching for a model for a new kind of energy politics, Germany would offer a reasonable starting point. Since 1990, Europe's eco-

nomic muscleman has been transforming an energy economy once dominated by coal and nuclear power into one where efficiency and renewable
technologies play a rapidly growing role. Today, the number of German
wind farms, solar systems, biofuel power plants, and other new energy systems is climbing so quickly that the country has already blown past its own
growth objectives and by 2010 will be producing nearly one-seventh of its
electricity without carbon emissions. Germany "is the locomotive pulling
renewable-energy development in Europe," boasts Arthouros Zervos, president of the European Renewable Energy Council, and the sentiment is
widely shared. Today, European companies, many of them German, dominate the world wind turbine market — a booming business that employs
more than two hundred thousand people and boasts annual sales of nearly
twelve billion dollars.[5]

What makes the German experience so useful is the degree to which it
demonstrates the capacity for political change. Before 1990, Germany was
the antithesis of a renewable-energy economy. German energy politics had
been ruled for decades by coal and nuclear companies, and the powerful
coal miners' unions. Utilities had a total monopoly over the power grid and
prohibited alternative-energy producers from selling their power, while the
very idea of "alternative" anything generated little interest among Germany's socially conservative majority.

All that began to change in the late 1980s. The 1986 Chernobyl nuclear
accident galvanized antinuclear opposition in Germany. New air-pollution
laws and declining domestic coal production made coal-fired power increasingly unattractive. Concerns about climate change had also begun to
resonate. German lawmakers, goaded by newly energized environmental
groups, had little choice but to cut the nation's reliance on nuclear power
and hydrocarbons — mainly through renewables and greater energy efficiency. In 1990, Germany passed the first of several laws to encourage carbon-free energy production.[6] Dubbed the Electricity Feed-in Law (EFL), it
required utilities to buy power from any renewables producer, no matter
how small. With the EFL in place, small investors began building tiny wind
power systems in Germany's windy north.

Germany's traditional power producers were so unconcerned by the
threat of renewables that they didn't oppose the law. But as wind farms began popping up across the German landscape, so did opposition, especially
from northern electrical utilities, which had to bear most of the costs of

connecting their grids to these widely scattered producers. In 1997, utilities and political conservatives amended the feed-in law to make sure that the share of renewable power could rise no higher than 5 percent of the total power market. With few prospects for growth, investors lost interest, and the nascent renewables industry stumbled.

In 1999, German energy politics changed again. The environmentalist Green party won enough seats in parliament to form a coalition government with Gerhard Schroeder's moderate Social Democrats. In 2000, the so-called red-green coalition government enacted a Renewable Energy Law (REL), repealing the 5 percent cap and offering generous subsidies to renewable energy producers, to be financed by a small tax on all energy sales. Utilities, arguing that the subsidies violated European Union free trade requirements, challenged the REL. The German government, however, countered that subsidies were intended simply to "internalize" the long-unpaid external costs of coal-fired and other polluting energy. The law stood.

With its profits all but guaranteed, the German renewables industry has been growing as quickly as did the cell phone market in the early 1990s. Each year, German wind-generating capacity is expanding by three thousand megawatts, or about 2 percent of total electricity use. That wind energy is coming into use faster than demand for electricity is growing means that wind energy is slowly but surely displacing fossil fuels.[7]

Other renewables are also growing rapidly. Biomass facilities — small, locally operated power plants that burn biofuels made from crop waste, wood chips, and other plant-based fuel — are being built so quickly that Germany is exhausting its supplies of crop wastes and will need to encourage farmers to grow fuel crops. Solar technology, though well behind wind and biomass, is also charging ahead. In 2004, Germany had 400 megawatts of solar capacity — more than the United States — and, more important, the increased sales have helped drive down photovoltaic production costs by more than a third. All told, says Volker Oschmann, a renewables expert with the federal Ministry for the Environment, Nature, and Nuclear Safety, Germany's renewables market is outpacing even optimistic projections. "Officially, the goal is to double the proportion provided by renewables by 2010," Oschmann told me, "but we are on track to reach that much sooner." By 2050, Germany intends to be producing fully half of its electricity from renewables technology.

German optimism goes beyond sheer numbers. The growing renewables industry is sending ripples throughout the German energy economy, encouraging new technologies, business ventures, and investment and, just as important, spawning new political alliances. The German wind industry is positioning itself as a major technology exporter, gaining increased acceptance among Germany's political powerful business community. Similarly, the rapid expansion of the biofuels market has created huge demand for specialized fuel crops at a time when the politically influential German farming sector is desperate for a new market. The result: political partnerships that can help the renewables industry counter the considerable resistance of traditional energy producers. Says Oschmann, "This is a perfect chance to gain partners for a renewables energy policy: politically powerful partners who wouldn't normally ally themselves with the Greens."

For all its apparent success, however, Germany's energy revolution faces serious challenges. Energy advocates worry that the success of renewable energy rests too heavily on its fashionableness among middle-class Germans who may lose interest after a year or two. Felix Christian Matthes, a veteran energy analyst with the Institute for Applied Ecology, says that rooftop solar panel installations have been growing at 75 percent a year in part because the equipment has become a status symbol. "It's mainly a case of image maintenance," says Matthes.[8] "If your third BMW isn't providing enough image growth, you have to do solar." And consumer energy attitudes can change. German researchers have been disturbed to discover that Germans are losing their tolerance for energy efficiency. As in America, the new popularity of larger cars and more energy-intensive homes has offset most of the gains made in reducing energy demand through improved efficiency.[9]

Political change could also shift the fortunes of renewable energy. An election loss by the Green party, for example, could leave the laws on renewables vulnerable to attack by conservative lawmakers, many of whom believe that German subsidies for renewables are far too generous. Even beyond such immediate challenges, advocates of a new German energy economy must confront the more basic political challenge of pushing renewables from their niche into the mainstream. The current growth rates of 50 to 75 percent a year are unsustainable, and as a result the country will face the classic twenty-first-century dilemma: how to meet growing energy needs while honoring Germany's commitment to reduce CO_2 emissions.

Many German policymakers are urging a move to gas-fired power as a transitional step, much as is happening in the United States and elsewhere. But some German energy experts fear that a major push toward gas will simply commit the nation and the industry to billions of dollars in infrastructure that takes fifty years to pay off and yet is obsolete after twenty. Although a gas economy would be considerably cleaner than a coal economy, Matthes says, it would still leave Germany's CO_2 emissions above the nation's Kyoto carbon budget.

Instead, says Daniel Becker, who is with the German Energy Agency, maybe Germany should consider a more cost-effective approach, such as continuing to use "coal power for ten to fifteen years, until we see what technology develops. We need to wait for a truly renewable technology to develop."[10] What that technology will be is far from clear. Some German energy advocates are betting on biomass and envision a large-scale farming industry growing nothing but fuel crops. Others are hoping for breakthroughs in nuclear technology that would improve the economics and safety of nuclear power plants.

Interestingly, support for a hydrogen economy here seems lukewarm, quite in contrast to the attitude in the United States. Although German automakers were among the first to demonstrate fuel cell cars, many policymakers here believe that fuels cells remain decades away from commercial viability, and some experts are even less hopeful about a new hydrogen fueling system. Instead, they want the German government to devote its research dollars to exploring energy technologies that can be deployed sooner, like biofuels, or even clean coal. In a pointed reference to the questionable hydrogen enthusiasms of U.S. president George Bush, many energy advocates here say that hydrogen has become political cover for energy companies. As Hermann Scheer, a Social Democratic politician and leading advocate of alternative energy, told me, oil companies "support hydrogen because it gives them time to work through their existing assets."

⟨₰⟩

Germany's successes in energy are often held up as proof of what a modern industrial economy is capable of. In most cases, the "modern industrial economy" being referred to is the United States, regularly pilloried for regressive energy policies and a refusal to address climate concerns. Such criticisms aren't entirely fair — Germany's aggressive commitments to CO_2 reductions, for example, have as much to do with that country's flat economic growth and declining coal industry as with any inherent Germanic altruism.[11] Yet the progressive policies of Germany and other European countries, as well as Japan, do tend to throw the shortcomings of American energy policy into harsher relief and, in the end, help explain why global energy policy has been so slow to evolve.

In theory, American energy policy harnesses the forces of a largely free energy market to ensure sufficient long-term energy supplies in a manner compatible with the nation's other social and environmental goals. In practice, American energy policy is incoherent and fragmented, without anything resembling a long-term strategy. In writing energy legislation, American lawmakers tend to be parochial, as interested in rewarding, or punishing, various states, regions, or industries, as in advancing some overarching national energy strategy. The resulting energy laws are frequently wish lists aimed at protecting regional interests, such as those of oil producers in Texas or Alaska or coal-mining companies in the East and in Wyoming or the big utilities in the Midwest and South, or ethanol producers in the Corn Belt, or the political interests of a particular lawmaker or committee chairperson.

If there are any unifying themes in American energy policy, they are the steady move away from the heavy regulation of the twentieth century and a steady movement toward greater supply. These have had the mostly beneficial impact of keeping energy prices lower than in more regulated economies, such as Europe's, but have also fostered an environment that gives the energy industry great influence over energy laws and policies. For example, U.S. lawmakers and presidents have historically favored policies that promote production of conventional energy sources — not just because vibrant industrial economies always need energy, but because energy producers make large campaign contributions. Since 1990, the oil and gas industry has given more than $159 million to American politicians; of that, 73 percent has gone to Republican candidates, who, not surprisingly, tend most often to side with the industry. In the 2000 election cycle alone, oil and gas companies gave $34 million, more than three-quarters of which went to Republicans.[12] By contrast, there is no industry built around using *less* energy, and thus few campaign contributions flow from backers of efficiency. And while the United States has a renewables industry, it is nowhere near as large as Europe's, and hardly in a position yet to play the political-contributions game.

Exactly how the steady flow of hydrocarbon campaign money affects U.S. energy policy is hard to know, but it's probably no coincidence that American energy laws and policies tend to favor not only increased production generally, but production of older, more traditional energy, as opposed to alternatives. This is one reason that the ancient American fleet of polluting coal-fired power plants was exempted from clean-air legislation: dozens of U.S. senators and representatives had power plants or coal mines in their states and districts and feared anything that might hurt coal and utility jobs — or stanch the flow of campaign contributions from coal and utility interests. This may also explain why the Republican-controlled House of Representatives refused in the fall of 2003 to approve a law that would have required that renewable energy make up a modest 10 percent of the national energy mix by 2020.

The steady flood of campaign contributions has also helped produce another industry advantage: a federal tax code that favors traditional energy companies and subsidizes hydrocarbon production. The so-called depletion allowance, for example, reduces the federal taxes an oil company pays, because in theory the company's assets are being exhausted with each barrel produced. Oil companies can also deduct most of their so-called intangible drilling costs, such as labor, materials, repairs, and supplies. All told, whereas the tax rate for non-oil industries is 18 percent on average, the oil industry is effectively taxed at just 11 percent, a sweet deal that amounted to tax savings for oil companies of $1.5 billion in 2000 and more than $140 billion since 1968.[13]

Nowhere has this reciprocal energy politics shown up more clearly than in the yearly fight over automotive fuel efficiency standards. By any reasonable standard, the most important step the United States could take to simultaneously improve energy security, cut CO_2 emission, boost urban air quality, and deprive Middle Eastern terrorists of financing would be to raise fuel efficiency requirements. American cars and trucks burn two of every three barrels of oil used in the United States — and one of every seven barrels used worldwide — a figure that is hardly surprising, given that economy standards have been frozen since 1988. Today, American cars need to achieve an average fuel economy of just 27.5 miles per gallon, while "light trucks," that hugely popular category that includes pickups and SUVs, need achieve only 20.5 miles per gallon. Even a modest improvement in fuel-economy standards — say, thirty-two miles per gallon for cars and twenty-four miles per gallon for light trucks — would by 2010 be saving 2.7 million barrels per day — or nearly twice as much as could be pumped every day from the Arctic National Wildlife Refuge.

Yet so far, even that small change has proved to be a political impossibility. Although such efficiency improvements are already technically feasible — Ford's Escape SUV, a gas-electric hybrid, gets 36 miles per gallon in the city — U.S. automakers and the big automotive unions have persuaded Congress not to raise fuel-efficiency standards since the late 1980s. Why? Among other reasons, because any regulations requiring greater fuel efficiency will initially favor Japanese and German automakers, whose fleets are already more fuel-efficient — thereby costing U.S. companies more of their market share and U.S. auto workers more of their jobs. And such losses are not inconsequential to American politicians. Since 1990, the U.S. transportation industry has made more than $256 million in campaign contributions. Whereas nearly 70 percent has wound up with Republicans, Democrats haven't been shy about asking for auto dollars, especially from the auto workers' unions. No surprise that CAFE has never come close to being updated.

American energy policies not only help preserve existing patterns of production and use, but indirectly discourage the development of newer and potentially better technologies. Government funding for research into nuclear and hydrocarbon technologies has run to tens of billions of dollars — an order of magnitude more than funding for solar, wind, and other renewables combined — at a time when the renewables industry desperately needs the kind of technical breakthroughs that government funding can help provide. Since 1947, for example, the U.S. government has spent $145 billion on nuclear R & D, as compared with around $5 billion for renewables.[14] In recent years, almost 65 percent of all federal production tax incentives, used to encourage certain energy industries, has gone toward gas production, as compared with 1 percent for renewables.[15]

It is not merely the lack of research funding that hurts newer energy technologies and industries. Whereas Japan and Europe have a tradition of long-term, government-dictated energy strategies, the fragmented, favor-granting nature of U.S. energy policymaking means that policies tend to change with every election cycle and business surge. "When Japan and Europe come to an energy policy, they tend to stick to it," says Merwin Brown, an energy economist and forecaster who helps develop U.S. renewables strategy; "but in the United States, it's much more helter-skelter, much more short-term, much more, 'Hey, let's try this — no, let's try this.'" The result, say Brown and other observers, is a degree of uncertainty that adds huge risk for energy investors, especially those looking to invest in unproven or alternative energy. Wind farm owners, for example, have lived in constant fear that U.S. lawmakers will cut the Production Tax Credit (PTC), the modest production subsidy that currently makes wind competitive with gas and coal — and their fear, apparently, is well founded. In the fall of 2003, Congress killed an energy bill containing the PTC, and the tax credits expired December 31. Lawmakers may resurrect the PTC in 2004, but in the meantime, energy analyst Janet Sawin told me, "many wind projects have already been put on hold." Despite rapid growth in the wind industry in 2003, "there will probably be relatively little development [of new wind power] in 2004."[16] Investors in other, more speculative technologies are even more gun-shy. Brown recalls a hydrogen seminar where investors were debating whether to bet on big stationary fuel cells, used to power buildings, or automotive fuel cells. "Most of the investors were ready to pursue stationary fuel cells because they are already selling and have some chance of being driven by the market, whereas automotive fuel cells are still policy-driven, in that they need a government policy to be competitive. So I asked them, Suppose government were to enact a policy that subsidized automotive fuel cells and made them cost-competitive, would you invest? And every one of them said, No, because they wouldn't trust the policy to last."

The inability of the United States to get behind a serious alternative-energy policy has crucial implications not simply for the future of the U.S. energy economy, but for the rest of the world as well. Many developing countries rely entirely on outside aid to help them modernize their energy systems. In China, India, and Russia — countries with huge coal resources and little incentive to use them cleanly — targeted investments to promote energy efficiency or clean-energy technologies could have a tremendous, positive impact on the way these countries' energy economies develop, with significant implications for future energy security and climate emissions. Yet whereas the United States spends billions of dollars each year promoting the production of hydrocarbons worldwide, especially in the developing world, U.S. efforts to help the rest of the world use *fewer* hydrocarbons are almost nonexistent. As one U.S. foreign aid official told me, "if you add up all the money the United States spends on U.S.-China energy cooperation, you'd have to stretch to get it over five hundred thousand dollars a year."

If American energy politics has always been dysfunctional, a new standard may have been set with the election of George W. Bush. The Texas Republican floated into office on a wave of campaign contributions from the energy and auto industries ($2.4 million from carmakers alone), and proceeded to assemble a White House that was closely aligned with both industries. Vice President Dick Cheney had been not only an oil executive at an energy company (Halliburton) but a congressman from Wyoming, one of the largest coal producers in the nation. National Security Adviser Condoleeza Rice had been a director at Chevron Oil. Commerce Secretary Donald Evans had run an oil exploration company. Energy Secretary Spencer Abraham had been a U.S. senator from Michigan, where he was known as the senator from Detroit and as a steady backer of the auto industry's political agenda — and a dependable opponent of higher fuel-efficiency standards.

Predictably, perhaps, the Bush energy policy has focused mainly on increasing energy production both at home and abroad, while assiduously avoiding domestic energy initiatives that threaten the economy, the president's political allies, or constituents that are critical to his and his party's retention of power. To be fair, Bush's policies are at least partly driven by a bona fide conservative political view on energy. Bush and his advisers have long hewn to a strain in conservative thought that regards environmental protection, energy conservation, and climate policy mainly as misguided liberal efforts to "save the planet" by weakening the economic and political power of the business community. In Bush's view, only a strong business community can keep the economy healthy enough (and America powerful enough) to take care of environmental concerns. "The Bush administration's implicit energy policy is that energy makes the world go around, oil and gas are the easiest, and the most sensible, way to maintain American economic and political might, and the energy industry, not government, knows best what it should do," says one energy analyst who has worked on U.S. climate policy for years. "As far as the Bush administration is concerned, the main challenges facing the U.S. energy system are Arabs and environmentalists."

It is also the case, however, that Bush's approach to energy transcends any conservative ideology and that it is often dictated by a political pragmatism and a cynicism that put even the very pro-business Clinton administration to shame. Consider the Bush administration's long-standing drive for increased domestic oil production, most notably in the Arctic National Wildlife Refuge (ANWR). Ostensibly, such production is meant to increase American "energy independence" and reduce our reliance on "foreign oil" — two glorious objectives that American politicians, and most recently Vice President Dick Cheney, trot out whenever gasoline prices get too high.

But in fact, though the White House may genuinely want more domestic oil production, administration officials know full well that U.S. "energy independence," at least in the short term, is largely a fantasy. The United States, lest we forget, is a "mature" oil province: we may consume 25 percent of the world's oil, but after a century's heavy production (accelerated, it should be noted, by billions of dollars in tax breaks), we now have less than 2 percent of the world's in-the-ground reserves. Since 1972, U.S. production has been declining steadily, and no amount of encouragement through tax incentives is going to change that. In even the most optimistic scenarios, U.S. oil production could be boosted from the current 9.5 million barrels a day to perhaps as much as 10.1 million barrels a day in 2020. Opening the Arctic to drilling will make little difference. Although some federal agencies, like the ever-optimistic U.S. Energy Information Administration, contend that the extensive coastal preserve holds tens of billions of barrels of crude, most oil company experts believe the number is far lower, and that, in at best ten years' time, ANWR could boost American production by 600,000 barrels a day. As Joseph Romm, the former Clinton energy official, told me, "you open up ANWR, and the only difference is that by 2020 we're importing 62 percent of our oil instead of 64 percent."

Such facts are not unknown to the White House: these are, after all, former oil company executives. As it turns out, though, the Bush administration is pushing ANWR not in order to improve American energy security, but because ANWR can be used as a bargaining chip in an energy debate with far larger political stakes: automotive fuel standards. Like Clinton before him, Bush knows that if he is forced into tightening CAFE standards, he will alienate the all-important auto lobby and the auto workers' unions — two constituencies no national politician can afford to lose (especially one who, like Bush, has trouble winning elections the traditional way). Like Clinton, however, Bush realizes he has an out: ANWR. Political strategists have long known that the Arctic wilderness carries a far higher emotional impact among voters than does fuel efficiency: even environmentally minded Americans would much rather save polar bears than conserve gallons of gasoline, and this is true even among the membership of big national environmental groups. Political strategists also know that many moderate members of Congress — the so-called swing block — feel they can vote "green" on perhaps one big issue a year without offending their more conservative constituents and colleagues.

In short, Bush strategists know that by effectively forcing a choice between imposing stricter fuel standards and saving ANWR, lobbyists for big groups like Sierra Club and the Wilderness Society, if they want to appease their own constituents, must choose ANWR — even though, privately, some Green activists admit that they would willingly sacrifice ANWR for higher fuel standards. As one somewhat embittered environmental lobbyist told me, when Greens try to lobby Congress, "lawmakers tell them, 'I've got one environmental vote I can give you this year: which do you want — CAFE or ANWR?' And most have asked for ANWR." Thus, as long as the White House can get ANWR on the table, it knows that CAFE won't be changed. "It's a huge win for the administration, because it's getting Greens to use up all their political ammo on something that doesn't matter, instead of dealing with fuel economy, which does."

Indeed, one of the most striking and discouraging facets of the Bush energy strategy has been the brazen attempt to maintain an obsolete fiction about energy and the United States: namely, that this country can keep ignoring fundamental weaknesses in the existing energy order, downplay the need to reduce demand for hydrocarbons, and simply drill its way to greater energy security — even though the White House knows full well that domestic oil production is in gradual but permanent decline and that "energy independence" is a sop to blue-collar voters and irate motorists.

Where Bush's political pragmatism may be most damaging, however, is on climate. While Europe and parts of Asia are already well along on implementing programs to reduce emissions, the United States remains locked in policy denial. True, there are many solid reasons for Americans to be skeptical of the Kyoto process — mainly, the too-hurried deadlines for CO_2 reductions. Bush's reluctance on climate, however, comes from other, less defensible motives as well — including the fear of any policy that might offend the powerful "coal vote" — coal companies, power utilities, and coal-mining states were all critical to his 2000 victory. "This administration's entire take on climate is rooted in the fact that Bush won the coal states essentially by being 'not Al Gore' on climate," says one Republican energy analyst, "and there is no way that Karl Rove is going to let Bush do anything to alienate those states this time around."

Above all, this administration, like others before it, fears that a truly honest climate policy could put long-term U.S. economic and political strength at risk. Under most climate policies, by the end of the century, the United States will need to have cut carbon emissions by 70 percent. This is a reduction that few believe can be achieved solely by switching to carbon-free fuels; it may require the mammoth U.S. economy actually to reduce its overall and per capita consumption — perhaps significantly. And while some energy efficiency experts believe that this energy reduction can be achieved through "transparent," painless efficiency measures, others believe that it will require significant changes in lifestyle, consumption patterns, and, perhaps, overall economic activity.

Whether such changes will truly be required is impossible to know at this point. Needless to say, given that American political power derives from its economy, and given that the U.S. economic strength depends entirely on rapid growth and ready access to cheap energy, the Bush administration shows an understandable lack of interest in any policy that carries the slightest hint of a reduction in either energy use or economic growth.

As a result, observers say, not only has the Bush administration simply refused to advance a substantive climate policy, or a serious alternative to the Kyoto treaty, but it is actively working to keep even the *idea* of a climate policy out of the public view. In the summer of 2003, under orders from White House staff, the Environmental Protection Agency (EPA) deleted most references to climate change from a long-awaited report on the state of the global environment. According to a story in the *New York Times*, the deleted sections included the EPA's conclusions about the role that human activity is playing in climate change, as drawn from a 2001 report by the National Research Council that the White House itself had requested and "which President Bush had endorsed in speeches that year."

Administration officials also cut any mention of a 1999 study showing how global temperatures had risen more sharply between 1990 and 2000 than at any time in the previous thousand years. "In its place," according to the *Times*, "administration officials added a reference to a new study, partly financed by the American Petroleum Institute, questioning that conclusion." Some EPA officials were so appalled by the revisions that they sent out a memo saying the revised report "no longer accurately represents scientific consensus on climate change." As one climate policy analyst told the *Times*, "this is like the White House directing the secretary of labor to alter unemployment data to paint a rosy economic picture."[17]

There is even evidence that the Bush White House helped coordinate a smear campaign against the EPA's report. In 2003, reporters for the London *Observer* gained access to an e-mail from the Competitive Enterprise Institute (CEI), a conservative, pro-business lobby group heavily funded by ExxonMobil, to Phil Cooney, chief of staff at the White House Council on Environmental Quality. In the June 3, 2002, e-mail, according to the *Observer*, CEI director Myron Ebell thanks Cooney for asking for CEI's help, then goes on to suggest a number of ways to discredit the EPA climate report and even remove some of its top officials, including former EPA director Christine Todd Whitman. "It seems to me that the folks at the EPA are the obvious fall guys and we would only hope that the fall guy (or gal) should be as high up as possible," Ebell writes. "Perhaps tomorrow we will call for Whitman to be fired." Whitman has since left the administration. The CEI also launched a lawsuit against another federal agency that does climate research, according to the *Observer*. Notes Richard Blumenthal, attorney general of Connecticut and a supporter of state and federal CO_2 reduction programs, "This email indicates a secret initiative by the administration to invite and orchestrate a lawsuit against itself seeking to discredit an official U.S. government report on global warming dangers."[18] Both White House officials and the CEI have denied any wrongdoing, but critics say the charges fit the administration's pattern when it comes to climate.

It is tempting, in light of such behavior, to blame *all* American energy problems on the current administration, with its links to the oil industry and its somewhat misplaced faith in an oil-driven energy economy. The more discouraging reality, however, is that neither Bush nor any of his predecessors could even *think* about advancing so risky and shortsighted an energy policy if American voters had any objections — and by and large, they do not. Americans are, in general, the least energy-conscious people on the planet. We are not only profoundly ignorant about what energy is, and the critical role it has played and continues to play in economics and politics, but most of us simply don't care about energy.

Some of this complacency is economic: for most Americans, apart from the periodic spikes in the price of gasoline or heating oil, or the occasional power outages, energy is so cheap and reliable that there seems little reason to give it a second thought. Then, too, some of this complacency stems from the near-religious confidence in the power of our technology: so successful have Americans been at innovation that most of us are quite content to leave the business of transforming the energy economy to those who have always handled it — namely, the energy companies and the government — and assume that *something* will work out.

Yet the truth is that there is more to our complacency than low prices or techno-confidence. In fact, most of us have a nagging suspicion that energy is more than a simple economic proposition — a mere question of whether we can afford to pay for a large car or a huge home or an expensive air vacation; rather, most of us know that energy is a much broader and subtler issue, with connections to all aspects of modern life. We may not all agree that filling up our Dodge Durango at an Orange County BP station is in any way linked to the scramble for drilling rights in Alaska or the brutal civil war in Nigeria or even the speed with which U.S special forces secured Iraq's offshore oil-export terminals. But even the most resolutely apolitical American knows at some level, or at least suspects, that energy is an intensely political commodity, and that our enormous demand for energy, our energy-intensive lifestyle, and our preference of ever-larger cars and trucks give energy companies greater political power, ensure that oil is central to nearly all domestic and foreign policy, and, in general, constitute one of the primary forces that keeps the existing energy order from changing.

Yet despite such awareness and suspicions, most of us continue to live in a state of denial, a willful and often quite creative ignorance about energy and its impact. As was true of Victorians and their famous refusal to acknowledge the sexual realities and tensions of their times, Americans have become prudes — but about energy. Most of us know that our energy consumption is excessive, that this excess is linked to myriad problems, ranging from air pollution and climate change to geopolitical chaos, and that it may even be the main obstacle to a more sensible and sustainable energy system. But we simply cannot bring ourselves to acknowledge these downsides, because doing so would force us to recognize a great many other problems with our energy economy, and our own role in it. In fact, when it comes to fundamental issues of contemporary energy, Americans have become so uninterested in and even hostile to the subject that many energy advocacy groups have simply ceased any public campaigning. "In the 1990s, people in my community essentially stopped talking about conservation," admits David Nemtzow, former president of the Washington-based Alliance to Save Energy and one of the founding lights of the American conservation movement. "We made a cold-hearted calculation that most Americans couldn't be bothered any more."

Thus it is that we Americans (and most of our media) are largely untroubled by Secretary Rumsfeld's absurd claim that the Iraqi war was "not about oil." We're not upset that the White House has steadily refused to disclose the names of the energy companies that helped write U.S. energy policy. We don't think it odd that the White House Energy Task Force was studying maps of Iraqi oil fields and pipelines as early as March 2000 — more than eighteen months *before* the September 11 attacks — or that the vice president's former oil company, Halliburton, won a multibillion-dollar U.S. government contract to repair Iraqi oil fields even before the second Iraqi war was under way. Or that one of the first actions by U.S. military forces in Iraq was to establish a tight security perimeter around the Iraqi Oil Ministry in downtown Baghdad, while hospitals, schools, utilities, and other critical elements in the infrastructure were left to be burned and looted. We refuse to be troubled by facts like these because even to look closely at them might force us to see them as extensions of an out-of-control energy system that begins at home, in our own cars and houses.

Americans' rising energy obliviousness is not, on the whole, good for a democratic system, nor is it particularly favorable to the making of smart energy policy. For all the criticism thrown at U.S. politicians by the media and other elites, the American people have been largely silent on questions of energy, thus giving lawmakers little incentive to make anything but a token effort at changing the system. As one veteran lobbyist for fuel efficiency told me, "when you try to make a case for a long-term policy on energy or climate, the common response from the Hill is, 'Well, that's all well and good, but I don't hear about this from my constituents back home.'"

In general, the only time U.S. politicians hear from voters about energy is during power outages or price spikes — that is, when people think they're paying too much for energy — circumstances not entirely conducive to development of a national energy policy aimed at *reducing* energy demand. Quite the contrary: when lawmakers hear complaints that gasoline or power prices are too high, they tend to vote for initiatives that expand supply and depress energy prices, not for carbon taxes. As another alternative-energy advocate notes, "if what you're trying to do is reflect the price of carbon on the costs of fuels in a way that will have the same effect as a new fuel tax, you may have the merits on your side, but the idea is flying in the face of everything politicians hear about energy back home. On energy issues, the public does not speak, which leaves a huge political vacuum in Washington — a vacuum the politicians are only too happy to fill." And what they will fill it with will be policies that continue to emphasize supply.

Whether we blame American energy politics on the people or their politicians, it's plain that U.S. energy policies will have a tremendous impact on the evolution of the global energy economy. Even the most devoted of "free-marketers" recognize that the rise of a truly new energy economy will be more an act of political transformation than a response to some technological or economic development. Though the market will play a fundamental role in creating any new energy economy, the notion that the market alone can produce a revolutionary and sustainable approach to energy production and consumption, and do it swiftly enough to avoid serious damage to the climate, becomes more dubious by the year.

In theory, the United States could lead that political transformation. Instead, U.S. energy policy remains captive to the politics of supply: before it stalled out in Congress in late 2004, the latest proposal for a U.S. energy policy was larded with still more tax breaks and other costly incentives for hydrocarbon fuels and new oil production — even though the decline in American prospects for oil production are undisputed.[19] Abroad, the United States is still scrambling for new oil allies, and as recently as July 2003 was lobbying the government of Nigeria to leave OPEC and begin pumping at maximum. Despite growing evidence of high-level Saudi complicity in the September 11 attacks, the White House seems to be doing everything it can to keep from offending the House of Saud.

In short, instead of leading the global energy revolution, the United States seems to be holding that revolution back — a delay that will surely cost us more in the long run than it can possibly save us today. The longer we wait to start moving toward a new energy system, the harder it will be to make any kind of orderly, progressive transition. Carbon dioxide emissions will be higher. Overall energy demand will be sharper. Energy poverty will be more profound and more volatile. Alternative-energy technologies will lag further behind. "There is no question that the day the United States got serious about a new energy system, we could have a huge impact in ten years," says one former top-level Energy Department official. "We could take action that would benefit the economy, the environment, and would improve energy security. That's the good news. The bad news is, we never will. Absent some crisis — something that would not only raise the price of oil but lead people to believe that prices would be high for a long time — the chances of the United States' acting proactively are nil."

In fact, the more we delay action, the more plausible such a crisis becomes — and not necessarily a neat, self-contained crisis, like the 1974 Arab oil embargo, that taught us an important lesson about waste, yet left our energy economy and the world it supports largely intact. Instead, the longer the world continues to rely on the current energy system, and the greater the demands we place on it, the more likely we are to see the kinds of serious system lapses that are only hinted at in media stories: nationwide blackouts; sabotage of critical infrastructure; yearlong, economy-sapping price spikes; violent instability in energy-producing states; even political or military conflict between big energy importers — any one of which could happen not in ten years or twenty-five years, but right now. Comforting as it might be to imagine the decline of our energy economy as a long-term process — with oil supplies peaking in 2025, say, or sea levels rising by 2050 — there are fewer and fewer reasons to believe that our overtaxed energy system won't have begun to collapse long before then.

13

HOW DO WE
GET THERE?

EACH AUTUMN SINCE 1999, a group of experts on energy and international relations has met with a handful of U.S. intelligence officials at a large, Georgian-style brick hotel just outside Washington for a closed-door conference entitled the Geopolitics of Energy in 2015. Organized by a government brain trust called the National Intelligence Council, "2015" is part of an effort by the U.S. intelligence community to imagine the global energy system of the future — not because the Central Intelligence Agency, the National Security Agency, or the Defense Department cares one way or another about hydrogen fuel cells or compact fluorescent light bulbs, but because energy affects global stability and global stability is key to American security.

At the 2002 session, held in mid-October, discussions revolved around a set of four scenarios, developed by the National Intelligence Council, which mapped out four different pathways to a new energy economy. The first two scenarios had fairly optimistic outcomes. One, entitled "Green as Green Can Be," began with what its authors called a "headline-grabbing environmental disaster" that "galvanizes public opinion" and causes the United States, Europe, and Japan to pursue "aggressively environmental policies," including heavy new gasoline taxes and stricter pollution regulations. The policies cut oil demand so significantly that by 2020, the world is using thirteen million barrels a day less than under most baseline forecasts. Scenario number two was even rosier: this time, a series of technology breakthroughs in everything from wind and solar energy to fuel cells and energy efficiency has led to substantial declines in energy intensity worldwide.

As the day wore on, the scenarios became bleaker — or perhaps more realistic. In the third scenario, participants discussed the consequences of a "peak" in conventional oil production, occurring sometime between 2010 and 2015. In this scenario, declining output from fields in the North Sea, Alaska, Venezuela, and Iran push oil prices to forty dollars. Higher energy prices begin eating into national economies, pushing the global economy toward recession.

The last scenario, entitled "A Darker Middle East," was the grimmest. Here, the U.S. defeat of Saddam Hussein has backfired, alienating many in the Arab world and leading to the overthrow of Saudi Arabia, Kuwait, and other "relatively friendly Arab governments by nationalist Islamic regimes." With U.S. forces tied up in Iraq and Afghanistan, the United States is reluctant to intervene militarily. Meanwhile, new nationalist governments in Saudi Arabia and Kuwait cut oil production by 20 percent for three years, and 10 percent thereafter. The crimp in supply, coupled with terrorist attacks on international oil shipments, pushes prices for crude to fifty dollars a barrel for five years — thereby setting the stage for the end to a modern energy economy based on cheap oil.

For each scenario, participants had been invited to assess coolly the geopolitical implications: which countries would benefit, say, or how international alliances might be affected. But some in attendance found the exercise unsettling — less because of the gory details of war and disruption in the grim scenarios than because of the implausibility of the optimistic story lines. In today's political climate, the idea of an energy future created proactively, by thoughtful policy or a technological breakthrough, struck some as highly unlikely. "I don't see anything really changing without some monumental event that *forces* change," Robert Ebel, a well-known energy analyst at the Center for Strategic and International Studies, told me later. Other participants were downright cynical. "No one takes these exercises seriously enough," remarked a gas industry analyst with long experience working with government. "Sure, a few people there seemed genuinely worried, but most were pretty complacent, which is pretty much how government and industry are about energy. They think that the energy future isn't going to be a whole lot different from the energy past, and to the extent that things are going to change, you can't predict it, or do anything about it, so why bother trying?"

⚬§§⚬

Such cynicism is no longer surprising. The more we learn about the history and character of our energy system, the harder it becomes to see how the world can escape some kind of wrenching disruption, given current trends. We know, for example, that our energy demand will eventually exceed our capability to supply it safely, especially in the developing world. Competition for energy resources will increasingly drive international relations, and produce conflict. Energy markets will continue to ignore the external costs of fossil fuels, thus confining new technologies to the margins and gradually poisoning the thin layer of soil, water, and air that supports all life.

And yet, the alternative scenarios — in which the world shifts gears and revolutionizes the way we make and use energy — are often more depressing, in part because we are increasingly aware of just how difficult change will be. In the simplest terms, the energy challenge of the twenty-first century will be to satisfy a dramatically larger demand for energy, while producing dramatically less carbon. Yet the availability of carbon-free energy on a mass scale — whether produced from renewable sources, like solar and wind, or from decarbonized fossil fuels — will not happen without significant technological developments. And such breakthroughs aren't likely until the market regards carbon as a cost to be avoided — not just in "progressive" enclaves like Germany or England, but in the big economies of Russia, China, and, above all, the United States.

Yet this, too, is increasingly difficult to imagine. Whereas European policymakers have finally begun shifting, however haltingly, toward a low-carbon energy regime, the United States, the one nation whose participation in any worldwide energy revolution is essential, seems unable to move without being pushed. The last American energy revolution came only in response to crisis — the 1974 Arab oil embargo — and since then, U.S. energy policy has become even more fractured and obsessed with supply. One possible opportunity for change — post–September 11, when the links between massive U.S. energy consumption and its high-risk foreign policy were starkly evident — came and went with little apparent impact on U.S energy policy. If anything, American political leaders have since then become even more hostile to the notion of a "new" energy economy, and more persuaded that the heart of American energy policy is, and always will be, defending "security of supply."

This is why, for many energy experts, true change in the global energy system is virtually impossible, except in response to some serious shock. In this somewhat pessimistic view, the question is not whether the world can avoid some kind of energy-related disaster, but whether our response will be reactionary and short-term, or constructive and long-term.

What kind of disruption is it likely to be? Nearly all the energy experts, oil company officials, and political analysts I've interviewed over the last several years believe that the most likely scenario involves upheaval in the Middle East. If the aging Saudi crown prince were to die (a commonly cited possibility), analysts say a succession battle could ensue between powerful Wahhabi clerics, who want to push Saudi Arabia toward a conservative Islamic theocracy à la Iran, and Saudi moderates, who hope for a more progressive, pro-Western regime. Experts say the struggle, even if not violent, could easily slow or halt Saudi export operations and cut world production by nearly 12 percent. In such dire circumstances world leaders, especially in the United States, would be forced into a series of tough choices that could alter the course of the energy future.

If such a disruption occurred today, analysts say, given how critical oil is to the global economy, and given the current political environment in the United States, there would be extraordinary pressure for military intervention — particularly if Saudi Arabia appeared to be tilting toward fundamentalism. As one foreign policy analyst who works closely with the CIA told me, "there is simply no way the United States would allow an Osama bin Laden to control the world's largest oil reserves."

That would be a grim dilemma indeed. If America launched a second military action, it might restore world oil supplies, at least temporarily, but the move would surely fuel Islamic rage, further destabilizing the Middle East and almost guaranteeing future supply disruptions. Yet if America declined to strike — if, for example, domestic political opposition halted a second Middle Eastern venture — and if Saudi oil were not immediately restored, importing nations would face an equally stark prospect — and none more so than the United States. Because it has made so little progress toward diversifying away from oil, a Saudi-centered disruption would be economically devastating. The closest precedent we have is the Iranian revolution, which took five million barrels out of production and sent prices up to forty dollars a barrel, or a hundred current dollars. Losing the Saudis' ten million barrels of daily production, though world emergency reserves would initially soften the blow, would be easily as destructive. Fuel-sensitive businesses, like airlines and trucking companies, would be affected immediately and drastically. Layoffs would ripple through the economy, sowing panic and causing companies to delay investments and expansions, and leading to more layoffs. And because energy costs affect the costs of producing goods and services but also hurt consumer buying power, higher oil prices would lead to the recession-inflation mix known as stagflation.

As the damage mounted, policymakers would be increasingly likely to take defensive and short-term actions, which, though necessary under the circumstances, could end any move toward a more progressive energy economy. To ease high energy prices, for example, regulators might waive emission requirements for coal-fired power plants. American policymakers would try to increase domestic oil production, by opening off-limits areas to drilling. They would also encourage additional production of "unconventional" oils, from the tar sands in Alberta, for example, and probably waive emission-control requirements there, too. "You can imagine a really ugly future where we're making massive quantities of synthetic fuels from coal and heavy oils and seeing huge increases in our carbon emissions," says Dan Lashof, an energy analyst with the environmentalist group Natural Resources Defense Council.[1] As recession set in, Congress would drop any plans to require automakers to raise efficiency standards. Lawmakers would also cut funding for nonessential energy programs, including R & D for hydrogen and subsidies for wind power and other renewable industries.

Such a defensive energy strategy would have catastrophic long-term impacts. Were the United States to move deeper into a traditional hydrocarbon economy, and further away from even a pretense of reducing CO_2 emissions, analysts fear that European governments might be pressured into delaying their own aggressive CO_2 reduction goals. As one U.S. climate policy expert put it, "any new U.S. move away from a climate policy could easily delay policy action in other countries, both by mobilizing the forces within those countries which are opposed to climate action, such as business lobbies, and also by giving cover to any leaders unwilling to take on climate change." China and India, too, could feel less pressure from the West to modernize their own energy economies and might resume the rapid expansion of conventional coal-fired power plants. If these developments occurred, energy analysts say, keeping atmospheric concentrations of CO_2 below the 550 parts-per-million threshold would prove impossible and catastrophic warming would become all but inevitable.

In the meantime, high oil prices would have encouraged a frenzy of new oil exploration and production — both in OPEC countries and in remote and hard-to-reach non-OPEC fields previously written off as too costly to develop. This new oil might reach the market within two to three years, bringing some price relief. Ultimately, however, this surge of high-priced oil production would only speed the depletion of non-OPEC oil reserves — and hasten the day when OPEC, even with a weakened Saudi Arabia, gains true pricing power over world oil markets.

Now let's consider a more optimistic scenario. Let's look at a future narrative in which the United States and the rest of the energy economy don't react defensively to a crisis, retreating deeper into the hydrocarbon economy, but rather move in an entirely different direction. Let's start by supposing that our oil disruption takes place under different circumstances. Suppose that our Saudi succession struggle occurs not today, but a few years from now — say, 2008 — and is slightly less severe, provoking a loss of just five million barrels of production. More important, suppose that, in the meantime, American confidence in energy has been badly shaken by a series of smaller, almost preparatory energy crises, and that consumers and politicians alike no longer find comfort in the energy status quo. Suppose that things have gone badly in Iraq, and that Americans are in no mood for any more oil interventions. Suppose that a progression of blackouts and natural gas price spikes have persuaded consumers that traditional U.S. energy policy has failed and that energy is too critical to be left entirely to the "free market." Suppose that, despite U.S. efforts to undercut OPEC, the cartel has kept oil prices above thirty dollars a barrel, and that these higher prices have eroded the economy, while spurring interest in efficiency and non-oil alternatives. Suppose that "energy security" and "volatility" have become nightly news topics, and that stories about civil strife in places like Nigeria and Bolivia, or the pipeline "wars" between China and Japan, are routine fare on front pages and Sunday talk shows.

Suppose, further, that the data on climate have become irrefutable. Imagine that the effects of global warming that we've already begun to see today — the heavier rainstorms and killer flash floods, the more intense summer droughts and forest fires, the steady declines in the mountain snow pack that most of the western United States depends on for water — start happening so frequently and with such great intensity and damage that we begin to register an atmosphere of crisis. Suppose that a few big states grow even more frustrated than they are now over federal energy policy, to the point where they begin acting independently: upping emissions requirements for cars and trucks (as has already happened in California and New York), phasing in a carbon tax or cap-and-trade system (as is under consideration by northeastern states), or launching programs to replace the unreliable regional grids with new "distributed generation" microgrids. "You could easily picture a situation where states start doing things on their own," one climate policy expert at a U.S. environmental group told me, "and pretty soon, you have a patchwork of different and sometimes conflicting state regulations, which annoys the hell out of industry because it's having to adapt to all these different laws, and pretty soon industry is actually *asking* Congress to adopt some kind of uniform carbon tax."

In such a political environment, analysts suggest, the United States might respond quite differently to a disruption or some other energy "event" than it would today. Rather than struggling to defend the energy status quo — say, by invading some oil-rich region — U.S. lawmakers might be willing to risk a more progressive and interventionist energy policy — one intended to balance the necessary focus on increased supply with a new emphasis on energy efficiency and low- and no-carbon fuels and energy technologies.

Such a sweeping policy, were it to be enacted, would probably be built around a core of long-term goals — among them, staying within a hundred-year carbon budget, and moving toward a hydrogen economy. Significantly, this new policy would emphasize the concept of a "bridge" economy, a transitional phase designed to arrest the worst of the current energy trends, while giving us more flexibility in eventually creating a new energy system. To encourage this transitional stage, the policy would focus on three near-term objectives designed to jump-start the process: first, an immediate move to expand natural gas imports; second, the rapid deployment of a carbon tax; and third, dramatically improved automotive fuel efficiency.

Just as important, whereas past energy policies have been centralized, top-down efforts — with government picking the winning technology and forcing compliance — analysts say that this policy would have to be a blend of incentives and constraints. First, the United States would commit to spending substantial sums — as much as twenty billion dollars a year — for the long-term energy research that is critical to achieving core break-throughs, but that private companies are typically unwilling to fund themselves. At the same time, government would set specific targets — such as emission levels or miles per gallon — complete with penalties for failure, but would allow the market considerable leeway in meeting those goals. Thus, if federal (or state) governments required that the fleet average for new vehicles sold in America be boosted to forty miles per gallon by 2020, for example, automakers would be largely free to choose how to achieve that goal. Similarly, where the government might set specific targets for a carbon budget or emissions levels, utilities would choose how they hit those targets.

⁂

Because the bridge economy will be fueled initially by gas, the first step for U.S. policymakers will be to dramatically increase the availability of gas as quickly as possible. Government will move immediately and aggressively to boost gas supplies — partly by increasing domestic drilling, but mainly through a rapid expansion of gas imports. American officials could accelerate construction of the long-delayed gas pipeline from Alaska and Canada to the lower forty-eight states by offering price supports and "soft financing" (low-interest loans with long payoff periods), to encourage skittish energy companies to make the necessary investments. As one industry analyst told me, "if the Feds asked companies to map out the cheapest pipeline route and then offered loan guarantees or soft financing, we could have that pipeline built in three years, max." Longer term, the United States would step up the approval process for construction of dozens of new sites for LNG regas terminals along the U.S. coasts and would work with Mexico to build up regas capacity in Baja.

Most of this new gas supply would be sucked up by the burgeoning gas power market, as coal-fired power plants were rapidly replaced by cleaner gas facilities, including smaller microturbines in office buildings and distributed power systems. As gas supplies gradually climbed, surplus gas would go toward other transitional energy uses. Natural gas could fuel fleets of specially converted buses, taxis, and other vehicles, especially government fleets, and slowly replace oil's share of the transportation market with a cleaner-burning fuel. As special "gas-to-liquids" technology developed, gas could be refined directly into synthetic gasoline and diesel for transportation — although this probably wouldn't be a major fuel source until after 2010. One important use of gas would be as a transitional feedstock to make hydrogen for fuel cells. Gas-derived hydrogen would itself serve as a transitional fuel, powering government fleets of fuel cell cars, as well as stationary fuel cells in experimental distributed power grids — thus helping push fuel cells from the research lab into the market.

In the long term, the gas bridge economy might run for two to three decades. During this time, the rate of emissions growth would begin to slow. A new model for a more distributed power system would be operating in places around the country. Fuel cells would be slowly but steadily penetrating both the automotive markets and the stationary power markets and laying the groundwork for the eventual emergence of a hydrogen economy once technologies to make hydrogen from renewable technologies, like solar or wind, or from clean coal, became cost-competitive.

The second component in an energy bridging strategy would be the adoption of a carbon penalty. Even the Bush administration recognizes the importance of this "market mechanism" as a catalyst for long-term climate and energy policy, but it has feared antagonizing the politically powerful coal interests, which believe such a penalty would put them out of business. A more progressive future administration could neutralize that resistance by making the carbon penalty part of an aggressive government campaign to develop clean coal as a long-term, carbon-free fuel.

First, analysts say, the White House would need to "de-Kyoto-ize" the debate over climate and CO_2 emissions. Rather than concentrating on the 1997 treaty (which is impossible for the United States to comply with today, and would be less achievable by 2008), the United States would embark unilaterally on its own domestic emission reduction campaign, possibly with the idea of rejoining a modified international effort at some point in the future. The government would create a carbon budget for each industrial sector, starting with the worst offender — power generation — and would assess a modest penalty for each ton of carbon. Rather than impose an actual carbon tax, the government, borrowing any useful lessons from the European systems, would encourage a carbon-trading system.

To ease the economic pain of a carbon penalty and deflect political opposition, the new American carbon cap-and-trade regime could be delayed — some analysts suggest starting it in 2018. The carbon penalty itself would also begin low — perhaps five or ten dollars per ton. It would rise gradually over the next two decades to a maximum of around one hundred dollars per ton, but with provisions for flexibility as new information emerges about carbon costs, or if the economic burdens of the tax are found to be too high. With the delayed schedule and low starting costs, utilities and other big emitters could plan ahead for the additional costs, phasing out high-emission assets slowly enough to avoid costly premature retirement.

Equally critically, by pairing the carbon penalty with a well-funded R & D program to develop technologies for coal gasification and carbon capture and sequestration — the essence of clean coal — the administration could create a long-term, carbon-free fuel solution while coopting coal interests. After years of being told that coal and climate policy don't mix, coal companies, unions, and utilities — as well as their allies in Congress — could become climate policy champions. "Right now, the coal industry sees itself as going out of business," says Reid Detchon, executive director of the Energy Future Coalition. Any policy that "can give them a brighter future, where coal is not the hated fuel, would be a huge winning scenario."

The cost of such a "winning scenario" would be relatively cheap. By some estimates, to have a near-zero-emissions clean-coal power plant, with carbon capture capabilities, up and running by 2020, the federal government will need to spend around twenty billion dollars, or a little more than a billion a year — a sizable sum, but far less than the government now spends subsidizing the oil and nuclear industries. The payoffs, if clean-coal technology succeeds, would be huge. An aggressive research and development program, coupled with the construction of a fleet of demonstration units, would drive down the costs of the gasification and carbon capture technologies, making these state-of-the-art power plants competitive on the U.S. power market and letting U.S. manufacturers begin selling this critical technology on the world market.

Over the longer term, a successful coal gasification industry could become a cornerstone of a hydrogen economy. Coal gas is rich in hydrogen, which means that, as coal gasification technology becomes more widespread and cost-effective, these high-tech factories could be used to convert coal, one of the cheapest and most plentiful fuels in the world, into low-cost hydrogen for use in large stationary fuel cells and, ultimately, in fuel cells for automobiles.

The third and final piece of the U.S. bridging strategy would be a no-holds-barred, multifront campaign to cut Americans' high consumption of oil and other energy. The White House would encourage stepped-up production of oil alternatives, such as biofuels from farm waste and, if CO_2 emissions could be controlled, synthetic crudes from Alberta tar sands. More importantly, the administration would move immediately to improve energy efficiency across the American energy economy — in homes and businesses, at industrial sites, in manufacturing processes, and above all in the transportation sector.

Boosting automotive fuel efficiency would have a huge, positive impact on long-term U.S. emissions and energy consumption — but could also be the most politically challenging to achieve. Although dramatic improvements in fuel efficiency are already technically feasible, there are few incentives to bring these kinds of cars to market. Since 1987, U.S. automakers, arguing that making cars more efficient would be too expensive and that American consumers don't care about fuel efficiency anyway, have stalled efforts to raise efficiency standards. While two-dollar-a-gallon fuel taxes in Europe and Japan have discouraged large cars there, that is not politically tenable in the United States; a more modest tax — say fifty cents a gallon — might fly, but it would have little effect: below a certain threshold, consumers don't take fuel costs into account when choosing a new car.

But many efficiency advocates and energy experts say that a U.S. president could break this decades-long impasse with a market-based "bridging" strategy that helps Detroit move away from existing automotive technologies and toward a low- or no-carbon vehicle. The key would be timing. Rather than forcing automakers to jump immediately to some new technology, like fuel cells, government would aim to improve the efficiency and emissions of existing internal-combustion technology. Whereas past policies have put the onus on Detroit to build fewer gas guzzlers, this policy would simply make gas guzzlers less attractive to consumers. Through a mechanism known as a "feebate," consumers choosing a vehicle that gets twenty miles per gallon or less would have to pay a stiff penalty, or fee — many advocates recommend as much as five thousand dollars. In contrast, someone buying a vehicle getting forty miles per gallon or better would receive a rebate in the same amount.

The feebate logic is simple: when buying a car, American consumers pay far more attention to up-front costs, such as sticker price, than to "life-cycle" costs, such as fuel or maintenance. Adding five thousand dollars to the sticker price, auto industry analysts say, would be enough to persuade most buyers to look at more efficient models. These buyers' purchase choices would then send a clear signal to automakers that greater fuel efficiency is a key selling point — and essentially harness market forces to improve fleet efficiency. Better still, because a feebate system is "revenue-neutral" — that is, fees collected pay for the rebates — it wouldn't require a budget battle in Congress.

American automakers and unions would probably fight feebates. Because Japanese and German automakers currently build and sell more fuel-efficient cars than American companies do today, foreign companies would have an advantage under the feebate program, at least until American companies could shift their own fleets toward greater fuel efficiency. But an activist U.S president could dissolve that resistance by offering two things. First, the White House would promote a market-based system that capped fleet emissions for each year but allowed carmakers to trade efficiency credits with one another, much as utilities would trade carbon emissions.[2] Second, the White House would offer Detroit a gold-plated olive branch: substantial subsidies or financial incentives to help American automakers and the unions make the transition to more fuel-efficient cars.

Such a bailout would be expensive, costing hundreds of million dollars, at least, and would provoke attacks from free-market proponents and fiscal conservatives. Even some environmental groups would be furious, charging that the White House was rewarding Detroit for decades of obfuscation and delay. But the deal could reasonably be pitched to the public as a national security issue — a kind of automotive version of the multibillion-dollar airline bailout enacted after the September 11 attacks. More to the point, a bailout may be the only way Detroit can be persuaded to improve efficiency without a costly political fight that could delay other elements of the U.S. energy program. "Basically, we're talking about a deal, a grand quid quo pro," concedes one former White House energy policy adviser who is now helping to craft an innovative automotive strategy. "In return for a promise of significant improvement in fuel economy, Detroit would basically get a massive bailout to pay for it to retool."

What is important about such a policy is that it recognizes the risks of forcing automakers to adopt untried, unproven technology. Fuel cells may indeed become the dominant power source for automobiles, and a critical part of long-term climate and energy strategies. But a self-sustaining fuel cell car industry is, at best, at least two decades away, and probably more — too long to wait to begin reducing our automotive emissions or energy use. Instead, under this plan, automakers would start cutting emissions immediately, through existing technologies, most probably gasoline-electric and diesel-electric hybrids, and perhaps direct-injection diesels and lean-burn ICEs. Within a decade, half the new cars on the road could be getting forty to sixty miles per gallon. This would cut CO_2 emissions dramatically and reduce oil demand by as much as five million barrels a day — in effect, we'd be improving energy security while buying ourselves another decade or two to develop a no-carbon transportation alternative. "It's the old adage that the perfect is the enemy of the good," says one energy analyst who works with the U.S. government. "You don't need to get to zero-emission fuel cells right away. If you can get to fifty miles to the gallon with a gasoline-electric hybrid, it means you've cut emissions and you can still use oil for a long time."

Eventually, of course, any bridging strategy would need to produce vehicles that are emission-free — which at this point probably means cars and trucks that run on some kind of fuel cell. The federal government could hasten this along in two ways. First, heavy investment in basic fuel cell research would speed solutions to such critical engineering obstacles as reliability, materials costs (especially platinum catalysts), and fuel storage issues. Second, a well-funded commitment to begin using fuel cells in fleets of government vehicles, such as the U.S. Postal Service trucks, or in cars for federal or state carpools, would create a small but important market for fuel cell vehicles. Last, just as the government now heavily subsidizes oil and gas production through tax breaks, during the bridge phase of the energy transition, energy companies and other producers, including homeowners with microelectrolyzers, would receive incentives for producing hydrogen.

❦

As the bridging strategy takes hold, the outlines of the transitional energy economy would emerge. Traditional hydrocarbon systems would be gradually replaced or upgraded with a diverse mix of fuels and technologies. Energy production and distribution would become more and more decentral-ized. Each development would bring immediate benefits, such as lower emissions or improved efficiency, but would also help foster a longer-term change in the way we make and use energy.

In the auto industry, for example, the combination of consumer fee-bates, government R & D funding, and fleet rollouts would give carmakers and fuel producers a low-risk environment in which to launch new auto-motive technologies and fueling systems, including a hydrogen system, while building critical public awareness. Joan Ogden, at the University of California, Davis, notes that yearly fleet vehicle sales total 750,000 cars and trucks. If just a fifth of those sales involved fuel cell vehicles each year, she says, within a decade, the United States could have 3,000,000 vehicles on the road — a number that many experts say would provide a kind of criti-cal mass for the new industry. At that level, volumes would be large enough to begin to make mass production cost-competitive, while providing en-ergy companies with enough of a market for hydrogen fuel to justify add-ing refining capacity.

We would see a similar ripple effect from a carbon tax. As carbon be-came a rising cost, power companies would accelerate efforts to avoid it — either by using emission credits, or, as credits become too expensive, by moving to other, less carbon-intensive ways of generating power. At first, utilities would probably begin replacing older, coal-fired plants with gas-fired turbines (a likelihood that makes increased gas supply all the more critical). As carbon costs continued to climb, utilities would have an incen-tive to look at other options. Companies might consider building state-of-the-art power plants using IGCC if the technology were cost-competitive. As carbon costs climbed still higher, utilities might add on the more expen-sive carbon capture and sequestration technology — which, in theory, could be available and cost-competitive by then.

Utilities would also invest heavily in renewable power. Wind farms are already approaching cost competitiveness in some regions and would quickly become competitive in nearly any market as coal-fired power be-came more heavily penalized. As demand for renewable energy climbs, ris-ing economies of scale would steadily bring down manufacturing costs for wind turbines as well as solar panels, allowing wind and solar farms to compete in more segments of the power market. For example, as wind power became more cost-competitive, it could be used more often as a base load, with more expensive gas-fired power being brought in during peak periods. Greater manufacturing volumes for solar panels would also intensify the pace of research and technical breakthroughs, including cheap, relatively efficient photovoltaic film that would, when applied to all building surfaces, dramatically improve the power-generating capacity of a specific site.

Granted, renewables still won't be as dependable as conventional power. Intermittency will remain a challenge to utilities and power managers, and there are clearly limits to the share of the market wind power could capture, regardless of carbon's rising cost. Yet many energy analysts believe that theoretical limits to renewables — for example, that wind can provide no more than a fifth of a market's power needs without causing disruptions — are based on the older grid model, with its huge central power plants and inefficient transmission lines and routing technologies. In the more decentralized models now under consideration, renewables will be able to play a much larger role.

In one experimental model, for example, utilities would use a combination of gas turbines and wind power to effectively invert the traditional base-and-peak-load scheduling model. First, a "smart" grid using hyperefficient switches and computerized power routing would be built to link several dozen regional wind farms, allowing operators to draw on power quickly from anywhere on the grid. If a scheduled delivery from a particular wind farm failed because of lack of wind, operators could quickly take a delivery of surplus power from any other wind farms with power to spare, thus smoothing out the supply curve. If shortfalls — due to unexpected wind deficits or periods of peak demand — were unavoidable, quick-starting gas-fired generators could be ramped up to fill in gaps in supply.

Such a model would require an overbuilding of wind-generating capacity, at additional expense, and would be dependent on careful site selection and a regionwide configuration to maximize the chances that wind was blowing somewhere in the system. Energy analysts say, though, that such a system, if carefully designed, would allow wind energy to take over a greater share of the base-load supply and cut the need for gas- and coal-fired power plants.

Overcoming the lower power density of renewable energy technology would be far harder — at least until hydrogen electrolysis became cost-effective. By as early as 2035, however, according to scenarios developed by the Pew Center on Global Climate Change, electricity from huge wind farms in the sparsely populated Midwest could be cost-effectively converted into batches of hydrogen. The hydrogen could then be piped to distant cities, much as natural gas is piped today, where it could be either used to make electricity in huge fuel cell power plants or sent to fueling stations for fuel cell cars.

Hybrid systems like these would dovetail with the larger movement, already under way, to revamp the overburdened U.S. power system. Even now, federal regulators are reassessing the impact of the 1990s deregulation trend, and many analysts expect that the government will reregulate parts of the power system, especially power lines and other pieces of the outdated transmission infrastructure that private utilities were reluctant to upgrade. As part of that movement, advocates say, the government could easily push a new grid structure, designed for maximum efficiency and flexibility, with new "smart" technologies that allow for faster, more efficient power scheduling and a wider range of power sources, renewable and conventional. A key feature would be the building of "microgrids," smaller, stand-alone systems that would let communities and businesses generate their own power. These systems would use whatever mix of renewable and traditional technologies was most economical. And with "net metering" (which is already available in thirty states) these systems could sell any power surplus back to the main grid.

This configuration would allow individual consumers to become power generators, as we saw in Chapter 8. It could also help address some of the obstacles to the development of a hydrogen economy; for example, until unit costs drop on automotive fuel cells, few consumers will be able to justify buying a fuel cell car simply for personal transportation. But keep in mind that fuel cell cars are not simply transportation devices: a fuel cell is a rolling generator — it doesn't "care" whether the electricity it produces goes to its own wheels or is sent elsewhere. In theory, when the owners of fuel cell cars parked at home or at work, they could plug their vehicles into the local microgrid and sell the fuel cells' power either to their employers or to utilities at daytime, high-demand rates.

Of course, this wouldn't be a money-making deal: as mini–power plants, automotive fuel cells couldn't compete with conventional power generators or even large stationary fuel cells. But as a way to partly offset the high capital costs of buying a fuel cell car, this plug-in approach could help the automotive fuel cell become economical sooner — while providing the power system with an additional source of power. This, say hydrogen enthusiasts, would be the first step toward integrating two markets that have long been separate — cars and power. "The beauty of hydrogen," says renewables researcher John Turner, "is that it blurs the differences between transportation and power, because it can be used for either."

It is important to note that the impact of an American bridging strategy would go well beyond the U.S. energy economy. Because the United States is so large a market for world energy products, a U.S. energy revolution would function as a catalyst in the transformation of the global energy economy, initiating a "domino effect" in energy that could ultimately change everything from emissions and energy use in the developing world to our oil-dominated geopolitical order.

The last time the United States got really serious about energy efficiency — after the 1974 oil price shocks — U.S. oil use fell so low that OPEC was nearly wiped out. A more permanent reduction — even if partly offset by rising demand in the fast-growing Asian economies — would completely change the global oil order. As oil prices fell — to as low as fifteen dollars a barrel, some analysts say — many big oil states would see their geopolitical status tumble. Some, like Russia, Venezuela, Iran, and Qatar, which have enormous gas reserves, could compensate by stepping up efforts to sell gas, especially to gas-hungry markets like China, India, and the United States. Other petrostates — like Mexico and Algeria, for instance — might be pushed into bankruptcy and would then require a massive, and inevitably United States–led, bailout.

Falling oil prices would also splinter OPEC. As Saudi Arabia, Kuwait, the United Arab Emirates, and Nigeria all tried to compensate for lower prices by boosting oil production, analysts say the inevitable glut would drive prices down further. Oil revenues would fall so sharply that many OPEC countries would suffer profound civil unrest. Some analysts believe unstable countries like Saudi Arabia would collapse. Others, however, argue that such lender nations as the United States, Europe, and Japan would step in quickly with financing packages — but would condition any loan on a commitment to economic and political reform. In either case, OPEC's power over the oil market would decline dramatically — as would petrostates' ability to finance terrorism.

Even as U.S. policies were undermining the existing energy order, they would be encouraging the development of a more sustainable one. A U.S. initiative to develop clean-coal technology, for example, would dramatically change the significance of an Asian economy powered by coal. If American companies can bring down the costs of IGCC and carbon capture technology sufficiently, China and India might find themselves able to burn their coal without dooming the climate to catastrophic warming.

In fact, many energy experts believe that the United States should not wait until the Chinese and Indians can afford clean-coal technology but should offer the technology as soon as it becomes available and should even subsidize the purchases, simply to avoid the catastrophe of an Asian energy economy based on dirty coal. Such energy charity would not be cheap: by one estimate, subsidies of that kind could run the United States at least ten billion dollars for the first hundred plants — a cost that conservative policymakers would oppose. But advocates of such clean-technology exports counter with three points. First, because China and India have little choice but to burn coal, if the United States hopes to avoid climate change, it has little choice but to help the Chinese and Indians adopt clean-coal technology. As one climate expert put it: "America is going to pay for climate, one way or another. It can either pay now to try to mitigate some of the effects, or it can pay later, when droughts and floods start decimating the developing world."

Second, advocates say that the United States could attach strings to its technology, making the offer contingent, for example, on a promise from Beijing to stop undercutting U.S. currency or dumping products on the U.S. market. Third, China and India will not be the only markets for U.S.-built clean-coal technology: many experts believe that the technology, once costs have been driven down, could give rise to a lucrative American export business — and reverse a depressing trend in which the United States has lost the lead in wind technology to the Danes and in solar technology to the Japanese. "We have to start looking at this less as a climate policy than as an economic stimulus for the U.S. industrial sector," argues Detchon. "We should be approaching this at scale, not as one-off R & D projects, but in a way that will make these units competitive overseas, where the bulk of the growth is. This is going to be a growth market, and the United States needs to build up a real manufacturing strength."[3]

Not every technology export will be so lucrative, particularly where the poorest countries are concerned. By lobbying multilateral lending insti-

tutions, such as the World Bank and the International Financial Corporation, to coordinate financing for the massive LNG projects, the United
States and other Western governments could help accelerate the laudable
effort to bring gas into China, India, and other "emerging markets." For
the desperately energy-poor countries in Asia, Africa, and South America,
the U.S. energy policy would need to be part of a broader development policy that included aid for basic energy purchases, transfers of U.S. energy
technology, especially power systems, and, over the longer term, initiatives
that encouraged third-world economic growth. Indeed, one of the most
straightforward strategies all industrialized nations could pursue, experts
say, would be to open up their own home markets to third-world agricultural exports, so that developing nations could begin earning hard currency to buy their own new energy technologies.

Politically, a new U.S. energy policy would send a powerful message to
the rest of the players in the global energy economy. Just as a carbon tax
would signal the markets that a new competition had begun, so a progressive, aggressive American energy policy would give a warning to international businesses, many of which now regard the United States as a lucrative dumping ground for older high-carbon technology. It would signal
energy producers — companies and states — that they would need to start
making investments for a new energy business, with differing demands and
product requirements. Above all, a progressive energy policy would not
only show trade partners in Japan and Europe that the United States is serious about climate but would give the United States the leverage it needs to
force much-needed changes in the Kyoto treaty. With a carbon program
and a serious commitment to improve efficiency and develop clean-energy
technologies, says one U.S. climate expert, "the United States could really
shape a global climate policy. We could basically say to Europe, 'Here is an
American answer to climate that is far better than Kyoto. Here are the practical steps we're going to take to reduce emissions, far more effectively than
your cockamamie Kyoto protocol.'"

Similarly, the United States would finally have the moral credibility to
win promises of cooperation from India and China. As James MacKenzie,
the former White House energy analyst who now works on climate issues
for the Washington-based World Resources Institute, told me, Chinese climate researchers and policymakers know precisely what China must do to
begin to deal with emissions but have thus far been able to use U.S. intran-

sigence as an excuse for their own inaction. "Whenever you bring up the question of what the Chinese should be doing about climate, they just smile. They ask, 'Why should we in China listen to the United States and take all these steps to protect the climate, when the United States won't take the same steps itself?'"

With a nudge from the United States, argues Chris Flavin, the renewables optimist at World Watch Institute, China could move away from its "destiny" as a dirty coal energy economy. Indeed, given China's urgent air quality problems, a growing middle class that will demand environmental quality, and a strategic desire to become a high-tech economy, Flavin says, Beijing is essentially already under great domestic pressure to look beyond coal and is already turning toward alternatives — gas, which is in short supply, but also renewables, especially wind, a resource China has in abundance. Once China's growing expertise in technology and manufacturing and its cheap labor costs are factored in, Flavin says, it has the basis for a large-scale wind industry — something the right push from the West could set in motion. "As China moves forward," asks Flavin, "is it really likely to do something that no other country has ever done: run a modern, high-tech, postindustrial economy on a hundred-year-old energy source?"

Flavin, for one, thinks not. During a visit two years ago to lobby reluctant Chinese government officials to invest in renewable energy, Flavin was pleasantly surprised to find in his hotel parking lot a truck owned by NEG Micon, a Danish company that is one of the world's largest wind turbine manufacturers. Flavin was elated: "At least one leading renewable-energy company, located halfway around the world, is confident enough of its business prospects in China that it now has its own vehicles in Beijing."

There is, of course, a real danger in relying on such hopeful scenarios. As important as optimistic forecasts may be in reminding us what is *possible,* they can also distract us from what is *probable* — namely, that the transition to a new energy system will be enormously challenging and the outcome almost completely uncertain. We may know, for example, that the energy economy of 2030 will be a hybrid of sorts, meeting demand with alternatives fuels and improved efficiencies, yet still heavily reliant on hydrocarbons — but we have little idea how large a share each energy source will be providing. We know that oil will have ceded some of its share of the transportation sector to some combinations of alternative fuels or energy technologies, but again, it is unclear which alternatives, or at what price. We know that our climate will be warmer, and that our various environments will be changed, perhaps forever, but we don't know how dramatically or fatally. Above all, we know that our energy lifestyle — how much we use and how we use it — will have changed, perhaps radically, but we don't know whether these changes will have been proactive and considered or reactionary and shortsighted.

In this sense, envisioning an energy future is as much about knowing what *can't* happen as imagining what can. We know, for example, that although the transformation of the energy economy will be market-driven, we won't get to the future we need without some degree of government intervention. We know that although the world must use energy differently than it does now — more efficiently and more thoughtfully — we can't realistically expect individuals, organizations, or nations to use *less* energy willingly if doing so means accepting lower living standards. We also know that a real energy revolution cannot happen without the involvement of Japan, Europe, and the United States — the only nations with the economic strength and technological know-how to bring this future into being, but also the countries with the most to lose should this transition fail.

We know that for all the importance that technology will have, we cannot expect a technological magic bullet. Of course, breakthroughs in some core technology can radically alter the course of the energy economy — the invention of the gasoline engine was a critical impetus for the early development of the oil-based energy economy, and some new breakthrough (a cheaper automotive fuel cell, for example, or a dramatically more efficient solar panel) could completely change the path of our energy future.

Yet we have also seen enough to realize that what technology gives us it can also take away. Human history is littered with brilliant mistakes — promising innovations that through inherent weakness, or poor timing, or simple bad luck, failed to deliver on their promise. Nuclear power was regarded for decades as the energy of the future, a clean, quiet power source that seemed well on its way to a dominant share of the energy market — until the accidents at Three Mile Island and Chernobyl. Almost instantly, "nukes" became a huge liability, opening a massive gap in the global power supply that nations are still struggling to fill with other energy sources.

"Failures in technology can be just as big a disruption to the energy economy as any kind of geopolitical crisis," says Gerald Stokes, director of the Joint Global Change Research Institute in Maryland. "What happens when the first hydrogen fueling station in Germany blows up, or you get a massive leak from a CO_2 pipeline that suffocates a bunch of people? An accident like that could terminate a technology and eliminate an entire energy option overnight."

The only way to minimize such risks is by hedging our bets — by putting in place policies that encourage countries and companies to innovate, but also by aggressively pursuing as many technologies as we can afford to. We may, for example, end up using a single fuel, like hydrogen, for our transportation and power, yet employ a broad range of technologies for generating that hydrogen: solar, wind, and tidal, perhaps, as well as others we have yet to imagine. In short, we need to have more options, not fewer. This means both avoiding the tendency to back a single technological horse — fuel cells over gasoline-electric hybrids, for instance — and avoiding the impulse to ostracize technologies out of hand. Making synthetic crude from tar sands may be unfeasible now, but with some future breakthrough in carbon capture technologies, it might play an important role in the transition to a post-oil economy.

Likewise, whereas nuclear energy seems untenable today for a host of technological, economic, and political reasons, breakthroughs in design, manufacturing, and waste storage could resuscitate "nukes" as a viable energy option. The "demonization of nuclear power is not helpful in a world which, for better or worse, gets nearly a fifth of its electricity from fissioning uranium and where many countries would find themselves in a precarious situation in regard to their electricity supply if they were to opt for a rapid closure of their reactors," says Vaclav Smil, an expert in energy and economics at the University of Manitoba. For all its risks and flaws, Smil says, nuclear power has been essential in keeping CO_2 emissions from being far worse than they are today. If all electricity that is currently produced by nuclear plants were to be produced by coal-burning plants, Smil says, global CO_2 emissions would be about a third higher — 2.3 billion tons — than they are today. "Curiously," Smil notes, "this impressive total of avoided emissions is almost never mentioned in current debates about the management of greenhouse gases."

Yet like most of those who devote their waking moments to imagining our shift to a new energy economy, Smil is far less concerned with the tech-nology that we embrace than with our ability to see these technologies as mere elements in a larger, proactive, and very long-term energy strategy. As Smil puts it, the "critical ingredients of an eventual success are straightfor-ward: beginning the quest immediately, progressing from small steps to grander solutions, persevering not just for years but for generations — and always keeping in mind that our blunders may accelerate the demise of modern, high-energy civilizations, while our successes may extend its life-span for centuries, perhaps even for millennia."[4]

Which path will we take? When I began work on this book, it was with a profound sense of pessimism. Given what I knew of the problems associ-ated with the modern energy economy — from pollution and declining production to mounting carbon emissions — I was dubious that the pro-cess could be turned around in time to make a difference. Experts talked endlessly of disruptions so savage that they would push the world into a kind of permanent energy crisis, a "forever war" in which concerns about future consequences like energy poverty or climate were sacrificed for near-term "security of supply." Or perhaps we could manage to avoid a disrup-tion for a few years or decades, meanwhile letting our "business-as-usual" energy economy move forward, its various instabilities and volatilities pro-liferating by the year, setting us up for a megadisruption from which civili-zation simply would not recover.

Today, my perspective is far more complex. I've come to see that the energy business is so innovative and fast to react and has proven so capable of overcoming obstacles in the past that I no longer doubt that companies themselves will be able to cope with the coming challenges — provided they get the right signals from government.

Here, too, the world is changing. For every reluctant policymaker in Congress and the White House, in Beijing and Moscow, in Riyadh and Lagos, there are leaders who are either brave enough to push for a new en-ergy order or smart enough to see the political or economic advantages to moving forward. Iceland is launching the world's first hydrogen economy. Germany, Denmark, and Holland are adding renewables capacity at rates that exceed even their own optimistic projections.

And Europe isn't the only place where politicians are seeing green. In

October 2003, Republicans in the U.S. Senate only narrowly defeated a bill that would have capped CO_2 emissions and created a national system for emissions trading. Despite heavy lobbying by mining, automotive, and utility interests, forty-three senators voted for the Climate Stewardship Act — the best performance for any climate-related federal legislation in the U.S. Senate — and its main sponsors, Democrat Joe Lieberman and maverick Republican John McCain, say they may reintroduce the bill in spring of 2004.[5] Lieberman and McCain aren't alone. Only a few weeks earlier, California's new governor, the cigar-chomping, Humvee-driving, archconservative Arnold Schwarzenegger, floored energy advocates when, less than forty-eight hours after being elected, he rolled out a plan to accelerate California's already aggressive targets for energy efficiency and renewable energy. If Schwarzenegger follows through — and the jury is still out — not only will California rival Germany and Denmark in the ambitiousness of its energy plans, but the Golden State will create an important and badly needed domestic market for new energy technologies, as well as a model for other states and even the federal government. One can even imagine a wave of miniature, state-led energy revolutions — not because other U.S. politicians will want to emulate Arnold Schwarzenegger, but because the evidence of the failing energy system will gradually become more conspicuous and thus harder for even the less daring among our politicians to ignore.

Perhaps most encouragingly, we may have more time than we think to overcome the current political inertia. In today's political environment, for example, pushing a carbon tax seems almost impossible. But as John Holdren, the former White House climate adviser, has pointed out, an entire climate policy need *not* be formulated and put in place today. Indeed, it might be preferable to put off some elements of a climate strategy. Allocations for per capita CO_2 emissions, for example, which many climate experts see as inevitable, would be strongly opposed in Washington today because those allocations favor developing countries and disadvantage wealthy states. But, says Holdren, such a policy "does not need to be politically feasible today, because [per capita CO_2 emissions allocations] would not need to begin being phased in before 2015 or 2020, by which time people's everyday experience of the impacts of climate change is likely to have stretched considerably the scope of what domestic and international politics will allow."[6] This is the perverse benefit of a slow-motion calamity: for a few decades, at least, the tougher decisions can wait.

Frankly, though, the thought of any kind of delay, no matter how rationally justified, terrifies me. No matter how successful or diverse our technology portfolio is, and no matter what kind of time frame we are working with, the sheer magnitude and complexity and unpredictability of the task at hand gives us little choice but to start transforming our energy system *now*. Energy poverty is not some future problem that may or may not materialize, but one that is occurring right now and will generate widespread instability and conflict if it is not immediately addressed. Even the long-term energy problems, like the decline of cheap oil or rising CO_2 concentrations, call for immediate action. It may be true that we can take two or even three decades to deploy carbon-free technologies and policies without seriously exceeding our 550ppm carbon budget. The point to remember here, though, is that to have those technologies ready by 2030, we need to start working on them today.

Starting now dramatically improves our chances of success, because it means we have more options, more freedom in how we deal with our energy problems. Starting now will allow our solutions more time to work, which means that we could take the cheaper, low-intensity routes — the incremental improvements in energy efficiency, for example, or the gradual improvements from low- to no-emission cars, or the cost-effective phasing out of coal-fired power plants — rather than having to make a last-minute, potentially ruinous leap to fuel cells. Starting now means we can test a fuller range of energy technologies and develop a full range of energy tools and methods and policies that give us an energy economy that is more diverse, more flexible, and, we hope, more effective.

Conversely, the costs of inaction are significant. Each year that we fail to commit to serious energy research and development or fail to begin slowing the growth of energy demand through fuel efficiency, each year that we allow the markets to continue treating carbon as cost-free, is another year in which our already unstable energy economy moves so much closer to the point of no return. Every delay means that our various energy gaps, when we finally get around to addressing them, will be wider and costlier to fill. By then, it will be too late for low-cost solutions and diverse portfolios and smooth, incremental transitions. Instead, we will need large-scale solutions that can be deployed rapidly. Little room will remain for concerns about sustainability or efficiency or equity, and our chances for long-term success will be seriously impaired.

The implications are stark. If we are to have any chance at building the kind of energy future we want, rather than having one foisted upon us, we need to begin constructing that future today — not in 2010, when the political atmosphere has perhaps become more favorable; not in 2020, when non-OPEC oil may have plateaued and rising oil prices are pounding our energy economy into a new, not altogether desirable shape; and certainly not after some supply disruption or energy war makes us even more defensive and reactionary and xenophobic, and thus even less inclined to save the world. In other words, we no longer have the luxury of simply waiting to see how the energy economy evolves and hoping for the best. From now on, we must take a proactive role in building our energy future, first by understanding why and how our energy system must be transformed, and then by working to ensure that the shift takes place. For, ultimately, the question facing us isn't whether our energy systems will change — indeed, the process is already well under way — but whether we can live with the outcome.

AFTERWORD

In the twelve short months since *The End of Oil* first appeared, the issues it raised about the health of the global energy economy have only become more pronounced. You can't turn on the TV, pick up a newspaper, or even drive past a gas station without being reminded that our system for producing and consuming energy worldwide is under enormous strain and that the very notion of "energy security" has become a thin fiction.

The most visible signs, of course, can be found on the world oil markets. In 2004 the price per barrel of crude topped fifty-five dollars — almost triple the three-decade average and nearly twice what it was before U.S. forces "secured" the oil fields in Iraq. True, today's oil prices haven't matched those during the energy crisis of the 1970s; in current dollars, the price then was well over eighty dollars a barrel. Yet the present cost of crude is having a profound impact nonetheless. The robust economic growth of the last several decades was possible in part because energy, especially oil, was so cheap. And if our modern economy is less affected by energy costs than its predecessors were, today's price spikes have managed to arouse that almost forgotten beast, inflation, provoking fears that the economic recovery we've all been counting on may take a long time to arrive. If we factor in the price of oil's cousin, natural gas, which has climbed recently to more than triple its historic average, it looks as if the long, blissful era of cheap energy is over.

Interestingly, even as oil companies rack up record revenues, the energy industry itself seems to have lost some of its old confidence. Although the United States and Saudi Arabia — the two bookends of the global oil order — continued to officially insist that high oil prices are an anomaly and that there is plenty of oil to go around, energy companies and the energy market as a whole seem to feel otherwise. The market is acutely aware that demand for oil is now growing far faster than anyone had anticipated — especially in countries like China, which now imports more oil than any other nation except the United States. The market knows that production is straining to keep up with this demand and that the Saudis, the Russians, and other formerly flush oil states are pumping crude at very close to their maximum. Above all, the market knows that the balance between supply and demand is now so tight that the slightest hiccup — a riot in Venezuela, a terrorist bomb in Baghdad — will send not just ripples but sharp spikes through the entire economy. Volatility has become the new market reality — not a reassuring scenario for a world that still depends on oil for 40 percent of its energy and more than 95 percent of its transportation fuel, a world that still has no real alternative to oil.

And keep in mind that this grim scenario is the short-term picture. In *The End of Oil*, I argued that in the longer term our oil-dominated global energy system was vulnerable to three major threats — depletion, environmental degradation, and geopolitical instability. If anything, these threats have become more acute during the last year. Today, for example, we find even more evidence that oil production is likely to hit a peak sooner rather than later. Western oil companies struggle to find enough new oil to replace the oil they are selling on the world markets. State-owned oil operations are also in trouble. According to a study last fall in the respected *Petroleum Review*, in Indonesia, the United Kingdom, Gabon, and fifteen other nations that together supply nearly a third of the world's daily oil needs, production is now falling by 5 percent a year — more than twice the rate of decline of the year before. Effectively, this means that other oil producers, like Saudi Arabia, Russia, and Venezuela, must pump extra oil just to keep global supplies steady, to say nothing of *raising* production to meet soaring demand. Chris Skrebowski, editor of *Petroleum Review* (and a former analyst for BP and the Saudi national oil company), put it this way: "Those producers still with expansion potential are having to work harder and harder just to make up for the accelerating losses of the large number that have clearly peaked and are now in continuous decline."

Optimists insist that the problem here is financial, not geological: the only reason production is falling is that oil states haven't invested enough money to increase their capacity. It is true that investment has not kept pace and that the global oil industry will need to spend several trillion dollars by 2020 to meet projected oil demand, but it now seems likely that no matter how much they invest, those oil states simply do not have the massive reserves they've always claimed — and upon which our energy optimism has largely rested. At an energy conference in Houston last spring, Saudi oil officials admitted that production at their largest fields was being maintained only by the injection of massive volumes of seawater to force the oil to keep flowing out. They also admitted that Ghawar, the largest oil field ever discovered and a mainstay of the world oil business, was more than half depleted and that reserves in parts of Ghawar were down to just 40 percent of their original volume.

As stark as such admissions may seem, some U.S. experts say this new candor doesn't go nearly far enough. At the same conference, Matt Simmons, a Houston energy investor and Bush administration energy adviser who has studied trends in world oil production, made the case that Ghawar is actually closer to 90 percent depleted and that the Saudi oil kingdom is much nearer its production peak than anyone in Riyadh — or Washington — wants to believe. The Saudis quickly rejected Simmons's thesis, insisting, as they always have, that Saudi Arabia, with its wealth of untapped fields, has enough oil to fuel the world for decades to come. But many oil analysts I've spoken to since the conference are far less confident about the "big" Saudi oil reserves. Simmons makes no apologies. "We could be on the verge of seeing a collapse of 30 or 40 percent of [the Saudis'] production in the imminent future," he was later quoted as saying. "And imminent means sometime in the next three to five years — but it could even be tomorrow."[1]

Depletion is, of course, only one threat to the modern energy system that depends on hydrocarbons. Levels of atmospheric carbon dioxide, the primary driver for climate change, continue to climb rapidly, as do signs that climate change is occurring. Polar ice is melting, as are glaciers around the world. And while parts of the world caught a break in 2004 — in the eastern United States, for example, temperatures were slightly cooler than the historic average — the warming trend continues, as do weather phenomena thought to be related to climate change. The eight named tropical storms that appeared in the Atlantic basin during August 2004 broke another record. New studies also show that crop yields are declining worldwide as temperatures rise — a trend that will be dangerously destabilizing in countries already suffering from poverty and unrest. In fact, the political ramifications of climate change are now gaining attention outside of activist circles. Early this year the U.S. Defense Department released a report stating that rising temperatures could be so destabilizing to world governments that "disruption and conflict will be endemic" and "warfare would define human life."[2]

Just as disturbing, the prospects for relief seem even dimmer now than they did a year ago. Although Europe has embarked on a relatively aggressive policy to cut CO_2 emissions, America remains as reticent as ever. The Bush administration did concede, a few months before the election, that climate change might indeed be linked to human-caused emissions. For all that, the White House hasn't shown much interest in enacting a climate policy that could actually cut those emissions. And our climate problems will not be solved simply by persuading U.S. lawmakers to act. For better and for worse, the future of the world's climate lies more and more in the hands of China and India, both of which will probably fuel a large percentage of their much-needed economic growth with climate-killing coal.

Sadly, these concerns about oil and climate may be overtaken by developments in the third arena: energy geopolitics. Oil has always been central to global power and has been at the heart of many of the most important political events of the twentieth century. But in the last year, oil's capacity to influence and upset the balance of global power has become more starkly apparent. Continued unrest in Venezuela and armed rebellion in Nigeria have demonstrated graphically just how vulnerable the global oil system is to the smallest political perturbations. In Russia, Vladimir Putin's steady consolidation of political power — and his recent battles with oil tycoon and political rival Mikhail Khodorkovsky — have thrown that country's oil sector into chaos and dampened Washington's hopes that a flood of new Russian oil would finally break the OPEC monopoly. Next door, China and Japan continue to spar diplomatically over access to Siberian oil. China is now so desperate for oil imports that it has stepped up efforts to forge a petro-alliance with the Middle Eastern producers that the United States is courting, effectively putting Beijing on a diplomatic collision course with Washington.

Yet perhaps the most disturbing developments in oil geopolitics have centered on Iraq, where the American-led war has already cost more than $100 billion, caused thousands of military and civilian casualties, and given radical Muslims more grist for their anti-West mill. The Bush administration continues to insist that the war wasn't "about oil." Yet in the past year, a series of high-level government investigations have steadily discredited the president's public rationales for invading Iraq — namely, that Saddam Hussein had weapons of mass destruction and that he maintained ties with Al Qaeda.

Ironically, even if the United States had hoped that a post-Saddam Iraq could be a new source of oil and a solution to America's growing oil anxieties, such hopes have proven false. The inability of the U.S. occupation forces to keep oil pipelines from being blown up, to protect civilian workers from being killed or kidnapped, or to maintain anything resembling civil order have undercut efforts to restore Iraq's oil exports, much less raise them. If anything, the war has actually made American energy security *worse:* not only is Iraq itself exporting less oil than before, but its neighbors are now less politically stable and thus less reliable as oil suppliers than they were before the war.

In fact, the greatest casualty of the Iraq war may be the very idea of energy security. Before the war, it was generally accepted by world leaders — and oil traders — that if global oil production truly did become threatened by political instability or terrorism, the United States could restore order, and exports, through some measure of diplomatic or military intervention. That confidence led George W. Bush to assure us he could use diplomatic leverage — "jawboning" — to persuade OPEC to raise production. More to the point, that confidence has given rise to an American energy strategy of projecting military force throughout the oil-producing world in order to guarantee access to oil — a Cold War–like doctrine that has seen the United States building up a military presence not only in the Middle East but in Africa, Central Asia, and South America, oil-rich regions whose exports might otherwise be vulnerable to interruption by terrorists or political unrest.

But with the continuing fiasco in Iraq, it is now clear that even the most powerful military entity in world history cannot stabilize a country at will or "make" it produce oil simply by sending in soldiers and tanks. In other words, since the Iraq invasion, the oil market now understands that the United States *cannot* guarantee the security of oil supplies — for itself or for anyone else. That new and chilling knowledge, as much as anything else, explains the high price of oil.

And what of efforts to move *beyond* oil? In *The End of Oil* I was highly skeptical of many of the more prominent energy alternatives. I was especially harsh on hydrogen fuel cells, a technology that its proponents, including the Bush administration, claim is all but imminent, yet whose real potential is still blocked by a number of technical, economic, and political obstacles. I also challenged the popular faith in renewable technologies like wind and solar power, which show great promise but which still lack the power density to easily and cost-effectively replace hydrocarbons. Many critics have taken me to task as too pessimistic; some have insisted that these and other energy alternatives are actually much closer to being cost-effective than I had suggested.

Have the past twelve months changed my mind? Perhaps. I concede that these technologies are improving rapidly — all the more because oil and natural gas prices are so high — and that governments and private companies should redouble their efforts to bring these and other energy technologies to the market. Yet I remain convinced that their time is still to come, and the most prudent course in the meantime is not to wait for the technology of tomorrow to sweep down and save us but to push *existing* technologies toward greater efficiency and lower emissions. In fact, no matter what energy technologies we end up using twenty or thirty years from now, we still won't have enough energy for everyone if we haven't found ways to use much less of it. Efficiency remains our greatest hope.

In this context, the most encouraging story of the last year has been the rising acceptance of gas-electric hybrids. These vehicles cut fuel use and emissions roughly in half, use a current technology, and have already earned respect from auto analysts and, more to the point, consumers. After the surprising success of hybrids from Toyota and Honda (waiting times are still more than a year in some markets), Ford has launched a hybrid SUV, the Escape, which gets nearly forty miles per gallon in the city, and other U.S. companies will have hybrids out soon. To be sure, hybrids will account for only a tiny piece of the enormous American auto market for some time. But given how quickly attitudes about hybrids have changed among consumers *and* automakers — and given that gasoline prices won't be coming back down very much for years — we shouldn't be surprised to see hybrids rapidly penetrate the U.S. auto market.

If the auto industry is ripe for an efficiency revolution, it's not clear whether that revolution can spread to other sectors. As *The End of Oil* argued, industrial nations currently waste an extraordinary amount of en-

ergy through poorly designed homes, office buildings, and factories — all
of which could be redesigned for dramatic energy savings. Yet the daunting
and hugely expensive task of reengineering such large pieces of infrastructure will require more than the kind of snappy ad campaign that has
worked for hybrid cars. Improving efficiency outside of the auto sector
must begin in the political sphere with a new consensus by policymakers
that the energy system must change in fundamental ways — and, above all,
real leadership to ensure that such change actually happens.

On this count, the prospects are still mixed. In America, local and
state policymakers have begun acting as if energy mattered. California, long
the leader in activist energy policy, has enacted tough new auto emission
standards and is again at the forefront of policymaking and technology research. Other states are also moving toward energy and emissions policies
that could eventually remake the American energy system. And, after decades of willful obliviousness, consumers seem to waking up to the possibility that the age of cheap energy is over. Not only are sales of the biggest
SUVs, such as the Hummer and Ford's obscene Excursion, flagging but, according to some surveys, a growing number of consumers say that if energy
prices stay high, they will seriously consider buying a more efficient car,
taking mass transit, or even moving to a home that shortens their com-
mute.

In other ways, however, energy attitudes remain unchanged. Even as
businesses and consumers scramble to adjust to higher energy prices, many
governmental leaders insist that the status quo needs no help. Indeed, for
anyone hoping for an American energy revolution, the reelection of President Bush last fall was hardly encouraging. Bush's energy policy during his
first term centered mainly on the outmoded assumption that the key to
U.S. energy security was simply to find *more* hydrocarbon energy, either at
home or abroad. As for creating a *new* energy economy, the president has
no vision. Early on, White House officials mocked energy "alternatives"
and the idea of cutting energy demand. When that approach played poorly,
the White House offered up vague promises of a "hydrogen economy," yet
committed only a fraction of the funding needed to bring such an economy
into being and said nothing about the obvious need to aggressively cut energy demand now while we wait for the development of a brave new energy
system.

The notion that the second Bush administration will be any different seems rather implausible. Although the president has the political "capital" and the credibility with the energy industry to launch a new energy policy, he seems far more likely to continue the policy of defending the status quo while denying that the current system is in need of change. Last fall, as crude oil prices sailed toward fifty-five dollars a barrel, White House officials insisted that the price rise was a "short-term phenomenon" driven by "an anomalous set of circumstances" including "the geopolitics of oil" — as if the geopolitics of oil were something that happens only once every few decades.

In many respects, the debate over energy is coming down to a single question: can the market deliver the kind of new energy economy we need, or must government step in? For the past few years the free-market argument has been dominant. Proponents, such as Alan Greenspan, the chair of the all-powerful Federal Reserve, insist that high oil prices will actually force whatever "adjustments" to the economy are necessary, without clumsy government intrusion. According to Greenspan, we will not only use less oil, thus reducing demand (and eventually the price), but just as important, the market will be spurred to develop alternatives to oil — everything from ethanol to hydrogen — which will begin to look quite affordable compared to oil. That's what happened in the 1970s, after the Arab oil embargo. And, says Greenspan, that's what is happening today. In other words, the current high prices are simply the beginning of a natural economic process, a historic transition that will eventually produce a new, smarter, cleaner, more efficient energy economy *without* government intervention.

But if anything, the last twelve months have pointed up the flaw in this otherwise comforting bit of classical economic theory. Yes, high oil and natural gas prices will change our behavior and our technologies. And if we could count on a nice smooth price rise from now to, say, 2030, chances are that the market could indeed build us a new energy economy while avoiding any really nasty side effects, such as economic dislocation. But as we've seen during the last year, the chances of an orderly price rise are slim. Gone are the days when countries like Saudi Arabia and Russia and Nigeria had limitless reserves and scads of extra production capacity. Markets are tight and are expected to stay that way for years, which all but guarantees that we'll see not just price spikes but severe ones. In fact, many analysts say that the oil markets see the market of the future as resembling a kind of permanent wave, with per-barrel prices oscillating between thirty dollars and eighty dollars or more. And that's not even considering the possibility that oil production will hit a peak and begin to decline — a possibility that seems less and less fantastic with each passing week.

In *The End of Oil* I laid out a scenario that might follow such a production peak — a grim worst case featuring global recession, worldwide unemployment, economic chaos, and, perhaps, a dangerous and escalating competition among the big oil-importing nations over the remaining reserves in the Middle East. Nothing in the last year has made me think such a scenario is less possible. What has changed, I would argue, is our awareness. More people and policymakers now seem to understand that the energy system is in serious and growing trouble and that without a fundamentally new approach we are almost assured of a catastrophic failure. What our new awareness actually means is hard to say. It may be the first tentative step toward building a more sustainable energy economy. Or it may simply mean that when our energy system does begin to fail, and we begin to lose everything that energy once supplied, we won't be so surprised.

Notes

Chapter 1. Lighting the Fire

1 Smil, *Energy in World History.*
2 Ibid.; Beaudreau, *Energy and the Rise and Fall of Political Economy.*
3 Smil, *Energy in World History,* 147.
4 Around A.D. 1100, for example, the superpower of North America was a people known as the Cahokia. The seat of empire, just east of present-day St. Louis, was a walled city enclosing houses, warehouses, workshops, plazas, and enormous burial mounds. Outside this complex lay an urbanized area, with homes and gardens, surrounded by well-tended fields of corn, squash, and other crops. At its peak, Cahokia ruled several hundred square miles and perhaps one hundred thousand citizens — the largest New World empire north of Mexico. But the Cahokia had at least one weak point: fuel. Early on, they appear to have developed a system for harvesting firewood from outlying forests and transporting it, via a network of swamps and creeks, into the city to feed fires for cooking, heating, and pottery making, as well as to supply building materials. Yet inevitably, as the Cahokia expanded, so did their fuel requirements. Wood-gathering parties were forced to travel farther and farther from home, until at last the distance became prohibitive. By A.D. 1250, the Cahokia had faded into obscurity, victims, it seems, of America's first energy crisis.
5 Te Brake, "Air Pollution and Fuel Crises," 83.
6 In 1653, the English Parliament authorized the logging of most of the remaining Crown forests, including Sherwood, but to little avail. When London burned in 1666, the city was forced to rebuild almost entirely with imported timber. Te Brake, ibid.
7 Schurr, *Energy in the American Economy,* 69.
8 Ibid., 70
9 Buxton, *The Economic Development of the British Coal Industry from Industrial Revolution to the Present Day.*

10 Ibid.

11 Ibid.

12 Ibid., 85.

13 Natural gas, too, after reaching the surface, escapes in natural jets that, if ignited, can burn for years and have doubtless been the source of numerous legends.

14 Schurr, *Energy in the American Economy*, 116.

15 Alliance of Automobile Manufacturers, "Economic Facts," *www.autoalliance .org/ecofacts.htm.*

16 British Petroleum was originally founded by the British government as Anglo-Persian Oil.

17 If a large refiner feared that prices for his raw material — crude oil — were trending up, he could buy a futures contract locking in a good price for a cargo of oil to be delivered and paid for at some future date. Likewise, producers learned to insulate themselves from falling prices by locking in future sales. Futures quickly became lucrative business, and not an altogether untainted one, either. In a foretaste of the Enron scandals, unscrupulous oil traders would agreed to buy (or sell) oil at a given price, then attempt to manipulate the market, through rumor or worse, in order to drive the price higher (or lower) by the time of settlement.

18 Williamson and Andreano, *The American Petroleum Industry,* 1:354.

19 O'Connor, *Empire of Oil,* 259. Some observers even feared that another war would break out between the two oil powers, England and the United States.

20 Williamson and Andreano, *The American Petroleum Industry,* 2:763.

21 Yergin, *The Prize.*

22 Schurr, *Energy in the American Economy,* 119.

23 From 5.2 million barrels a day to 8.45 million barrels a day: Williamson and Andreano, *The American Petroleum Industry,* 2:805.

24 In fact, U.S. fields could have produced another two million barrels a day, thereby forestalling imports for years, but the oil companies were held back by overzealous state authorities intent on avoiding a glutted domestic market. The resulting gap was quickly filled with foreign oil, much of it produced and imported by large U.S. companies with foreign holdings. Small independent oil companies, which depended entirely on regulated domestic production, complained bitterly that they were being killed by cheap foreign oil.

25 O'Connor, *Empire of Oil,* 252.

26 South Coast Air Quality Management District, www.aqmd.gov/smog/ inhealth.html#historical.

CHAPTER 2. THE LAST OF THE EASY OIL

1 Personal communication.

2 Personal communication.

3   This estimate includes about one trillion barrels in known reserves, plus another seven hundred billion barrels in "reserve additions" — basically, new oil discovered in existing or even abandoned fields. International Energy Agency, *World Energy Outlook,* 47.

4   Future demand depends on a variety of factors, from how fast the economy grows to how energy-efficient we become; oil price, too, is key: higher prices tend to lower demand and thus delay reaching a peak.

5   They engaged in underreporting mainly to avoid being compelled by their host governments to produce more oil and thus to pay additional taxes and royalties.

6   Nor is it merely the OPEC countries whose numbers are questionable. During Mexico's financial collapse in the mid-1990s, oil officials there reportedly exaggerated the reserves to enhance the country's collateral and borrowing power: Campbell, *The Coming Oil Crisis,* 73.

7   Riva, *World Oil Production After Year 2000: Business as Usual or Crises?*

8   In fact, this "front-loaded" pattern of discovery shows up across oil fields of all sizes. According to an analysis by USGS scientists, in any given geographical area — whether one is looking at a single oil field or a single oil-producing country, a region or an entire hemisphere — the bulk of the oil is almost always discovered during the first third of the exploration period.

9   Personal communication, October 20, 2003.

10 Personal communication, October 20, 2003.

11 Practically the only optimistic forecasts came from adherents of a somewhat bizarre Soviet theory that oil is created continuously deep in the bowels of the earth.

12 Recovery rates depend on a wealth of factors, from the viscosity of the oil — thicker oil tends to be harder to extract — to the size of the pores in the reservoir rock to how much gas pressure remains in the field.

13 Personal communication.

14 Personal communication, January 2002.

15 It is thus off-limits to Western companies, which, had they been allowed to look, would probably have found it decades ago, when most megagiants were discovered.

16 "BP Gushes over Treasure in Deep Water of Texas Coast," *Dallas Morning News,* August 6, 2002.

17 "Oil Majors Wonder," *Financial Times,* April 25, 2002.

18 Personal communication.

19 Reuters, "UK Hails Crop of Small North Sea Oil, Gas Fields," August 5, 2002; Urquhart, "North Sea Oil Output to Peak This Year."

20 Ruppert, "Interview with Matt Simmons."

21 Production, or "lifting" costs, in the Middle East are around $1.50 a barrel, compared with $4 to $6 a barrel for non-OPEC oil. Add $7 to $8 for shipping and a comfortable profit and you have a price of around $10 a barrel, which is what some analysts say oil would cost in a free market. OPEC members,

however, base their oil price not on actual costs of production but on the very high level of revenue they need to keep their corrupt and inefficient governments in the black — around $22 to $28 a barrel. "The higher costs of non-OPEC oil are a classic indication that oil reserves on a global scale are being depleted, just as should be expected," says energy economist-consultant Alfred Cavallo (unpublished article).

22 Personal communication, October 20, 2003.

23 Greenland, for example, is a hot prospect not because big oil has actually been found there, but, mostly, because Greenland has geological features similar to those beneath the oil-rich North Sea.

24 Personal communication.

25 Agence France Press, "ExxonMobil to Plug Azerbaijan Well," February 25, 2002.

26 Ibid.

27 Reuters, "Oil Majors Cool on Kazakh Offshore Plans," October 3, 2002.

28 Environmentalists could also delay development of a host of "unconventional" oil projects, including plans to mine the huge tar-sand fields in Alberta. Technological advances now make it possible to refine tar sands into usable crude at a cost that is competitive with that of "conventional" oil, but because the refining process itself produces so many pollutants, a full-scale tar-sands industry must first overcome substantial political hurdles.

29 Personal communication.

30 Ruppert, "Interview with Matt Simmons."

31 Personal communication.

32 Personal communication.

CHAPTER 3. THE FUTURE'S SO BRIGHT

1 Motavalli, "Harnessing Hydrogen," 34.

2 Ibid.

3 Scott, "Fuel Cells Power Toward the Mass Market," 41.

4 Personal communication from John Turner, National Renewable Energy Laboratory.

5 "Buried Losses: The Journey from Plant to Coal."

6 This is why pure hydrogen is often used to fuel rockets, which need to release a maximum amount of energy.

7 Energy is the capacity to do work — such as moving a car or creating heat. The more energy you have, the more work you can do. Fuels like gasoline or carbohydrates or hydrogen contain varying amounts of energy, which can be liberated via oxidation and other processes and put to work. To use a standard measure, one *calorie* of energy is the amount of energy required to raise one cubic centimeter of water one degree Celsius.

8 "My first three months," recalls David Watkins, a chemical engineer involved in the early days, "I read all this crap on [fuel cells] and decided, 'Hell, no one's done any engineering.'" In other words, the possibilities for innovation and improvement were unlimited.

9 Koppel, *Powering the Future.*

10 Ibid.

11 Ibid.

12 Automakers fought on two fronts, lobbying and litigating to overturn the ZEV laws, but meanwhile slowly making preparations to meet those laws through new programs for electric cars, electric-gasoline hybrids, and fuel cell vehicles. Shrewdly, carmakers also cheerily supported the Partnership for a New Generation of Vehicles, or PNGV, a 1993 federal program to help U.S. automakers develop a super-fuel-efficient, eighty-mile-per-gallon car by 2004. Although PNGV did advance fuel cell technology marginally, its goals were nonbinding: automakers did not have to sell a single fuel-efficient car; they had supported PNGV mainly to distract lawmakers and consumers from demanding tougher fuel-efficiency requirements. By 1996, after industry lobbyists finally killed California's ZEV law and throttled congressional attempts to raise fuel efficiency, alternative automotives seemed at a standstill.

13 Initially, Daimler had wanted to build cars with hydrogen-fueled internal-combustion engines and had regarded fuel cells mainly as a means to generate onboard electricity for accessories. But after seeing Ballard's technology up close, the Germans realized that the fuel cell itself was the key and, switching strategies, announced ambitious plans for a fuel cell–powered vehicle within the decade.

14 As cited in Koppel, *Powering the Future,* 219 and 225.

15 Ibid., 218.

16 Hoffmann, "Hydrogen," p. 146; Motavalli, *Forward Drive,* p. 157; Koppel, *Powering the Future,* p. 222.

17 Koppel, *Powering the Future,* p. 222.

18 Hoffman, *Tomorrow's Energy,* viii.

19 Personal communication from Karen Miller, November 2002.

20 As far back as 1997, Ballard's new chairman, Firoz Rasul, had warned, "Offering zero or low emissions levels will not be enough" to break into the consumer market. No one doubted that issues of cost and fueling could be solved eventually; but the fixes would take time — far longer than many investors appeared willing to wait.

21 Personal communication.

22 Personal communication, December 12, 2002.

23 Personal communication from Peter Hoffmann, December 12, 2002.

24 One company estimates refitting costs of four hundred thousand to five hundred thousand dollars per station. The United States has 200,000 service

stations, a third of which must be refitted before consumers will be comfortable buying a hydrogen-powered car.

25 Quoted by Alec Brooks, Evworld.com.
26 Personal communication.
27 Personal communication.

CHAPTER 4. ENERGY IS POWER

1   Personal communication.
2   Ecuador and the minuscule Gabon were briefly OPEC members but have since withdrawn.
3   Mitchell et al., *The New Economy of Oil*, 136.
4   The natural decline in American production was exacerbated by a series of inflation-fighting measures undertaken in 1971 that capped the price of domestic oil and thereby discouraged U.S. oil companies from expanding domestic production. (A similarly misguided policy in the power markets contributed to the California energy crisis of 2000.) By the time Libya raised prices, the U.S. domestic market was already under serious pressure.
5   The Venezuelan dictator Marcos Perez, when he wasn't torturing and killing thousands of political opponents and establishing himself as the archetype for all South American dictators, squandered hundreds of millions of oil dollars on lavish monuments, including a replica of New York's Rockefeller Center.
6   Haggerty, *VENEZUELA — A Country Study*.
7   Ibid.
8   Mitchell et al., *The New Economy of Oil*, 136.
9   Personal communication.
10 Verleger, "A Collaborative Policy to Neutralize the Economic Impact," 1.
11 Personal communication.
12 In January 1998, Rumsfeld, Wolfowitz, Perle, and other members of the conservative think tank Project for the New American Century (PNAC) wrote a letter to then-president Clinton, stating, "It hardly needs to be added that if Saddam does acquire the capability to deliver weapons of mass destruction . . . a significant portion of the world's supply of oil will all be put at hazard . . . The only acceptable strategy is . . . to undertake military action as diplomacy is clearly failing. In the long term, it means removing Saddam Hussein and his regime from power. That now needs to become the aim of American foreign policy." From PNAC's Letter to President Clinton, as cited on PNAC's website. *www.newamericancentury.org/iraqclintonletter.htm*.
13 Quoted in Diebel, "Oil War."
14 Burbach, "Bush Ideologues Trump Big Oil Interests in Iraq," *Redress Information & Analysis*, www.redress.btinternet.co.uk/rburbch21.htm, September 30, 2003.

15 Personal communication.
16 "OPEC's Attempt to Flex May Fold," *Chicago Tribune*, September 28, 2003.

CHAPTER 5. TOO HOT

1   "Global Energy Technology Strategy," 22.
2   The remainder comes mainly from the intestines of farm animals; for example, half of New Zealand's CO_2 emissions derive from its fifty million sheep and cattle — Reuters, October 18, 2002.
3   To help people visualize a ton of carbon, Gerry Stokes, director of the U.S. Joint Global Change Research Institute in Washington, likes to carry a five-pound bag of charcoal briquettes around. Each bag contains at least one hundred briquettes, which means that a car that gets twenty-five miles to the gallon produces carbon emissions the equivalent of throwing a briquette out the window every quarter of a mile. "Since the average American drives ten thousand miles a year, that's forty thousand briquettes per car, year in and year out," says Stokes. "Imagine the roadside."
4   Griscom, "the Thrill of Defeat."
5   And this despite the fact that the 1991 eruption of Mount Pinatubo cooled average temperatures during two of those years.
6   As reported in Gelbspan, *The Heat Is On*, 148.
7   See www.british-energy.com/environment/BE/academic/key_studies.htm.
8   Ibid., 162.
9   Sawin, "Interview with Rajendra Pachauri," 13.
10 Gelbspan, *The Heat Is On*, 159.
11 Even today, atmospheric CO_2 declines each spring as new plants absorb more carbon dioxide, then rises each fall, as plants die and decompose.
12 Concentrations over the last fifty thousand years have ranged from 200 to 270ppm.
13 "Special Report: CO_2 Sequestration Adds New Dimension."
14 Emissions of CO_2 are measured in terms of the "contained carbon" in the CO_2 molecule, a measurement that excludes the oxygen. If we include the oxygen, the emissions are substantially larger: every ton of coal or oil in fact produces *three* tons of CO_2 — a discrepancy in weight explained by the weight of the oxygen.
15 "Special Report: CO_2 Sequestration Adds New Dimension."
16 Or sooner: climate scientists fear that the trend could be accelerated by the release of methane, a potent greenhouse gas, which is now trapped beneath the permafrost and which, if released into the atmosphere, could cause so-called runaway warming.
17 Holdren, "The Energy-Climate Challenge," 26.
18 Personal communication.
19 Mintzer, Leonard, and Schwartz, *U.S. Energy Scenarios for the 21st Century*.

20 See www.puaf.umd.edu/CISSM/Projects/NIC/Geopol-Tech&Environment
.htm.

21 "Global Energy Technology Strategy."

22 Titus, "The Costs of Climate Change to the United States."

23 Europeans had also hoped to exploit the fact that their CO_2 emissions were
already down, for reasons having nothing to do with climate policy. East
Germany's heavily polluting power industry had been replaced with cleaner
power; Great Britain, in an effort to crush the coal unions, had shifted its
coal-fired electrical industry to gas. As a result, E.U. emissions had not
grown significantly since 1990, and Europeans states could cut emissions far
more easily and cheaply than the Americans — a serious economic advantage that would be undermined if the Americans won their emission exemp-
tions.

24 Indeed, the E.U. states were sufficiently unified in their disgust for American
climate policy that they were able to resolve remaining conflicts and ratify
Kyoto — something Clinton had been unable to do.

25 Goulder, "U.S. Climate-Change Policy."

26 O'Connor, *Empire of Oil*, 252.

CHAPTER 6. GIVE THE PEOPLE WHAT THEY WANT

1 "PetroChina and Sinopec Endure Violent Clashes over Petrol Stations," *Financial Times*, April 25, 2002.

2 Reuters, "China's Car Sales Hit One Million for First Time," December 16,
2002.

3 As reported in "China's Boom Adds to Global Warming Problem," *New York
Times*, October 22, 2003.

4 Calder, "Asia's Empty Tank," *Foreign Affairs*, March/April 1996, 56, as quoted
in Manning, *The Asian Energy Factor*.

5 Manning, *The Asian Energy Factor*, 60.

6 U.S. Energy Information Administration, *International Energy Outlook, 2001*,
199, 201, 204, 207.

7 I am indebted to Mark Hertsgaard's excellent *Earth Odyssey*, 158, for this account.

8 Ibid.

9 Mitchell et al., *The New Economy of Oil*, 9.

10 In December 1952, London suffered coal "smog" so thick that visibility was
cut to fifty meters. More than four thousand people died; the magnitude of
the disaster was not recognized for days, until undertakers began running
out of coffins and florists ran out of flowers. The ensuing outcry led eventually to laws regulating coal-fired emissions: www.portfolio.mvm.ed.ac.uk/
studentwebs/session4/27/greatsmog52.htm.

11 European Wind Energy Association, "Record growth for global wind," press release, March 3, 2003.
12 Lave, "A New CAFE," 2.
13 U.S. Energy Information Administration, *International Energy Outlook, 2001*, 179.
14 Hertsgaard, *Earth Odyssey*, 167.
15 Manning, *The Asian Energy Factor*, 70
16 Mitchell et al., *The New Economy of Oil*, 79.
17 Ibid. (tripled); U.S. Energy Information Administration (will double), *International Energy Outlook, 2001*, 199; Manning, *The Asian Energy Factor*.
18 Zhou et al., *Transportation in Developing Countries*, 7.
19 "Car Makers Prepare Profit Road," *South China Morning Post*, June 11, 2002.
20 In a textbook example of depletion, the forty-year-old Shengli peaked in 1987 at six hundred thousand barrels a day, plateaued for eight years under increased production efforts, and then, despite those efforts, went into an acute decline in 1996.
21 U.S. Energy Information Administration, *International Energy Outlook, 2001*, 179.
22 International Energy Agency, *World Energy Outlook, 2001*, 69.
23 Ibid.
24 Personal communication, February 7, 2003.
25 O'Ryan et al., *Transportation in Developing Countries*, iv.
26 Hertsgaard, *Earth Odyssey*, 244.
27 Xinhua News Agency, "Beijing Auto Show," June 13, 2002; *China Daily* (no headline or byline), June 13, 2002.

CHAPTER 7. BIG OIL GETS ANXIOUS

1 Because the lighter crudes yield more gallons of gasoline per barrel than heavier oils do and are thus more profitable to refine, companies preferred light crudes, especially in late winter, when they would begin building up their stocks of gasoline for the summer driving season.
2 Gheit is quoted in "BP Focuses on Water off the Texas Coast," *Dallas Morning News*, June 8, 2002, as reprinted on the Web site for Alexander's Gas & Oil Connections, www.gasandoil.com/goc/company/cnn23619/htm.
3 Banerjee, "For ExxonMobil."
4 Callus, "BP Says Will Miss Output Target in 2002."
5 Picerno, "If We Really Have the Oil."
6 "Shell Faces Lawsuit," *Financial Times*, January 26, 2004; Traynor, "Shell T & T."
7 Tar sand refining is so polluting that Canada recently had to exempt its tar sands operations from its otherwise aggressive policy to reduce CO_2 emissions.
8 www.power-technology.com/projects/mexicali/.

9   Ibid.

10 Personal communication, August 2003.

11 By one estimate, if nuclear power plants were banned and all current nuclear power replaced by coal-fired power, carbon emissions would increase from 6.5 billion tons a year to 9 billion tons.

12 World Nuclear Association (WNA), "World Nuclear Power Reactors, 2000–2001," www/world-nuclear.org/info/reactors.htm. I am indebted to Smil, *Energy at the Crossroads,* for this cite.

13 Of course, in industrializing countries, like China, where environmental regulations are few, coal power is much cheaper and is growing rapidly.

14 "The Strategic Role of Gas," speech delivered by Lord Browne, group chief executive of BP, to the 22nd World Gas Conference in Tokyo, June 2, 2003. www.bpgas.co.uk/perspectives/hot/060203.html.

15 Barrionuevo, "How Trinidad Became a Big Supplier."

16 According to one study, as much as 2 percent of gas piped through the United States system escapes into the air, a percentage that raises the climatic impact of a natural gas economy by a considerable margin. And the U.S. gas system is in far better shape than those in the rest of the world.

17 BP, *BP Statistical Review of World Energy,* 5.

18 Ruppert, "Interview with Matt Simmons."

19 Gas is measured by volume, in cubic feet.

20 Lance Van Anglen, Unocal Corp, as quoted in *Petroleum Finance Week.* www.hartenergynetwork.com/info.php?PETM.

21 Ryan Lance, ConocoPhillips, as quoted in *Petroleum Finance Week.*

22 Personal communication. www.bpgas.co.uk/perspectives/hot/060203.html.

23 Of course, companies wanted to avoid reacting too quickly, because higher prices were a handy source of profits, as long as they didn't reduce the demand or come to the attention of government regulators.

24 Weissman, "Days of Shock and Awe About to Hit the Natural Gas and Power Markets," Part 1, www.energypulse.net/centers/article/article_display.cfm?a_id=324.

25 Currie, "Natural Gas Supply."

26 Personal communication.

27 Personal communication. www.bpgas.co.uk/perspectives/hot/060203.html.

28 American Wind Energy Association press release, June 18, 2003.

29 Fee, "Russian and Iranian Gas and Future U.S. Security," *Middle East Economic Survey* on-line, September 15, 2003. www.mees.com/postedarticles/oped/a4637d01.htm.

30 This presented a particularly choice irony, since the Chinese probably would have taken Australian gas in any case, in the belief that Australia was more stable than either Qatar or Indonesia.

CHAPTER 8. AND NOW FOR SOMETHING
COMPLETELY DIFFERENT

1 As Maycock explains, in a California climate with two thousand hours of peak sun per year, the eight-dollar generator would produce two kilowatt-hours of electricity per year. The cost depends on the amount of capital used. For a twenty-year fixed-rate loan at 7 percent, with amortization of about 10 percent a year, costs would be about forty cents per kilowatt-hour.

2 Up to a point: the power curve tends to level off at around forty-five miles per hour, and at fifty-six miles per hour, most turbines automatically shut down.

3 According to a new study (Awerbuch, "Determining the Real Cost"), long-term gas price volatility actually drives up operating costs, and thus whole-sale electricity costs, far more than most utilities expect. When future volatility is correctly assessed, says Shimon Awerbuch, a former energy analyst at the International Energy Agency and author of the study, the long-term wholesale price of electricity from gas or coal is anywhere from one to three cents *higher* than industry analysts predict — whereas wind power's costs remain steady. "Because they ignore these risk differentials," Awerbuch argues, "traditional analyses incorrectly *overestimate* the cost of renewable-based electricity."

4 The remaining 10 percent is lost as a result of maintenance shutdowns.

5 Personal communication, June 26, 2003.

6 Smil, *Energy at the Crossroads,* 298.

7 Personal communication, June 23, 2003.

CHAPTER 9. LESS IS MORE

1 Lovins and Lovins, "Energy Forever," 2.

2 Because the American economy and energy consumption were both growing at the same rate — 3 percent a year — most experts believed economic growth and energy consumption were fundamentally linked, as if by some natural law, and that any attempt to conserve would curb American economic power. Interlaboratory Working Group, *Scenarios for a Clean Energy Future,* 2.6–2.8.

3 Rosenfeld, "The Art of Energy Efficiency," 37.

4 Personal communication.

5 Personal communication.

6 Personal communication.

7 Personal communication.

8 Smil, *Energy at the Crossroads.*

9 Unless they use the savings to buy an SUV.

10 Under the program, utilities that encouraged customers to reduce power by 1 percent could add a 1 percent charge to the customer's bill. Consumers saw no net increase in their bills, and utilities earned a small, but significant return for cutting power demand.

11 Personal communication.

12 Personal communication.

13 As mentioned earlier, it is now clear that California's energy "crisis" was a fiction, brought on mainly by dishonest utilities and rapacious energy traders. Recent reports have shown that utilities consciously chose to idle power plants in record numbers, in order to cut the supply of power and raise prices. As Lovins puts it, "the same system that met a fifty-three-billion-watt load in the summer of 1999 couldn't meet a twenty-nine-billion-watt load in January 2001 — not because half the capacity vanished, but because ten billion watts 'called in sick.'"

14 Smil, *Energy at the Crossroads*, 317.

15 See www.home.earthlink.net/~andrewrudin/index2.html.

CHAPTER 10. ENERGY SECURITY

1 Agence France Press, August 16, 2003.

2 Ibid.

3 Although several midlevel meetings were arranged between the White House and the Indian government (and although the Dhabol deal was, rather suspiciously, mentioned in the Bush energy policy), the White House, already under fire for its cozy relationship with U.S. energy companies, did little to help Enron.

4 Energy poverty takes a disproportionate toll on women: because women and especially girls do most of the fuel gathering, they have even less time for any activity remotely resembling self-improvement, such as learning to read or acquiring job skills, than do men.

5 World Energy Council, *The Challenge of Rural Energy Poverty*.

6 Personal communication.

7 "Energy Is Key Area," 29.

8 Arab OPEC nations are among the biggest donors to African countries.

9 Burn, "The Hunt for New Oil."

10 "China's Boom Adds to Global Warming Problem."

11 National Bureau of Statistics, as reported ibid.

12 Hertsgaard, *Earth Odyssey*, 181–82.

13 International Energy Agency, as reported in "China's Boom Adds to Global Warming Problem."

14 Manning, *The Asian Energy Factor*, 70.

15 The issue is mainly economic: whoever owns the land the pipeline traverses gets the construction dollars and the eventual transit fees.

16 "Higher Oil Prices Here to Stay," *Arab Oil & Gas Magazine,* August 31, 2003.

17 Ibid.

18 This may already be happening. Since 1999, under something called the Saudi Gas Initiative, Riyadh has invited bids from Western oil companies to develop the kingdom's vast and largely neglected gas resources for Saudi domestic use. True, this is gas, not oil, but many industry observers see in the Gas Initiative a face-saving first step toward eventually allowing the majors back into the Saudi Arabia oil fields — and a tacit admission that the Saudis lack the money to develop the gas reserves themselves.

19 "Higher Oil Prices Here to Stay."

20 McKillop, "Why Venezuela and World Oil Exporters Can Target U.S. $36–$45 per Barrel."

21 Ibid.

22 Baer, "The Fall of the House of Saud," 53.

23 Burn, "The Hunt for New Oil."

24 Russian Information Agency, "Japan Ready to Invest $14 Billion in Russia's Far Eastern Oil and Gas Projects."

25 Interfax New Agency, "Output Continues to Fall at Sinopec Shengli."

26 Obaid et al., "The Sino-Saudi Energy Rapprochement," 35.

CHAPTER 11. THE INVISIBLE HAND

1 Ford, for example, is pursuing a "hedging strategy," simultaneously developing several fuel options — including natural gas, ethanol, gasoline-electric hybrids, and hydrogen fuel cells, but says it won't roll out any new technology until it can "prove itself economically viable."

2 Hakim, "Cloaked in Green."

3 Chris Isadore, "Automakers Say Oil Spike Won't Stab Them," CNN Money on-line, www.money.cnn.com/2003/01/07/news/companies/autoshow_fuel, January 7, 2003.

4 So large are the sector's capital requirements that historically governments have either built the facilities themselves — as is the case with many hydroelectric dams — or allowed utilities to operate as monopolies, thus helping ensure recovery of the investment.

5 Smil, *Energy at the Crossroads.*

6 Personal communication.

7 Holt, personal communication.

8 An important exception is American Electric Power, a U.S. utility that believes coal can be decarbonized economically.

9 This assessment is even more the case in countries that actually encourage

the use of fossil fuel; Eastern Europe, Germany, Spain, and many developing
countries currently subsidize the use of coal.

10 Personal communication.

CHAPTER 12. DIGGING IN OUR HEELS

1 Swarns, "Compromise Brings Accord."
2 ExxonMobil, the largest private oil company in the world, had revenues of
$185 billion in 2000, and is actually relatively small potatoes in the oil world.
3 There is a good deal of variation in the politics and business practices of producers themselves. Companies like British Petroleum and Shell have all but
embraced the idea of climate change as real and are trying to position themselves as "early adopters," in hopes of profiting from any move toward a gas
or hydrogen economy — and both advertise themselves as renewables pioneers. By contrast, more conservative companies like ExxonMobil remain
largely committed to an oil-dominated energy economy and have used their
considerable political and financial resources to delay policies that put that
the oliocentric business model at risk.
4 Interview, Dahrhan, Saudi Arabia, May 2002.
5 "Germany Key to Overall European Renewables."
6 Other sectors began moving toward cleaner energy as well. German automakers, in a move to avoid new government regulations, voluntarily agreed
to reduce tailpipe emissions and develop zero-emission cars.
7 Personal communication from Felix Christian Matthes, December 2001.
8 Ibid.
9 According to a study by the Fraunhofer Institute, "larger, heavier automobiles are increasingly 'frittering away' the energy savings achieved by technical improvements to new vehicles. . . . Energy efficiency is suffering particularly from the space heating of larger premises at higher temperatures,
from stand-by losses from electrical equipment and from the use of powerful, high-performance appliances." "Little Progress in Energy Efficiency,"
www.isi.fhg.de/pr/2000engl/epro32000.htm.
10 Personal communication.
11 It should also be noted that German energy politics owe much to a host of
historical factors, ranging from decades of high taxes and energy scarcity to a
radicalized and competent Green movement, which simply do not exist in
America.
12 See www.opensecrets.org/industries/indus.asp?Ind=E01.
13 See, for non-oil industries, Lazarri, "Energy Tax Policy," 2; for the oil industry, Geller, *Energy Revolution*, 38.
14 Smil, *Energy at the Crossroads*, 84

15 U.S. Energy Information Administration, as cited in Smil, *Energy at the Crossroads*, 36.

16 Personal communication, December 16, 2003.

17 Revkin and Seelye, "Report by the E.P.A. Leaves Out Data."

18 Harris, "Bush Covers Up Climate Research."

19 To be fair, the proposed energy bill also contained similar incentives for wind generators, including a production tax credit, although these incentives were far smaller than their hydrocarbon counterparts.

CHAPTER 13. HOW DO WE GET THERE?

1 Personal communication.

2 For example, if Toyota's 2010 fleet had exceeded the fuel-efficiency targets or was well below its emissions cap for that year, the Japanese company would earn some kind of efficiency credit, based perhaps on total fuel saved, which could then be sold to another carmaker, like General Motors, whose 2010 fleet had fallen short of the fuel efficiency target. General Motors could then use the credits to avoid any penalties. This cap-and-trade system would effectively harness market forces, not regulations, to achieve higher fuel efficiency.

3 Personal communication.

4 Smil, *Energy at the Crossroads*, 357.

5 Griscom, "The Thrill of Defeat."

6 Holdren, "The Energy-Climate Challenge," 43.

AFTERWORD

1 Darley, "A Tale of Two Planets: A Report on the Conference 'Future of Global Oil Supply: Saudi Arabia,'" FromTheWilderness.com, March 17, 2004, www.fromthewilderness.com/free/ww3/031704_two_planets.html.

2 Townsend and Harris, "Now the Pentagon tells Bush: Climate change will destroy us," (London) *Observer*, February 22, 2004, a copy of which can be seen at www.notinourname.net/war/climate-22feb04.htm.

BIBLIOGRAPHY

Adams, Robert McC. *Paths of Fire: An Anthropologist's Inquiry into Western Technology.* Princeton, N.J.: Princeton University Press, 1996.

Agence France Press. "ExxonMobil to Plug Azerbaijan Well." *Economic Times,* February 25, 2002.

Anderson, Robert O. *Fundamentals of the Petroleum Industry.* Norman: University of Oklahoma Press, 1984.

Awerbuch, Shimon. "Determining the Real Cost: Why Renewable Power Is More Cost-Competitive Than Previously Believed." www.jxj.com/magsandj/rew/2003_02/real_cost.html, March–April 2003.

Baer, Robert. "The Fall of the House of Saud." *Atlantic Monthly,* May 2003, 53.

Ballard Power Systems. "Ballard announces plan and restructuring to extend cash to 2007," Ballard Power Systems company press release, December 9, 2002.

———. "Ballard, Coleman Powermate to collaborate on portable fuel cell power generators," Ballard Power Systems company press release, January 16, 2000.

———. "Ballard extends industry lead with unveiling of next-generation fuel cell." Press release, January 9, 2000.

———. "Ballard Power Systems Inc. announces pricing of public offering of U.S. $340,725,000 of common stock." Press release, February 23, 2000.

———. "Ballard Power Systems Inc. files registration statement and prospectus for public offering of common shares," Press release, January 28, 2000.

———. "Expanding Our Strength." Annual report, 2001.

———. "Ford presents FC5 prototype vehicle powered by Ballard's next-generation fuel cell," Ballard Power Systems company press release, January 11, 2000.

Banerjee, Neela. "For ExxonMobil, Size Is a Strength and a Weakness." *New York Times,* March 4, 2003.

Barrionuevo, Alexei. "How Trinidad Became a Big Supplier of Liquefied Natural Gas to the U.S." www.wallstreetjournal.com, March 13, 2001.

Bartlett, D. "The Great Energy Scam: How a Plan to Cut Oil Imports Turned into a Corporate Giveaway." *Time,* October 13, 2003, 61.

Beaudreau, Bernard C. *Energy and the Rise and Fall of Political Economy.* Westport, Conn: Greenwood Press, 1999.

BP. *BP Statistical Review of U.S. Energy, June 2001.*
———. *BP Statistical Review of World Energy, June 2001.*
———. "Investing in Renewables and Alternatives." Company materials, October 10, 2002.
Bradsher, Keith. *High and Mighty SUVs: The World's Most Dangerous Vehicles and How They Got That Way.* New York: Public Affairs, 2002.
Brown, Stuart F. "Gearing Up to Make Fuel Cells." *Fortune,* June 25, 2001.
"Buried Losses: The Journey from Plant to Coal Wastes a Lot of Energy." www.economist.com/science/displaystory.cfm?story_id=2155375, October 23, 2003.
Burn, Timothy. "The Hunt for New Oil." *Washington Times,* September 28, 2003.
Buxton, Neil K. *The Economic Development of the British Coal Industry from Industrial Revolution to the Present Day.* London: Batsford Academic, 1978.
Calder, K. "Asia's Empty Tank." *Foreign Affairs,* March/April 1996.
Callus, Andrew. "BP Says Will Miss Output Target in 2002." Reuters, www.zawya .com/oilgas/, September 4, 2002.
Camozzi, R. "Eugene, Ore., Companies Offer Biodiesel, a Soybean Oil-Petroleum Blend." *Eugene, Oregon, Register-Guard,* October 20, 2002.
Campbell, Colin. *The Coming Oil Crisis.* Essex, England: Multi-Science Publishing Company and Petroconsultants, 1988.
Cassidy, John. "Beneath the Sand: Can a Shattered Country Be Rebuilt with Oil?" *The New Yorker,* July 14 & 21, 2003, 64–75.
Chalabi, Fadhil J. "Oil and Development Policy in Saudi Arabia." A presentation for the Institute of Petroleum, U.K., February 14, 2000.
———. "OPEC: An Obituary." *Foreign Policy,* Winter 1997–98.
Cheney, Richard. *National Energy Policy: Report of the National Energy Policy Development Group.* Washington, D.C.: Office of the Vice President of the United States, May 16, 2001.
Chicago Tribune, "OPEC's Attempt to Flex May Fold," September 28, 2003.
Christian Science Monitor, "Can Clean Diesel Power Past Gasoline?" October 15, 2002.
Considine, J. "Battle for Market Share: World Oil Market Projections, 1995–2010." *Journal of Energy Literature* (Oxford Institute for Energy Studies), June 1996.
Cordesman, Anthony. *Saudi Arabia: The Broader Factors Driving the Need for Foreign Investment and Economic Diversity.* Washington, D.C.: Center for Strategic and International Studies, April 2002.
———. "Saudi Arabia Enters the Twenty-first Century: Shaping the Future of the Saudi Petroleum Sector." Review draft. Washington, D.C.: Center for Strategic and International Studies, January 31, 2002.
Currie, Jeffrey. "Natural Gas Supply and Demand Issues." Congressional testimony, the Committee on Energy and Commerce, June 10, 2003. Available online at *http://energycommerce.house.gov/108/Hearings/06102003hearing944/Currie 1524print.htm.*
Daily, Matt. "U.S. Seen Turning Abroad to Feed Natgas Appetite." Reuters,

www.planetark.com/dailynewsstory.cfm/newsid/19832/newsDate/14-Feb-2003/ story.htm, February 14, 2003.

Dallas Morning News, "BP Gushes over Treasure in Deep Water off Texas Coast," August 6, 2002.

Darley, Julian, "A Tale of Two Planets: A Report on the Conference 'Future of Global Oil Supply: Saudi Arabia.'" www.fromthewilderness.com/free/ww3/ 031704_two_planets_html. March 17, 2004.

Deffeyes, K. *Hubbert's Peak: The Impending World Oil Shortage.* Princeton, N.J.: Princeton University Press, 2001.

Deutsche Bank. "Global Energy Wire: Reservations About Reserves," January 23, 2004.

Diebel, Linda. "Oil War: 23 Years in the Making." *Toronto Star,* March 9, 2003. www.globalpolicy.org/security/oil/2003/0309oilwar.htm.

Easterbrook, G. "Axle of Evil: America's Twisted Love Affair with Sociopathic Cars." *New Republic,* January 20, 2003.

———. "Car Talk: Why Bush's H-Car Is Just Hot Air." *New Republic,* February 24, 2003.

Economides, M. *The Color of Oil: The History, the Money, and the Politics of the World's Biggest Business.* Katy, Tex.: Round Oak, 2000.

"Energy Report." *Oil & Gas Journal,* May 12, 2003.

European Commission. *The Report of the G8 Renewable Energy Task Force.* Brussels: European Commission, July 17, 2001.

European Wind Energy Association. "Record Growth for Global Wind." Press release, March 3, 2003.

"ExxonMobil to Cease Operations in Turkmenistan Newsbase News Services." www.gasandoil.com/goc/company/cnc21243.htm, March 21, 2002.

Feld, L. *OPEC Revenues Fact Sheet.* www.eia.doe.gov/emeu/cabs/opecrev.html (U.S. Energy Information Administration), December 2001.

———. *Venezuela Fact Sheet.* Washington, D.C.: U.S. Energy Information Administration, April 2002.

Financial Times, "Oil Majors Wonder," April 25, 2002.

———, "PetroChina and Sinopec Endure Violent Clashes over Petrol Stations," April 25, 2002.

———. "Shell Faces Lawsuit over Oil Reserves Cut," January 26, 2004.

Flavin, C., and N. Lensser. *Power Surge: Guide to the Coming Energy Revolution.* Worldwatch Institute Environmental Alsert Series. New York: Norton, 1994.

Fox, Justin. "A Gas-Guzzling Scapegoat." www.fortune.com, October 24, 2001.

Gantz, Rachel. "Success of Renewables Demands a Change in 'Petroleum Culture.'" *Octane Week,* July 15, 2002.

Gelbspan, Ross. *The Heat Is On: The Climate Crisis.* Reading, Mass.: Perseus, 1998.

Geller, Howard. *Energy Revolution: Policies for a Sustainable Future.* Washington, D.C.: Island Press, 2003.

"Germany Key to Overall European Renewables." www.solaraccess.com/ newsstory?storyid=5214, October 1, 2003.

Giusti, Luis. "*La Apertura:* The Opening of Venezuela's Oil Industry." *Journal of International Affairs,* Fall 1999.

"Global Energy Technology Strategy: Addressing Climate Change. Initial Findings from an International Public-Private Collaboration." Report prepared for the U.S. Global Energy Technology Strategy Program by the Battelle Memorial Institute, Washington.

Glunt, J. "High Volume of Renewables Needed to Displace Petroleum Fuels." *Octane Week,* October 21, 2002.

Goldberg, Marshall. "Federal Energy Subsidies: Not All Technologies Are Created Equal." Report by the Renewable Energy Policy Project. Washington, D.C.: Center for Rewnewable Energy and Sustainable Technology, July 2000.

Goulder, L. "U.S. Climate-Change Policy: The Bush Administration's Plan and Beyond." iis-db.stanford.edu/viewpub.lhtml?pid=20399&cntr=cesp (Stanford Institute for Economic Policy Research), February 2002.

Griscom, Amanda. "The Thrill of Defeat: The Climate Bill Lost Out, but the Environment May Yet Prove the Winner." www.gristmagazine.com/muck/muck110503.asp, November 5, 2003.

Haggerty, R. A. *VENEZUELA — A Country Study.* memory.loc.gov/frd/cs/vetoc .html (Federal Research Division, Library of Congress), December 1990.

Hakim, Danny. "Cloaked in Green, but Guzzling Gas, " *New York Times.* April 19, 2003.

Harris, Paul. "Bush Covers Up Climate Research: White House Officials Play Down Its Own Scientists' Evidence of Global Warming." *London Sunday Observer,* September 21, 2003.

Heinberg, Richard. *The Party's Over: Oil, War, and the Fate of Industrial Societies.* Gabriola Island, British Columbia: New Society Publishers, 2003.

Hertsgaard, Mark. *Earth Odyssey: Around the World in Search of Our Environmental Future.* New York: Broadway Books, 1998.

Hoffmann, Peter. "Hydrogen." *World & I Magazine,* October 2002.

———. *Tomorrow's Energy.* Cambridge, Mass.: MIT Press, 2001.

Holdren, John P. "The Energy-Climate Challenge." *Environment,* June 2001, 21–44.

———. "Energy Efficiency and Renewable Energy in the US Energy Future." Invited testimony before the House Science Committee, February 28, 2001.

———. "Meeting the Energy Challenge." *Science,* February 9, 2001.

———. "Searching for a National Energy Policy." *Issues in Science and Technology,* Spring 2001.

Hutzler, Mary, et al. *Analysis of Corporate Average Fuel Economy (CAFE) Standards for Light Trucks and Increased Alterative Fuel Use.* Washington, D.C.: U.S. Energy Information Administration, March 2002.

IBM Consulting Services. *Playing to Win in the Downstream: Looking Beyond Tomorrow.* Sommers, N.Y.: IBM Global Services, March 2003.

Imle, John F., Jr. "Multinationals and the New World of Energy Development: A Corporate Perspective." *Journal of International Affairs,* Fall 1999.

Interfax News Agency. "Output Continues to Fall at Sinopec Shengli." news .daylightonline.com/Hefei.html, September 9, 2003.

Intergovernmental Panel on Climate Change. *Climate Change 2001: Synthesis Report.* Geneva: United Nations / World Meteorological Organization, 2001.

Interlaboratory Working Group. *Scenarios for a Clean Energy Future.* www.msnbc .msn.com/id/3071521. Springfield, Va.: National Technical Information Service, U.S. Department of Commerce, November 2000.

International Energy Agency. *World Energy Outlook, 2001.* Paris: International Energy Agency/Organization for Economic Cooperation and Development, 2001.

Inter Press Service. "Automakers Dropping Electric Cars as Sales Sag." www.zawya .com/oilgas/, September 30, 2002.

James A. Baker III Institute for Public Policy, Rice University / Council on Foreign Relations. "Strategic Energy Policy: Challenges for the 21st Century." (Report of an independent task force, Edward Morse, chair.) Washington, D.C.: James A. Baker III Institute for Public Policy / Council on Foreign Relations, 2001.

Karl, Terry Lynn. "The Perils of the Petro-State: Reflections on the Paradox of Plenty." *Journal of International Affairs,* Fall 1999.

Kempton, Willet, J. Boster, and J. Hartley. *Environmental Values in American Culture.* Cambridge, Mass.: MIT Press, 1996.

"Key Studies Deriving Monetary Estimates." www.british-energy.cmo/environment/BE/academic/key_studies. htm.

Khalip, Andrei. "Oilmen's 'Green' Pledges Met with Disbelief." Reuters, www .planetark.com/dailynewsstory.cfm/newsid/17641/story.htm, September 4, 2002.

Kilgore, W. *Measuring Energy Efficiency in the United States Economy: A Beginning.* DOE/EIA 0555 (95)/2. Washington, D.C.: U.S. Energy Information Administration, 1995.

Knopman, Debra. "Assessing Natural Gas and Oil Resources." Testimony presented before the U.S. House Subcommittee on Energy and Mineral Resources, June 24, 2003.

Koppel, Tom. *Powering the Future: The Ballard Fuel Cell and the Race to Change the World.* Toronto: Wiley, 1999.

Kyodo News, "Toyota Brings Forward Release of Fuel Cell Car to Year-End," July 1, 2002.

Lave, Charles. "A New CAFE." *Access Magazine* (University of California, Berkeley, Transportation Center), Fall 2001.

Lavelle, Marianne. "Sand Dollars: New Technology Makes It Easier to Tap Canada's Oil Reserves." *U.S. News & World Report,* October 13, 2003.

Lazzari, S. "Energy Tax Policy." www.ncseonline.org/NLE/CRSreports/energy/eng-60.cfm (U.S. Congressional Research Service), September 19, 2001.

Lipschultz, David. "Solar Power Is Reaching Where Wires Can't." *New York Times,* September 9, 2001.

Lopez, Jose. "Richmond, Calif., to Be Home to Fueling Station That Extracts Hydrogen." *Contra Costa Times,* October 31, 2002.

Lovins, Amory, and L. H Lovins. "Energy Forever." *American Prospect,* February 11, 2002, 30–34.

———. "Fool's Gold in Alaska." *Foreign Affairs,* July/August 2001, 72–85.

———. "Mobilizing Energy Solutions." *American Prospect,* January 28, 2002, 18–21.

Loy, Frank, and Bruce Smart. *U.S. Policy on Climate Change: What Next?* Queenstown, Md.: Aspen Institute, 2002.

Lugar, Richard, and R. James Woolsey. "The New Petroleum." *Foreign Affairs,* January/February 1999, 88–102.

Mabro, Robert. "The World's Oil Supply, 1930–2050: A Review Article." *Journal of Energy Literature,* June 1996 (Oxford Institute for Energy Studies).

MacKenzie, James. "A History of Energy." In *The Energy Book: A Look at the Death Throes of One Energy Era and the Birth Pangs of Another,* ed. Wayne Hanley. Brattleboro, Vt.: Stephen Green Press.

Manning, Robert A. *The Asian Energy Factor: Myths and Dilemmas of Energy, Security, and the Pacific Future.* New York: Palgrave, 2000.

Matthes, Felix, and Christof Timpe. *Sustainability and the Future of European Electricity Policy.* Berlin: Heinrich Boell Foundation, 2000.

McKillop, Andrew. "Why Venezuela and World Oil Exporters Can Target US$36–$45 per Barrel." www.petroleumworld.com/SuF082403.htm, October 2, 2003.

Mintzer, I., J. A. Leonard, and P. Schwartz. *U.S. Energy Scenarios for the 21st Century.* Arlington, Va.: Pew Center on Global Climate Change, July 2003.

Mitchell, John, with K. Morita, N. Selley, and J. Stern. *The New Economy of Oil: Impacts on Business, Geopolitics, and Society.* London: Earthscan Publications, 2001.

Moran, Michael, and Alex Johnson. "Oil After Saddam: All Bets Are In." www.msnbc.msn.com, November 7, 2002.

Morse, Edward. "A New Political Economy of Oil?" *Journal of International Affairs,* Fall 1999, 1–29.

———. "The U.S. and the International Petroleum Sector: The Rogue Elephant in the Jungle of Geopolitics." Presentation at the Oxford Energy Seminar, August 31, 1995.

Morse, Edward, and James Richard. "The Battle for Energy Dominance." www.foreignaffairs.org/20020301faessay7969/edward-l-morse-james-richard/the-battle-for-energy-dominance.html, March/April 2002.

Motavalli, Jim. *Forward Drive.* San Francisco: Sierra Club Books, 2000.

———. "Harnessing Hydrogen." www.emagazine, March/April 2000.

National Environmental Trust. *America, Oil, and National Security: What Government and Industry Data Really Show.* Washington, D.C.: National Environmental Trust, 2002.

National Science Foundation. "Scientists Explore Large Gas Hydrate Field off Oregon Coast." Press release, www.eurekalert.org/pub_releases/2002-09/nsf-selo91002.php, September 11, 2002.

Noreng, Øystein. *Crude Power: Politics and the Oil Market.* London: Taurus, 2002.

New York Times, "China's Boom Adds to Global Warming Problem," October 22, 2003.

Obaid, Nawaf E., A. Jaffe, E. Morse, C. Gracia, and K. Bromley. "The Sino-Saudi Energy Rapprochement: Implications for U.S. National Security." Report prepared for the Office of the Secretary of Defense, U.S. Department of Defense, January 82002.

O'Connor, Harvey. *Empire of Oil*. New York: Monthly Review Press, 1955.

Ogden, Joan, R. Williams, and E. Larson. "Societal Lifecycle Costs of Cars with Alternative Fuels/Engines." *Energy Policy*, November 2003.

Olcott, Martha Brill. "Pipelines and Pipe Dreams: Energy Development and Caspian Society." *Journal of International Affairs*, Fall 1999.

"Oil and Gas: Long-Term Contribution Trends." www.opensecrets.org/industries/indus.asp?Ind=E01

O'Ryan, Raul, D. Sperling, M. Delucchi, and T. Turrentine. *Transportation in Developing Countries: Greenhouse Gas Scenarios for Chile*. Arlington, Va.: Pew Center on Global Climate Change, August 2002

Petroleum Finance Week, April 14, 2003.

Picerno, James. "If We Really Have the Oil." www.wealth.bloomburg.com, September 2002.

Pickler, Nedra. "Honda, FuelMaker Unveil At-Home Natural Gas Refueling System." Associated Press, www.automotiveindustrynews.com/auto33022958.htm, October 8, 2002.

Qing, Dai, and L. Sullivan. "The Three Gorges Dam and China's Energy Revolution." *Journal of International Affairs*, Fall 1999, 53–71.

Raloff, Janet. "Power Harvests: The Salvation of Many U.S. Farmers May Be Blowing in the Wind." *Science News*, July 21, 2001.

Reuters. "China's Car Sales Hit One Million for First Time." www.chinacars.com/english/content/business/62048.asp, December 16, 2002.

———. "Oil Majors Cool on Kazakh Offshore Plans." www.zawya.com/oilgas/, October 3, 2002.

———. "UK Hails Crop of Small North Sea Oil, Gas Fields." www.zawya.com/oilgas/, August 5, 2002.

Revkin, A., and K. Seelye. "Report by the E.P.A. Leaves Out Data on Climate Change." *New York Times*, June 19, 2003.

Rifkin, J. *The Hydrogen Economy*. New York: Jeremy Tarcher, 2002.

Riva, Joseph P., Jr. *World Oil Production After Year 2000: Business as Usual or Crises?* www.ncseonline.org/NLE/CRSreports/energy/eng-3.cfm?&CFID=11855416&CFTOKEN=86271523 (U.S. Congressional Research Service Report for Congress, 35–925 SPR 1995).

Rolt, L. T. C., et al. *The Steam Engine of Thomas Newcomen*. Ashbourne, England: Landmark Publishing, 1997.

Rosenfeld, Arthur. "The Art of Energy Efficiency: Protecting the Environment with Better Technology." *Annual Review of Energy and the Environment*, 1999.

Rudin, Andrew. "How Improved Efficiency Harms the Environment." www.home.earthlink.net/~andrewrudin/index2.html, August 1999.

Rudolph, Richard. *Power Struggle: The Hundred-Year War over Electricity.* New York: Harper & Row, 1986.

Ruppert, Mike. "Interview with Matt Simmons" (on-line article for From the Wilderness). www.copvcia.com/ftw/free/ww3/061203_simmons.html, June 12 2003.

Russian Information Agency. "Japan Ready to Invest $14 Billion in Russia's Far Eastern Oil and Gas Projects." www.eng.mineral.ru/Chapters/News/9218 .html (Mineral Information and Research Centre, Ministry of Natural Resources of the Russian Federation), August 13, 2003.

Sampson, Anthony. *The Seven Sisters: The Great Oil Companies and the World They Shaped.* New York: Bantam Books, 1991.

Sawin, Janet. "Interview with Rajendra Pachauri." *World Watch,* March/April 2003.

Schurr, Sam H., B. Netschert, V. Eliasberg, J. Lerner, and H. Landsberg. *Energy in the American Economy, 1850–1975: An Economic Study of Its History and Prospects.* Baltimore, Md.: Johns Hopkins University Press, 1960.

Schweizer, Peter. *Victory: The Reagan Administration's Secret Strategy That Hastened the Collapse of the Soviet Union.* New York: Atlantic Monthly Press, 1994.

Scott, Alex. "Fuel Cells Power Toward the Mass Market." *Chemical Week,* February 28, 2001, 41.

Smil, Vaclav. *Energy at the Crossroads: Global Perspectives and Uncertainties.* Cambridge, Mass.: MIT Press, 2003.

———. *Energy in World History.* Boulder, Colo.: Westview Press, 1994.

South Coast Air Quality Management District. "Historic Air Pollution Disasters." www.aqmd.gov/smog/inhealth.html#historical.

South China Morning Post, "Car Makers Prepare Profit Road," June 11, 2002.

"Special Report: CO_2 Sequestration Adds New Dimension to Oil, Gas Production." *Oil & Gas Journal,* March 3, 2003.

Stipp, David. "The Coming Hydrogen Economy." *Fortune,* November 12, 2001.

Swanson, Gary. "Alternative Energy a New Cash Crop for Colorado Agriculture." *Denver Post,* March 24, 2002.

Swarns, R. "Compromise Brings Accord on Renewable Energy Closer." *New York Times,* September 2, 2002.

Te Brake, William H. "Air Pollution and Fuel Crises in Preindustrial London, 1250–1650." In *Technology and the West: A Historical Anthology from Technology and Culture,* ed. Terry S. Reynolds and Stephen H. Cutcliffe. Chicago: University of Chicago Press, 1997.

"Termoelectrica de Mexicali." www.power-technology.com/projects/mexicali/.

Titus, James. "The Costs of Climate Change to the United States." yosemite .epa.gov/oar/globalwarming.nsf/content/ResourceCenterPublicationsSLR_ US_Costs.html (U.S. Environmental Protection Agency Global Warming Publications).

Townsend, Mark and Paul Harris, "Now the Pentagon tells Bush: Climate Change will destroy us," *Observer,* London, February 22, 2004.

Traynor, J. "Shell T & T: Sins of the Past." Deutsche Bank research report, January 9, 2004.

Urquhart, F. "North Sea Oil Output to Peak This Year." www.thescotsman.online, October 22, 2002.

U.S. Energy Information Administration. *International Energy Outlook, 2001.* Washington, D.C.: U.S. Government Printing Office, 2001.

U.S. Library of Congress. *Venezuela: A Century of Caudillismo.* December 1990.

Van Dyke, K. *Fundamentals of Petroleum,* 4th ed. Austin: University of Texas Press, 1997.

Verleger, Philip. "A Collaborative Policy to Neutralize the Economic Impact of Energy Price Fluctuations." Policy paper, June 10, 2003.

Victor, David. *The Collapse of the Kyoto Protocol and the Struggle to Avoid Global Warming.* Princeton, N.J.: Princeton University Press, 2001.

Victor, David, A. Hayes Jeffrey, and H. Huntington. *Geopolitics of Gas: An Analysis of Prospective Developments in the Natural Gas Trade and Geopolitical Implications.* Prospectus. Stanford, Calif.: Stanford University Program on Energy & Sustainable Development, 2003.

Williamson, Harold F., and R. Andreano. *The American Petroleum Industry.* Vols. 1 and 2. Evanston, Ill.: Northwestern University Press, 1959–1963.

World Bank. *Energy After the Financial Crisis: Energy and Development Report.* Washington, D.C.: World Bank Publications, 1999.

World Energy Council. *The Challenge of Rural Energy Poverty in Developing Countries.* London: World Energy Council, 1999–2003.

Xinhua News Agency. "Beijing Auto Show Wins International Plaudits." June 13, 2002.

Xu, Xiaojie. *Petro-Dragon's Rise: What It Means for China and the World.* Florence: European Academic Press, 2002.

Yergin, D. *The Prize: The Epic Quest for Oil, Money, and Power.* New York: Simon & Schuster, 1992.

Zhou, Hongchang, and D. Sperling. *Transportation in Developing Countries: Greenhouse Gas Scenarios for Shanghai, China.* Arlington, Va.: Pew Center on Global Climate Change, July 2001.

Acknowledgments

A book such as this is not written solo. Since I began work on *The End of Oil* in 2001, dozens of brilliant, dedicated people have helped me master a complex universe of material and ideas.

I received invaluable help from countless experts in the energy business — academics, analysts, and company executives — who took the time to share their insights into the largest and most important industry on the planet. David Victor, director of the Program on Energy and Sustainable Development at Stanford University; Edward Morse at Hess Energy Trading Company; and Vaclav Smil at the University of Manitoba were particularly generous with their time and wisdom.

I am also heavily indebted to Gerald Stokes and Bill Chandler at the Joint Global Change Research Institute; Ira Joseph at PIRA Energy; John Turner at the National Renewable Energy Laboratory; Joan Ogden at the University of California, Davis; Matt Simmons at Simmons and Company in Houston; Phil Verleger at PKVerleger; John Mitchell at the Royal Institute of International Affairs; Anthony Richter at the Open Society Institute; Reid Detchon at Energy Future Coalition; Lee Lynd at Dartmouth College; Tasios Melis at the University of California, Berkeley; Paul Maycock at PV Energy Systems; Willett Kempton at the University of Delaware; Ian Bremmer and Vitali Meschoulam at Eurasia Group; Julia Nanay at PFC Energy; Manouchehr Takin and Julian Lee at the Centre for Global Energy Studies; Janet Sawin and Chris Flavin at WorldWatch Institute; Amory Lovins at RMI; Denis Hayes at the Bullitt Foundation; Kalee Kreider at Fenton Communications in Washington; Eileen Clausen at the Pew Center on Global Climate Change; and Lawrence Goulder at Stanford University.

Thanks as well to Arthur Rosenfeld, Peter Hoffmann, Joseph Romm, Merwin Brown, Parag Khanna, David L. Stern, and, of course, Michael Lynch and Colin Campbell — the matter and antimatter of the oil-depletion debate.

The energy industry is not the most open or accessible subject for writers, especially those with a critical perspective. Having been demonized for more than a century by journalists, activists, and irate consumers — not to mention politicians needing a quick bounce in the polls before election day — few executives involved with energy or energy technologies are eager to talk frankly or publicly with reporters about the problems of the energy business. I did find important exceptions, though. Lauren Segal at BP (formerly British Petroleum); David Frowd and Douglas McKay at Royal Dutch–Shell; Xiaojie Xu at China National Petroleum Corporation; Abdulaziz Al-Khayyal and his colleagues at Saudi Aramco; and Ali Bin-Ibrahim al-Naimi, the Saudi minister of oil, were among those who were willing not only to talk but to share their hopes and concerns for an industry in the midst of an uncertain transition. In addition, dozens of other energy industry professionals spoke to me off the record.

Turning the raw data into a usable form was a joint effort. My editor at Houghton Mifflin, Anton Mueller, understood from the start the story I wanted to tell and guided its evolution from idea to finished book; it couldn't have happened without him. For their efforts, his colleagues Erica Avery, Gracie Doyle, and Susan Abel should all get extra vacation time. Clara Jeffery, my former editor at *Harper's Magazine*, encouraged my first story on energy trends, and my agent, Heather Schroder at ICM, persuaded me to consider a full-length book. I also want to thank the people who helped me get these ideas into the marketplace, including Amy Carlson, Pat Rutledge, Jenny Pederson, Jean Spalti, Ian Dunn, Jay Harris, Eileen Ellis, and Luke Mitchell at *Harper's*, who deserves a medal for forbearance.

Just as important are the friends and family who encouraged, inspired, and supported me during what seemed like an endless process, among them Molly Roberts, Lynn Roberts, Matt Roberts, and Ann Lister, who cheered me on; Pat and Byron Dickinson; Sue Dickinson and Rob Werth, who provided life support and lots of free meals; Randall Smith, who taught me to smile while suffering; and Richard Champion, who listened patiently for hours, his face never once betraying any sign of boredom.

Above all, I want to thank my children, Hannah and Isaac, who endured cheerfully, and especially my wife, Karen Dickinson, whose love kept me going and whose sharp eye and natural editor's touch kept me from going on too long.

Index

turbines as alternative to, 179

See also Cars: hybrid; Fuel efficiency; Transportation fuel

Gelbspan, Ross, 122

General Electric (GE), 75–76, 199, 200

General Motors Corporation (GM), 82, 83, 86

in China, 144, 149, 158, 164

Geopolitical issues (of energy). *See* Energy security

Geopolitics of Energy in 2015 conference, 307–8

Geothermal energy, 190, 192

Germany

alternative energy in, 192, 194–95, 198, 200, 285, 289–93, 329, 330

carbon dioxide emissions reduction in, 131, 135, 138–39

cars made in, 264, 296, 318, 358n. 6

as coal producer, 28, 29, 270, 290, 294

energy consumption in, 150

energy efficiency in, 14, 81

and Kyoto Protocol, 127

and oil in World War II, 38, 39–40

Ghadaban, Thamer al-, 238

Ghawar oil field (Saudi Arabia), 1–2, 12, 92

Gheit, Fadel, 58, 171, 265

Ghosn, Carlos, 159

Global warming. *See* Climate change

GM. *See* General Motors Corporation

Goldman Sachs, 183

Goldstein, David, 225

Gordon, Rick, 178, 265

Gore, Albert, 127, 128, 301

Goulder, Lawrence, 138

Government

and carbon capture, 208

and energy companies, 30, 38–39, 41, 94, 98, 109–13, 135–36, 191, 214, 240, 269–70, 282, 286–87, 294–302, 315–16, 357n. 4

and hydrocarbon research, 296

and lack of planning for future energy crises, 46, 65, 94, 133, 134, 305–6, 331–32

and natural gas, 186, 249, 258

and new energy economy, 279–80, 294–302, 313–27, 329–30

and oil security, 38–39, 41, 109–13, 255–56, 286–87

and power sector, 357n. 4

U.S. state opposition to federal, 313, 330

See also Energy security; *Names of specific countries*

Great Britain

alternative energy in, 285

coal in, 21–24, 27–30, 35, 124, 135

in Iraqi wars, 9–10

and Kyoto Protocol, 127

in Middle East, 98

as natural gas importer, 181

and oil, 35, 36, 38–41, 54, 103, 104, 251

SUVs in, 263

See also London; North Sea

Great Smoky Mountains National Park, 259

Greenhouse effect

from burning hydrocarbons, 6, 119–20, 179

in primordial times, 123, 124

See also Carbon dioxide emissions; Climate change

Greenland, 56, 64, 348n. 23

Greenpeace, 130, 287

Greenspan, Alan, 166, 183

Grove, William, 72

Guangdong Province (China), 185

Gulf of Mexico

natural gas in, 181

oil in, 51, 56, 58, 96

Gulf Oil Corporation, 40

Haldane, John, 74

Halliburton energy company, 298, 304

Hamer, Glenn, 195

Hammil, Al and Curt, 32–33, 35

Oil (*cont.*)

refining of, 36, 37, 170

reserves of, 4, 6, 46, 48–52, 54, 178, 251–55

search for more, 8, 9, 11–13, 36, 38–39, 42, 46, 49–53, 62–64, 173, 305, 312

self-pressurized, 2, 96

share of energy market dominated by, 5, 13, 166

and standard of living, 2–3, 35–36

surplus inventory of, 171, 202, 203, 253–55

transporting of, 7

"unconventional," 46, 55, 173, 208–9, 211, 311

U.S. dependence on foreign, 4, 9–10, 15, 38–39, 41–43, 54, 59, 94–95, 97, 105–6, 109–13, 152, 180, 204, 217–18, 237–38, 254, 285, 286

U.S. Iraqi wars linked to, 9–10, 13, 97, 105–13, 123, 219–20, 237–38, 252, 255, 257, 304, 308, 312, 350n. 12

"water cut," 2

See also Air pollution; Arab oil embargo; Carbon dioxide emissions; Cars; Energy; Fuel efficiency; Hydrocarbons; Non-OPEC oil; Oil companies; Oil depletion; Oil prices; OPEC; Price volatility; War

Oil companies

and alternative energy sources, 77–78, 81, 82, 85, 87–88, 193, 194–95, 285–87, 293

and carbon capture, 208

corporate model for, 37–38, 170–71

and electric trolleys, 78

government's ties to, 38–39, 41, 94, 98, 286–87, 294–302, 316, 347n. 4

independent, 40, 104, 346n. 24

influence of, on George W. Bush, 109–13, 135–36, 191, 214, 282, 298

infrastructure and technologies de-

veloped by, 36, 40, 45, 52, 54–57, 186–87, 213, 237–38, 251, 329

major, 40, 42, 98–100, 104, 170–74, 357n. 18

and Middle Eastern oil operations, 38–39, 49, 98–100, 111–12, 252–53

in natural gas business, 8, 166, 167–74, 176–78, 270, 357n. 18

and oil prices, 97, 104

search for oil by, 8, 9, 11, 13, 36, 38–39, 42, 46, 49–53, 62–64, 173, 305, 312

tax breaks for, in United States, 295

volatility among, 6

See also Energy security; Investments; Non-OPEC oil; Oil depletion; Oil prices; Price volatility; Profits; *Names of specific companies*

Oil depletion, 2, 4, 12–13, 43–65, 284

allowances for, in U.S. tax code, 295

in Azerbaijan, 44–45, 52–53, 62–64

as crisis, 46, 145, 253, 306, 331–32

lack of panic about, 59–62, 65, 271, 300

of non-OPEC oil, 57–60, 95, 114, 168, 173, 178, 239, 251–52, 312, 332

and oil prices, 173, 253, 312

planning and preparation for, 65, 123, 168, 308–9, 331–32

in supergigantic oil fields, 57–58

Oil market

anxiety in, 108

as not free, 60–61, 104–7, 111, 113–14

See also Energy security; Government; Markets; Oil companies; Oil prices; Price volatility; Profits

Oil optimists, 46–49, 54–56, 59, 62

Oil pessimists, 46, 49, 52, 54, 65

Oil prices

average, 60, 96

and economic growth, 108, 111, 113

high, 97, 100–105, 108–10, 162, 169,

Shaw, Bob, 89
Shayba oil field (Saudi Arabia), 1–2, 92
Shell Oil Company, 36, 54, 242
   and alternative energy sources, 82,
       87, 194–95, 358n. 3
   as major oil company, 40, 286
   in Mexico, 39, 166
   and natural gas, 166, 167, 177
   oil production by, 172, 244
   oil searches by, 13, 173
   profits of, 173
Shengli oil field, 160
Ships
   oil-powered, 35, 36, 38, 40
   steam-powered, 28, 29
   See also Tankers
Siberia
   climate changes in, 116–17
   natural gas in, 167
   oil in, 8, 48, 55, 56, 96, 116–17
Sierra Club, 300
Simmons, Matt, 60, 64, 65
Sims-Gallagher, Kelly, 162, 164
Singer, Fred, 119
Sinopec Corporation, 143–44
Siradjev, Sahib, 44, 52–53
Slingenberg, Yvon, 283
Smelting industries, 25–29
Smil, Vaclav, 206–7, 232, 328–29
Smog, 42, 144, 151, 259
   See also Air pollution; Respiratory
      diseases
Smoking, 275
SOCAR (State Oil Company of Azer-
     baijan Republic), 53, 62–63, 65
Solar energy, 5, 9, 192–95, 209
   amount of world's energy provided
     by, 191
   costs of, 191, 192, 194, 195, 201–2, 320–
     21
   and developing countries, 162
   Europe and Japan as leaders in, 15,
     291–92, 324

   growth of industry for, 190, 194–95,
     201, 320–21
   obstacles to use of, 192–94, 201–4,
     206, 216, 230, 321
   quality of energy provided by, 192,
     206, 216, 230, 321
   research on, 68, 76, 131, 137
   storing of, in hydrogen fuel cells, 75,
     76, 78, 87, 204–5
   See also Photosynthesis; Photovoltaic
     cells
Solar Energy Industries Association, 195
Source rock, 33, 34, 50, 65
South America
   energy demand in, 146
   natural gas in, 165, 167, 181
   oil in, 48, 56, 57, 244, 256
   political decisions about energy in,
     284
   See also Names of specific countries in
Southeast Asia, 241
   See also Names of specific countries in
South Korea, 150, 156–57, 160, 162, 175,
     185, 239
Southwest Powerlink, 174
Soviet Union. See Russia
Spain, 196
"Spark spread," 177, 240
Spindletop (Beaumont, Texas), 31–35,
     38, 50
Sport-utility vehicles. See SUVs
"Squeezing the market" tactic, 97, 183
Standard Oil Company, 36–39
   See also Esso; ExxonMobil Corpora-
     tion; Exxon Corporation
Standards of living
   amount of energy needed to raise,
     242–43
   relation of, to energy, 2–6, 11–12, 15,
     24–26, 29, 31, 36, 40–41, 71, 113,
     147, 155, 210, 230, 241–43, 327
Stanford Institute for Economic Policy
     Research, 138